ARITHMETIC FOR COLLEGE STUDENTS

ARITHMETIC FOR COLLEGE STUDENTS

FOURTH EDITION

D. FRANKLIN WRIGHT

CERRITOS COLLEGE

D. C. HEATH AND COMPANY

Lexington, Massachusetts   Toronto

This book is dedicated to the memory of
Sara M. Wright

PREFACE

Just as in the three previous editions, the basic philosophy of this text is to give students a chance to understand *why* arithmetic procedures work as they do as well as *how* to perform them. By giving brief but sound explanations, I have tried to eliminate most of the guesswork and "lost" feelings that produce fear of arithmetic for many students.

In particular, prime factorization still forms the foundation for understanding fractions; fractions form the foundation for proportions; proportions underlie percents; and these ideas all lead to practical applications.

CONTENT

With the exception of Chapter 1 on whole numbers and Chapter 7 on the metric system, the chapters are best covered in the order presented because many topics depend on previously discussed material. The text includes enough material for a three or four semester-hour course.

Chapter 1 (Whole Numbers) is review. Exponents are now introduced in Chapter 2 (Prime Numbers). An instructor could therefore start the course with Chapter 2 and have students review Chapter 1 on their own time.

Chapter 2 (Prime Numbers) forms a foundation for the text. It contains the basic ideas of exponents, order of operations, and prime numbers used throughout the text. To avoid confusion with the more important topic of least common multiple, the topic of greatest common divisor, with exercises, now appears in Appendix III. It can be taught with Chapter 2 if desired.

The approach to fractions in Chapter 3 (Rational Numbers) is based on using prime factorizations. Many students have told me that fractions "make more sense" with this approach, and that working with fractions is no longer a series of mysterious steps that they cannot remember.

Chapter 4 (Decimal Numbers and Proportions) emphasizes reading and writing decimal numbers and using ratios and proportions in solving word problems. These topics form the basis for the development of percent in Chapter 5 (Percent). The problems using percent in Chapter 5 have been upgraded, and students may use calculators at this time.

Chapter 6 (Applications with Calculators) is an entirely new chapter; it discusses simple interest, compound interest, installment purchasing, real estate, and reading graphs and charts. This chapter allows students to understand how to use a calculator, including its limitations in rounding off answers. The problems involve many calculations. They are quite reasonable and informative with the use of calculators but would be too time-consuming without them. Students also learn the importance of good organization of quantities of data.

By using calculators only in Chapter 6 and in portions of Chapter 5 (Percent), the instructor can emphasize that the calculator is a tool and not a substitute for skill and understanding. Students must know *what* buttons to push, *why* those particular buttons, and be able to recognize the reasonableness of the answers they get from a calculator.

Chapter 7 (Measurement: The Metric System) is particularly useful, both for initial learning and later as a reference source. The emphasis is on changing units within the metric system and using these units correctly in calculating length, area, mass, and volume of geometric figures. The chapter also discusses Celsius and Fahrenheit temperature equivalents and U.S. Customary and Metric equivalents.

Chapter 8 (Negative Numbers), Chapter 9 (Solving Equations), and Chapter 10 (Real Numbers and Graphing Linear Equations) have been completely rewritten, and form the pre-algebra part of the text. These chapters are particularly valuable for those students who will continue in mathematics. Chapter 8 introduces the basic operations with integers and negative rational numbers, and Chapter 9 discusses solving equations and word problems. Simplifying square roots, real number lines, working with inequalities, graphing lines in a plane, and the Pythagorean Theorem are in Chapter 10. The changes in these chapters are: Number lines introduce negative numbers; vectors and sets have been eliminated except for minimal use of braces to indicate sets of ordered pairs. New sections cover negative rational numbers and graphing straight lines. Because of the proliferation of calculators, the value of studying square root algorithms is questionable. Thus, in Chapter 10, one algorithm for square roots has been eliminated and the divide-and-average algorithm now appears in Appendix IV.

The appendices contain material and problems on ancient numeration systems, base two and base five, the greatest common divisor, and square roots. The inside back cover contains a table of powers, roots, and prime factorizations.

EXERCISES

There are more than 3600 exercises carefully chosen and graded, proceeding from easy exercises to more difficult ones, plus practice quizzes. Answers to all the exercises except multiples of 4 are in the back of the text. Answers to all Chapter Review questions and all Chapter Test questions are also in the back of the text. I have made a special effort to provide interesting and meaningful word problems.

SPECIAL FEATURES
1. Each chapter has
   - a Chapter Summary of definitions and key ideas
   - a Chapter Review of questions similar to the exercises
   - a Chapter Test similar to a test the students can expect to take
2. There are more examples, more exercises, and more word problems.
3. The explanations are shorter and easier to read.
4. Calculators are used in special sections on practical applications.
5. Most sections include Practice Quizzes of 3–5 exercises to provide immediate classroom feedback to the instructor.

ADDITIONAL AIDS
An *Instructor's Guide* contains brief discussions of each chapter, the answers to the exercises numbered as multiples of four, and sample tests for each chapter.

The *Student Workbook* presents explanations of common arithmetic operations, extensive examples of those operations, and exercises with completely worked-out solutions.

ACKNOWLEDGMENTS
I would like to thank the editorial staff of D. C. Heath, which has given great service to this fourth edition. Pat Wright did another marvelous job of typing and proofreading the manuscript. All reviewers were helpful with their constructive and critical comments: Lawrence Brenton, Wayne State University; Kathryn P. Caraway, Charles S. Mott Community College; Betty Lou Field, Glendale Community College; Michael Gallo, Monroe Community College; John T. Gordon, Georgia State University; Kay Hudspeth, Pennsylvania State University; Michael R. Karelius, American River College; William Keils, San Antonio College; Richard Kuuttila, Macomb County Community College; Pamela E. Matthews, Chabot College; Eileen P. Neely, Northern Virginia Community College; Mark Phillips, Cypress College; Joe Prater, University of Southern Colorado; Sister M. Geralda Schaefer, Pan American University.

D. FRANKLIN WRIGHT

CONTENTS

4

DECIMAL NUMBERS AND PROPORTIONS   114

5

PERCENT   149

6

APPLICATIONS WITH CALCULATORS   172

7

MEASUREMENT: THE METRIC SYSTEM   203

8

NEGATIVE NUMBERS   237

9
SOLVING EQUATIONS   262

10
REAL NUMBERS AND GRAPHING LINEAR EQUATIONS   283

Appendix I
ANCIENT NUMERATION SYSTEMS   313

Appendix II
BASE TWO AND BASE FIVE   321

Appendix III

Appendix IV

1
WHOLE NUMBERS

1.1   THE DECIMAL SYSTEM (BASE TEN)

The term **numeral** refers to the symbols used to represent the idea of **number.** For the Romans, the numerals X, L, and C represented the same numbers as the numerals 10, 50, and 100 do for us. We have adopted the Hindu-Arabic system, invented about A.D. 800, because it allows us to add, subtract, multiply, and divide easier and faster than any of the ancient number systems, including the Roman system. (A discussion of several ancient number systems is included in Appendix I.)

The **decimal system** (deci means ten in Latin) is a **place value system** that uses ten digits, and the value of each place is a **power of ten** (1, 10, 100, 1000, and so on). Each power of ten is found by multiplying 10 times the previous power. Since 10 is the basis of the system, the decimal system is also known as the **base ten system.** (The word **base** will be discussed in more detail in Chapter 2.)

OUR SYSTEM OF REPRESENTING NUMBERS DEPENDS ON THREE THINGS

1.   The ten digits $\{0, 1, 2, 3, 4, 5, 6, 7, 8, 9\}$.

2.   The placement of the digits.

3.   The value of each place.

Consider the two numerals 40 and 400. The digit 4 has a different meaning (or value) in each of the numerals. Why? The symbol 4 is the same in each numeral, so the difference must be in the placement of the 4. In the symbol 40, the digit 4 is in the second place from the right, and the second place has a value of ten. In the symbol 400, the digit 4 is in the third place, and the third place has a value of one hundred.

We start (see Figure 1.1) with a point called a **decimal point.** Each place to the left of the decimal point has ten times the value of the place on its right. When we write a digit in a place, we mean that the value of the

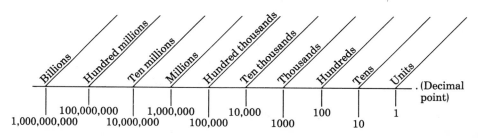

Figure 1.1

digit is to be multiplied by the value of the place. The value of any numeral is found by adding the results of multiplication of the digits by their place values. For example,

$$257 = 2(100) + 5(10) + 7(1) = 200 + 50 + 7 \qquad \text{(expanded notation)}$$

[NOTE: Writing a numeral next to a numeral in parentheses means to multiply.] Also, if the decimal point is not written, it is understood to be to the right of the rightmost digit.

To write a numeral in **expanded notation,** each digit is multiplied by the value of its place, and all the products (the result of multiplication) are added as in the following examples. The English word equivalents can then be read from the products. If a numeral has more than four digits, commas are placed to separate every three digits. Commas are used in the same manner in the word equivalents.

EXAMPLES

Write each of the following numerals in expanded notation and in its English word equivalent.

1. 573

$$573 = 5(100) + 7(10) + 3(1) \qquad \text{(expanded notation)}$$
$$= 500 + 70 + 3$$

five hundred seventy-three

2. 4862

$$4862 = 4(1000) + 8(100) + 6(10) + 2(1) \qquad \text{(expanded notation)}$$
$$= 4000 + 800 + 60 + 2$$

four thousand eight hundred sixty-two

3. 8007

$$8007 = 8(1000) + 0(100) + 0(10) + 7(1)$$
$$= 8000 + 0 + 0 + 7$$

eight thousand seven

4. 1,590,768

$$1,590,768 = 1(1,000,000) + 5(100,000) + 9(10,000) + 0(1000)$$
$$+ 7(100) + 6(10) + 8(1)$$
$$= 1,000,000 + 500,000 + 90,000 + 0 + 700 + 60 + 8$$

one million, five hundred ninety thousand, seven hundred sixty-eight

[NOTE: The word **and** does not appear in the English word equivalents, and it should **not** be said when reading a decimal numeral unless there are digits to the right of the decimal point. We will discuss this in more detail in Chapter 4.]

PRACTICE QUIZ	Write the following numerals in expanded notation and in their English word equivalents.	ANSWERS
	1. 512	**1.** 5(100) + 1(10) + 2(1) Five hundred twelve
	2. 6394	**2.** 6(1000) + 3(100) + 9(10) + 4(1) Six thousand three hundred ninety-four
	3. Write one hundred eighty thousand, five hundred forty-three as a decimal numeral.	**3.** 180,543

EXERCISES 1.1

Did you read the explanation and work through the examples before beginning these exercises?

Write the following decimal numerals in expanded notation and in their English word equivalents.

1. 37	**2.** 84	**3.** 98	**4.** 56
5. 122	**6.** 493	**7.** 821	**8.** 1976
9. 1892	**10.** 5496	**11.** 12,517	**12.** 42,100
13. 243,400	**14.** 891,540	**15.** 43,655	**16.** 99,999
17. 8,400,810	**18.** 5,663,701	**19.** 16,302,590	
20. 71,500,000	**21.** 83,000,605	**22.** 152,403,672	
23. 679,078,100	**24.** 4,830,231,010	**25.** 8,572,003,425	

Write the following numbers as decimal numerals.

26. seventy-six

27. one hundred thirty-two

28. five hundred eighty

29. three thousand eight hundred forty-two

30. two thousand five

31. one hundred ninety-two thousand, one hundred fifty-one

32. seventy-eight thousand, nine hundred two

33. twenty-one thousand, four hundred

34. thirty-three thousand, three hundred thirty-three

35. five million, forty-five thousand

36. five million, forty-five

37. ten million, six hundred thirty-nine thousand, five hundred eighty-two

38. two hundred eighty-one million, three hundred thousand, five hundred one

39. five hundred thirty million, seven hundred

40. seven hundred fifty-eight million, three hundred fifty thousand, sixty

41. ninety million, ninety thousand, ninety

42. eighty-two million, seven hundred thousand

43. one hundred seventy-five million, two

44. thirty-six   **45.** seven hundred fifty-seven

1.2 ADDITION

The letter W is used to represent all whole numbers: $0, 1, 2, 3, 4, 5, \ldots$. The three dots indicate that the pattern of numbers is to continue without end. Thus, 13, 92, and 10,000,000 are all whole numbers.

Addition with whole numbers is indicated either by writing the numbers horizontally with a plus sign ($+$) between them or by writing the numbers vertically in columns with instructions to add.

$$6 + 23 + 17 \qquad \text{Add} \quad \begin{array}{r} 6 \\ 23 \\ \underline{17} \end{array}$$

The numbers being added are called **addends,** and the result of the addition is called the **sum.**

$$\text{Add} \quad \begin{array}{r} 6 \\ 23 \\ \underline{17} \\ 46 \end{array} \quad \begin{array}{l} \text{addend} \\ \text{addend} \\ \text{addend} \\ \text{sum} \end{array}$$

Be sure to keep the digits aligned (in column form) so you will be adding units to units, tens to tens, and so on. Neatness is a necessity in mathematics.

TABLE 1.1   BASIC ADDITION FACTS

+	0	1	2	3	4	5	6	7	8	9
0	0	1	2	3	4	5	6	7	8	9
1	1	2	3	4	5	6	7	8	9	10
2	2	3	4	5	6	7	8	9	10	11
3	3	4	5	6	7	8	9	10	11	12
4	4	5	6	7	8	9	10	11	12	13
5	5	6	7	8	9	10	11	12	13	14
6	6	7	8	9	10	11	12	13	14	15
7	7	8	9	10	11	12	13	14	15	16
8	8	9	10	11	12	13	14	15	16	17
9	9	10	11	12	13	14	15	16	17	18

To be able to add with speed and accuracy, you must **memorize** the basic addition facts, which are given in Table 1.1. Practice all the combinations so you can give the answers immediately. Concentrate.

As an exercise to help find out which combinations you need special practice with, write, on a piece of paper in mixed order, all one hundred possible combinations to be added. Perform the operations as quickly as possible. Then, using the table, check to find the ones you missed. Study these frequently until you are confident that you know them as well as all the others.

Your adding speed can be increased if you learn to look for combinations of digits that total ten.

EXAMPLE

To add the following numbers, we note the combinations that total ten and find the sums quickly.

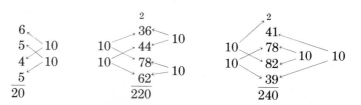

When two numbers are added, the order of the numbers does not matter. That is, $4 + 9 = 13$ and $9 + 4 = 13$. Also,

$$\frac{\begin{array}{r}5\\7\end{array}}{12} \quad \text{and} \quad \frac{\begin{array}{r}7\\5\end{array}}{12}$$

By looking at Table 1.1 again, we can see that reversing the order of any two addends will not change their sum. This fact is called the **commutative property of addition.** We can state this property using letters to represent whole numbers.

COMMUTATIVE
PROPERTY OF
ADDITION

If a and b are two whole numbers, then $a + b = b + a$.

To add three numbers, such as $6 + 3 + 5$, we can add only two at a time, then add the third. Which two should we add first? The answer is that the sum is the same either way:

$$6 + 3 + 5 = (6 + 3) + 5 = 9 + 5 = 14$$

and $$6 + 3 + 5 = 6 + (3 + 5) = 6 + 8 = 14$$

We can write

$$8 + 4 + 7 = (8 + 4) + 7 = 8 + (4 + 7)$$

This illustrates the **associative property of addition.**

ASSOCIATIVE
PROPERTY OF
ADDITION

If a, b, and c are whole numbers, then

$$a + b + c = (a + b) + c = a + (b + c)$$

Another property of addition with whole numbers is addition with 0. Whenever we add 0 to a number, the result is the original number:

$$8 + 0 = 8, \quad 0 + 13 = 13, \quad \text{and} \quad 38 + 0 = 38$$

Zero (0) is called the **additive identity** or the **identity element for addition.**

ADDITIVE
IDENTITY

If a is a whole number, then there is a unique whole number 0 with the property that $a + 0 = a$.

EXAMPLES

1. $(15 + 20) + 6 = 15 + (20 + 6)$ associative property

2. $(15 + 20) + 6 = (20 + 15) + 6$ commutative property

3. $17 + 0 = 17$ additive identity

Note that Example 1 shows a change in the **grouping** (association) of the numbers, while Example 2 shows a change in the **order** of two numbers.

PRACTICE QUIZ	Which property of addition is illustrated?	ANSWERS
	1. $12 + 0 = 12$	1. Additive identity
	2. $15 + 3 = 3 + 15$	2. Commutative property
	Find the following sums.	

	3. 57
	    98 4. 36
	           78
	           89
		3. 155 4. 203

EXERCISES 1.2

Did you read the explanation and work through the examples before beginning these exercises?

Do the following exercises mentally and write only the answers.

1. $6 + (3 + 7)$ 2. $(4 + 5) + 6$ 3. $(2 + 3) + 8$

4. $(2 + 6) + (4 + 5)$ 5. $8 + (3 + 4) + 6$ 6. $9 + (8 + 3)$

7. $9 + 2 + 8$ 8. $7 + (6 + 3)$ 9. $(2 + 3) + 7$

10. $8 + 7 + 2$ 11. $9 + 6 + 3$ 12. $4 + 3 + 6$

13. $4 + 4 + 4$ 14. $9 + 1 + 5 + 6$ 15. $5 + 3 + 7 + 2$

16. $3 + 6 + 2 + 8$ 17. $5 + 4 + 6 + 4 + 8$

18. $6 + 2 + 9 + 1$ 19. $4 + 3 + 6 + 6 + 2$

20. $8 + 8 + 7 + 6 + 3$

Show that the following statements are true by performing the addition. State which property of addition is being illustrated.

21. $9 + 3 = 3 + 9$ 22. $8 + 7 = 7 + 8$

23. $4 + (5 + 3) = (4 + 5) + 3$ 24. $4 + 8 = 8 + 4$

25. $2 + (1 + 6) = (2 + 1) + 6$ 26. $(8 + 7) + 3 = 8 + (7 + 3)$

27. $9 + 0 = 9$ 28. $(2 + 3) + 4 = (3 + 2) + 4$

29. $7 + (6 + 0) = 7 + 6$ 30. $8 + 20 + 1 = 8 + 21$

Copy the following problems and add.

31.	65	**32.**	24	**33.**	73	**34.**	165	**35.**	876
	43		78		68		276		279
	54		95		98		394		143

36.	268	**37.**	981	**38.**	2112	**39.**	114	**40.**	1403
	93		146		147		5402		7010
	74		92		904		710		622
	192		17		1005		643		29

41.	213,116	**42.**	21,442	**43.**	438,966	**44.**	123,456
	116,018		32,462		1,572,486		456,123
	722,988		564,792		327,462		879,282
	24,336		801,801		181,753		617,500
	526,968		43,433		90,000		740,765

45. Mr. Jones kept the mileage records indicated in the table shown here. How many miles did he drive during the six months?

MONTH	MILEAGE
Jan.	546
Feb.	378
Mar.	496
Apr.	357
May	503
June	482

46. The Modern Products Corp. showed profits as indicated in the table for the years 1974–1977. What were the company's total profits for the years 1974–1977?

YEAR	PROFITS
1974	$1,078,416
1975	1,270,842
1976	2,000,593
1977	1,963,472

47. During six years of college, including two years of graduate school, Fred estimated his expenses each year as $2035; $2842; $2786; $3300; $4000; $3500. What were his total expenses for six years of schooling? [NOTE: He had some financial aid.]

48. Apple County has the following items budgeted: highways, \$270,455; salaries, \$95,479; maintenance, \$127,220. What is the county's total budget for these three items?

49. The following numbers of students at South Junior College are enrolled in mathematics courses: 303 in arithmetic; 476 in algebra; 293 in trigonometry; 257 in college algebra; 189 in calculus. Find the total number of students taking math.

50. In one year, the High Price Manufacturing Co. made 2476 refrigerators; 4217 gas stoves; 3947 electric stoves; 9576 tractors; 11,872 electric fans; 1742 air conditioners. What was the total number of appliances High Price produced that year?

1.3   SUBTRACTION

A long-distance runner who is 25 years old had run 17 miles of his usual 19-mile workout when a storm forced him to quit for the day. How many miles short was he of his usual daily training?

What *thinking* did you do to answer the question? You may have thought something like this: "Well, I don't need to know how old the runner is to answer the question, so his age is just extra information. Since he had already run 17 miles, I need to know what to add to 17 miles to get 19 miles. Since $17 + 2 = 19$, he was 2 miles short of his usual workout."

In this problem, the sum of two addends was given, and only one of the addends was given. The other addend was the unknown quantity.

$$\underset{\text{addend}}{17} \; + \; \underset{\text{missing addend}}{\square} \; = \; \underset{\text{sum}}{19} \qquad 17 + \underline{} \; = 19$$

As you may know, this kind of addition problem is called **subtraction** and can be written:

$$\underset{\text{sum}}{19} \; - \; \underset{\text{addend}}{17} \; = \; \underset{\substack{\text{missing addend}\\\text{(or difference)}}}{\square} \qquad \text{(Read: "19 minus 17 equals blank.")}$$

Or
$$\begin{array}{r} 19 \\ -17 \\ \hline \square \end{array} \quad \begin{array}{l} \text{sum} \\ \text{addend} \\ \text{difference or missing addend} \end{array}$$

In other words, subtraction is a reverse addition, and the missing addend is called the **difference** between the sum and one addend.

Perform the following subtraction mentally: $17 - 8 = \square$. You should think, "8 plus what number gives 17? Since $8 + 9 = 17$, we have $17 - 8 = 9$."

<u>EXAMPLES</u> 1. Find the difference for

$$\begin{array}{r} \overset{6}{75} \\ -48 \\ \hline \mathit{27} \end{array}$$

Since thinking of a number to add to 48 to get 75 is difficult, we resort to place value and write the numbers in expanded form.

$$\begin{array}{r} 75 = 70 + 5 = 60 + 15 \\ -48 = 40 + 8 = 40 + 8 \\ \hline 20 + 7 = 27 \end{array}$$

(10 is "borrowed" from 70 since no whole number can be added to 8 to get 5.)

The above procedure is commonly written as

$$\begin{array}{r} \overset{6}{} \\ \overset{6}{7}{}^{1}5 \\ -48 \\ \hline 27 \end{array}$$

(Can you see that $\overset{6}{7}{}^{1}5$ is shorthand for the expanded form $60 + 15$?)

2. Find the difference for $482 - 195$ using place value.

$$\begin{array}{r} 482 = 400 + 80 + 2 = 400 + 70 + 12 = 300 + 170 + 12 \\ -195 = 100 + 90 + 5 = 100 + 90 + 5 = 100 + 90 + 5 \\ \hline 200 + 80 + 7 = 287 \end{array}$$

Notice that 10 is "borrowed" from 80, and then 100 is "borrowed" from 400. Do you know why these "borrowings" are necessary?
In shorthand,

$$\begin{array}{r} 3\,17 \\ \cancel{4}\,\cancel{8}{}^{}2 \\ -195 \\ \hline 287 \end{array}$$

(You should realize that the numbers are not written in expanded form here but should be visualized that way.)

3. Find the difference for $500 - 132$.

$$\begin{array}{r} 500 = 490 + 10 = 400 + 90 + 10 \\ -132 = 130 + 2 = 100 + 30 + 2 \\ \hline 300 + 60 + 8 = 368 \end{array}$$

Or

$$\begin{array}{r} 500 = 490 + 10 \\ -132 = 130 + 2 \\ \hline 360 + .8 = 368 \end{array}$$

Or

$$\begin{array}{r} 49 \\ \cancel{5}\,\cancel{0}{}^{1}0 \\ -132 \\ \hline 368 \end{array}$$

4. The cost of repairing Ed's used TV set was going to be $230 for parts (including tax) plus $50 for labor. To buy a new set, he was going to pay $400 plus $24 in tax and the dealer was going to pay him $75 for his old set. How much more would Ed have to pay for a new set than to have his old set repaired?

Solution:

Used Set	New Set	Difference
$230 parts	$400 cost	$349
+50 labor	+24 tax	−280
$280 total	$424	$ 69
	−75 trade-in	
	$349 total	

Ed would pay $69 more for the new set than for having his old set repaired. What would you do?

PRACTICE QUIZ	Find the following differences.	ANSWERS
	1. 83 −54	1. 29
	2. 600 −368	2. 232
	3. 7856 −6397	3. 1459

EXERCISES 1.3

Subtract. Do as many problems as you can mentally.

1. $8 - 5$	2. $19 - 6$	3. $14 - 14$	4. $17 - 9$
5. $20 - 11$	6. $17 - 0$	7. $17 - 8$	8. $16 - 16$
9. $11 - 6$		10. $13 - 7$	

11.   17
     −17
	12. 42
	     −31
		13. 89
		     −76
			14. 53
			     −33
				15. 47
				     −27

16.   96
     −27
	17. 23
	     −18
		18. 74
		     −29
			19. 61
			     −48
				20. 52
				     −27

21.   126
     −32
	22. 174
	     −48
		23. 347
		     −129
			24. 256
			     −118	

25.   692
     −217
	26. 543
	     −167
		27. 900
		     −307
			28. 603
			     −208	

29. $\begin{array}{r} 474 \\ -286 \end{array}$ 30. $\begin{array}{r} 657 \\ -179 \end{array}$ 31. $\begin{array}{r} 7843 \\ -6274 \end{array}$ 32. $\begin{array}{r} 6793 \\ -5827 \end{array}$

33. $\begin{array}{r} 4376 \\ -2808 \end{array}$ 34. $\begin{array}{r} 3275 \\ -1744 \end{array}$ 35. $\begin{array}{r} 3546 \\ -3546 \end{array}$ 36. $\begin{array}{r} 4900 \\ -3476 \end{array}$

37. $\begin{array}{r} 5070 \\ -4376 \end{array}$ 38. $\begin{array}{r} 8007 \\ -2136 \end{array}$ 39. $\begin{array}{r} 4065 \\ -1548 \end{array}$ 40. $\begin{array}{r} 7602 \\ -2985 \end{array}$

41. $\begin{array}{r} 7,085,076 \\ -4,278,432 \end{array}$ 42. $\begin{array}{r} 6,543,222 \\ -2,742,663 \end{array}$ 43. $\begin{array}{r} 4,000,000 \\ -2,993,042 \end{array}$

44. $\begin{array}{r} 8,000,000 \\ -647,561 \end{array}$ 45. $\begin{array}{r} 6,000,000 \\ -328,989 \end{array}$

46. What number should be added to 978 to get a sum of 1200?

47. What number should be added to 860 to get a sum of 1000?

48. If the sum of two numbers is 537 and one of the numbers is 139, what is the other number?

49. A man is 36 years old, and his wife is 34 years old. Including high school, they have each attended school for 16 years. How many total years of schooling have they had?

50. Basketball Team A has twelve players and won its first three games by the following scores: 84 to 73, 97 to 78, and 101 to 63. Team B has ten players and won its first three games by 76 to 75, 83 to 70, and 94 to 84. What is the difference between the total of the differences of Team A's scores and those of its opponents and the total of the differences of Team B's scores and those of its opponents?

51. The Kingston Construction Co. made a bid of $7,043,272 to build a stretch of freeway, but the Beach City Construction Co. made a lower bid of $6,792,868. How much lower was the Beach City bid?

52. Two landscaping companies made bids on the landscaping of a new apartment complex. Company A bid $550,000 for materials and plants and $225,000 for labor. Company B bid $600,000 for materials and plants and $182,000 for labor. Which company had the lower total bid? How much lower?

53. In June, Ms. White opened a checking account and deposited $1342, $238, $57, and $486. She also wrote checks for $132, $76, $25, $42, $480, $90, $17, and $327. What was her balance at the end of June?

54. A manufacturing company had assets of $5,027,479, which included $1,500,000 in real estate. The liabilities were $4,792,023. By how much was the company "in the black"?

55. A man and woman sold their house for $135,000. They paid the realtor $8100, and other expenses of the sale came to $800. If they owed the bank $87,000 for the mortgage, what were their net proceeds from the sale?

56. A woman bought a condominium for a price of $150,000. She also had to pay other expenses of $750. If the local Savings and Loan agreed to give her a mortgage loan of $105,500, how much cash did she need to make the purchase?

57. Junior bought a red sports car with a sticker price of $10,000. But the salesman added $1200 for taxes, license, and extras. The bank agreed to give Junior a loan of $7500 on the car. How much cash did Junior need to buy the car?

58. In pricing a four-door car, Pat found she would have to pay a base price of $9500 plus $570 in taxes and $250 for license fees. For a two-door of the same make, she would pay a base price of $8700 plus $522 in taxes and $230 for license fees. Including all expenses, how much cheaper was the two-door model?

1.4 BASIC FACTS OF MULTIPLICATION

Repeated addition of the same addend can be shortened considerably by learning to multiply. Here a raised dot between two numbers indicates multiplication. Thus,

$$7 + 7 + 7 + 7 = 4 \cdot 7 = 28$$
$$279 + 279 + 279 + 279 = 4 \cdot 279 = 1116$$

The repeated addend (279) and the number of times it is used (4) are both called **factors,** and the sum is called the **product** of the two factors.

$$279 + 279 + 279 + 279 = \underbrace{4}_{\text{factor}} \cdot \underbrace{279}_{\text{factor}} = \underbrace{1116}_{\text{product}}$$

Several notations can be used to indicate multiplication. In this text, we will use the raised dot and parentheses much of the time.

(a) $4 \cdot 279$ (b) $4(279)$ (c) $(4)279$

(d) $(4)(279)$ (e) 4×279 (f) $\begin{array}{r} 279 \\ \times 4 \\ \hline \end{array}$

(g) Directions are given:

Multiply $\begin{array}{r} 279 \\ 4 \\ \hline \end{array}$

To change a multiplication problem to a repeated addition problem every time we are to multiply two numbers would be ridiculous. For example, 48 · 137 would mean using 137 as an addend 48 times. The first step in learning the multiplication process is to **memorize** the basic multiplication facts in Table 1.2. The factors in the table are only the digits 0 through 9. Using other factors involves the place value concept, as we shall see.

TABLE. 1.2   BASIC MULTIPLICATION FACTS

·	0	1	2	3	4	5	6	7	8	9
0	**0**	0	0	0	0	0	0	0	0	0
1	0	1	2	3	4	5	6	7	8	9
2	0	2	4	6	8	10	12	14	16	18
3	0	3	6	9	12	15	18	21	24	27
4	0	4	8	12	**16**	20	24	28	32	36
5	0	5	10	15	20	**25**	30	35	40	45
6	0	6	12	18	24	30	**36**	42	48	54
7	0	7	14	21	28	35	42	**49**	56	63
8	0	8	16	24	32	40	48	56	**64**	72
9	0	9	18	27	36	45	54	63	72	**81**

If you have difficulty with **any** of the basic facts in the table, write all the possible combinations in a mixed-up order on a sheet of paper. Write the products down as quickly as you can and then find the ones you missed. Practice these in your spare time until you are sure you know them.

The table is a mirror image of itself on either side of the main diagonal (the numbers 0, 1, 4, 9, 16, 25, 36, 49, 64, 81). This indicates that multiplication is **commutative.** For example, we note that

$$5 \cdot 3 = 3 \cdot 5, \qquad 6 \cdot 2 = 2 \cdot 6, \qquad 7 \cdot 1 = 1 \cdot 7, \qquad 8 \cdot 0 = 0 \cdot 8$$

COMMUTATIVE
PROPERTY OF
MULTIPLICATION
    If a and b are whole numbers, then $a \cdot b = b \cdot a$.

A close look at Table 1.2 also shows that multiplication by 0 gives 0 and that multiplication by 1 gives the number being multiplied. These two properties are called the **zero factor law** and **multiplicative identity,** respectively.

EXAMPLES	Zero Factor Law	Multiplicative Identity
	$0 \cdot 7 = 0$	$1 \cdot 7 = 7$
	$9 \cdot 0 = 0$	$9 \cdot 1 = 9$
	$83 \cdot 0 = 0$	$83 \cdot 1 = 83$
	$0 \cdot 654 = 0$	$1 \cdot 654 = 654$

We state these two properties formally.

ZERO FACTOR LAW   If a is a whole number, then $a \cdot 0 = 0$.

MULTIPLICATIVE IDENTITY   If a is a whole number, then there is a unique whole number 1 with the property that $a \cdot 1 = a$.

Only two numbers can be multiplied at a time. Therefore, if three or more factors are to be multiplied, we have to decide which two to **group** or **associate** together first. Does it matter? The answer is no. For example,

$$2 \cdot 3 \cdot 7 = (2 \cdot 3) \cdot 7 = 6 \cdot 7 = 42$$

and

$$2 \cdot 3 \cdot 7 = 2 \cdot (3 \cdot 7) = 2 \cdot 21 = 42$$

Also,

$$9 \cdot 2 \cdot 5 = (9 \cdot 2) \cdot 5 = 18 \cdot 5 = 90$$

and

$$9 \cdot 2 \cdot 5 = 9 \cdot (2 \cdot 5) = 9 \cdot 10 = 90$$

These examples illustrate the **associative property of multiplication.**

ASSOCIATIVE PROPERTY OF MULTIPLICATION   If a, b, and c are whole numbers, then $a \cdot b \cdot c = a(b \cdot c) = (a \cdot b)c$.

PRACTICE QUIZ	Which property of multiplication is illustrated?	ANSWERS
	1. $5 \cdot (3 \cdot 7) = (5 \cdot 3) \cdot 7$	1. Associative property
	2. $32 \cdot 0 = 0$	2. Zero factor law
	3. $6 \cdot 8 = 8 \cdot 6$	3. Commutative property

EXERCISES 1.4

Did you read the material in the text and study the examples before starting the exercises?

Do the following problems mentally and write only the answers.

1. $8 \cdot 9$ 2. $7 \cdot 6$ 3. $8(6)$ 4. $9(7)$ 5. $6 \cdot 5$

6. $5 \cdot 9$ 7. $3(9)$ 8. $8(5)$ 9. $6(4)$ 10. $7(4)$

11. $8(7)$ 12. $3(6)$ 13. $0 \cdot 3$ 14. $5 \cdot 0$ 15. $1 \cdot 9$

16. $4 \cdot 1$ 17. $9 \cdot 0$ 18. $7 \cdot 0$ 19. $6 \cdot 1$ 20. $8 \cdot 1$

21. $2 \cdot 7 \cdot 3$ 22. $4 \cdot 3 \cdot 5$ 23. $2 \cdot 5 \cdot 6$ 24. $5 \cdot 7 \cdot 1$

25. $3 \cdot 2 \cdot 5$ 26. $6 \cdot 1 \cdot 4$ 27. $5 \cdot 1 \cdot 8$ 28. $3 \cdot 2 \cdot 9$

29. $8(6)(2)$ 30. $3(7)(5)$ 31. $4(3)(6)$ 32. $6(1)(9)$

33. $0 \cdot 3 \cdot 96$ 34. $5(0)(42)$ 35. $16(0)(82)(193)$

For each property listed, give two examples that illustrate the property with whole numbers.

36. Associative property of multiplication

37. Associative property of addition

38. Commutative property of addition

39. Commutative property of multiplication

40. Multiplicative identity

41. Zero factor law

42. Additive identity

43. In your own words, describe the meaning of the term **factor.**

44. Fill in the missing numbers in the chart. Five is added to the given number. The sum is then doubled, and ten is subtracted from the product. The answer is written in the last column. [HINT: To fill in the last two rows, you must think backwards.]

GIVEN NUMBER	ADD 5	DOUBLE	SUBTRACT 10
2	7	14	4
0	5	?	?
1	6	12	?
7	?	?	?
?	?	?	16
?	?	?	10

45. Fill in the missing numbers in the chart. Five is added to the given number. The sum is then tripled (multiplied by 3), and fifteen is sub-

tracted from the product. The answer is written in the last column. [HINT: To fill in the last two rows, you must think backwards.]

GIVEN NUMBER	ADD 5	TRIPLE	SUBTRACT 15
2	7	21	6
1	6	18	?
7	12	?	?
8	?	?	?
?	?	?	15
?	?	?	18

1.5   MULTIPLICATION BY POWERS OF TEN

A **power** of any number is either 1 or that number multiplied by itself one or more times. We will discuss this idea in more detail in Chapter 2. Some of the powers of ten are: 1, 10, 100, 1000, 10,000, 100,000, and so on.

$$1$$
$$10$$
$$10 \cdot 10 = 100$$
$$10 \cdot 10 \cdot 10 = 1000$$
$$10 \cdot 10 \cdot 10 \cdot 10 = 10,000$$
$$10 \cdot 10 \cdot 10 \cdot 10 \cdot 10 = 100,000$$

and so on

Multiplication by powers of ten is useful in explaining multiplication with whole numbers in general, as we will do in Section 1.6. Such multiplication should be done mentally and quickly. The following examples illustrate an important pattern:

$$6 \cdot 1 = 6$$
$$6 \cdot 10 = 60$$
$$6 \cdot 100 = 600$$
$$6 \cdot 1000 = 6000$$

If one of two whole number factors is 1000, the product will be the other factor with three zeros (000) written to the right of it. Two zeros (00) are written to the right of the other factor when multiplying by 100, and one zero (0) is written when multiplying by 10. Will multiplication by one million (1,000,000) result in writing six zeros to the right of the other factor? The answer is yes.

Many products can be found mentally by using the properties of multiplication and the techniques of multiplying by powers of ten. The processes are written out in the following examples, but they can easily be done mentally with practice.

EXAMPLES

1. $6 \cdot 90 = 6(9 \cdot 10) = (6 \cdot 9)10 = 54 \cdot 10 = 540$

2. $3 \cdot 400 = 3(4 \cdot 100) = (3 \cdot 4)100 = 12 \cdot 100 = 1200$

3. $2 \cdot 300 = 2(3 \cdot 100) = (2 \cdot 3)100 = 6 \cdot 100 = 600$

4. $6 \cdot 700 = 6(7 \cdot 100) = (6 \cdot 7)100 = 42 \cdot 100 = 4200$

5. $40 \cdot 30 = (4 \cdot 10)(3 \cdot 10) = (4 \cdot 3)(10 \cdot 10) = 12 \cdot 100 = 1200$

6. $50 \cdot 700 = (5 \cdot 10)(7 \cdot 100) = (5 \cdot 7)(10 \cdot 100) = 35 \cdot 1000$
      $= 35,000$

7. $200 \cdot 800 = (2 \cdot 100)(8 \cdot 100) = (2 \cdot 8)(100 \cdot 100) = 16 \cdot 10,000$
      $= 160,000$

8. $7000 \cdot 9000 = (7 \cdot 1000)(9 \cdot 1000) = (7 \cdot 9)(1000 \cdot 1000)$
      $= 63 \cdot 1,000,000 = 63,000,000$

[NOTE: To find the product in each example, the nonzero digits are multiplied, and the appropriate number of zeros is written.]

EXERCISES 1.5

Use the techniques of multiplying by the powers of ten to find the following products mentally.

1. $25 \cdot 10$	**2.** $76 \cdot 100$	**3.** $47 \cdot 1000$	**4.** $18 \cdot 10$
5. $72 \cdot 10$	**6.** $13 \cdot 1$	**7.** $50 \cdot 60$	**8.** $90 \cdot 80$
9. $20 \cdot 20$	**10.** $60 \cdot 60$	**11.** $30 \cdot 40$	**12.** $70 \cdot 80$
13. $90 \cdot 70$	**14.** $300 \cdot 30$	**15.** $200 \cdot 20$	**16.** $500 \cdot 70$
17. $500 \cdot 30$	**18.** $120 \cdot 30$	**19.** $130 \cdot 40$	**20.** $200 \cdot 60$
21. $200 \cdot 80$	**22.** $300 \cdot 600$	**23.** $100 \cdot 100$	**24.** $100 \cdot 50$
25. $3000 \cdot 20$	**26.** $500 \cdot 50$	**27.** $400 \cdot 30$	**28.** $50 \cdot 200$
29. $40 \cdot 6000$	**30.** $2000 \cdot 400$	**31.** $80 \cdot 600$	
32. $3000 \cdot 5000$	**33.** $20,000 \cdot 30$	**34.** $4000 \cdot 4000$	
35. $70 \cdot 9000$	**36.** $800 \cdot 4000$		

1.6   MULTIPLICATION OF WHOLE NUMBERS

To multiply $3(70 + 2)$, we can add first, then multiply

$$3(70 + 2) = 3(72) = 216$$

But we can also multiply first, then add in the following manner:

$$3(70 + 2) = 3 \cdot 70 + 3 \cdot 2$$
$$= 210 + 6$$
$$= 216$$

The result is the same. This is an example of the **distributive property of multiplication over addition.**

DISTRIBUTIVE
PROPERTY OF
MULTIPLICATION
OVER ADDITION

If a, b, and c are whole numbers, then $a(b + c) = a \cdot b + a \cdot c$.

We can use the distributive property and our skills with multiplication by powers of ten from Section 1.5 to explain the procedure for multiplying two whole numbers.

EXAMPLES

1. Multiply $4 \cdot 68$.

(a) $4 \cdot 68 = 4(60 + 8)$
$\qquad = 4 \cdot 60 + 4 \cdot 8$ distributive property
$\qquad = 240 + 32$
$\qquad = 272$

(b)   In another form,

$$
\begin{array}{cc}
60 + 8 & \quad 68 \\
\underline{\phantom{60 + {}}4} & \quad \underline{4} \\
\underline{240 + 32} = \underline{272} & \quad 32 \\
\text{partial products} \quad \text{product} & \quad 240 \\
& \quad \underline{272}
\end{array}
$$

$$240 + 32 = 272$$
partial products   product

$\left.\begin{array}{r}32\\240\end{array}\right\}$ partial products

272   product

or

2. Multiply $37 \cdot 42$.

(a) $37 \cdot 42 = (30 + 7)(40 + 2)$
$\qquad = (30 + 7) \cdot 40 + (30 + 7) \cdot 2$ distributive property
$\qquad = 30 \cdot 40 + 7 \cdot 40 + 30 \cdot 2 + 7 \cdot 2$ distributive property twice
$\qquad = 1200 + 280 + 60 + 14$
$\qquad = 1554$

(b)   In another form,

$$
\begin{array}{r}
40 + 2 \\
30 + 7 \\
\hline
280 + 14 \\
\end{array}
\qquad \text{or}
$$

$$
\begin{array}{r}
280 + 14 \\
1200 + \;\;60 \\
\hline
1200 + 340 + 14 = 1554 \\
\end{array}
$$

3(70+2)

3×(72)

42	factor
37	factor
14	$(7 \cdot 2 = 14)$
280	$(7 \cdot 40 = 280)$
60	$(30 \cdot 2 = 60)$
1200	$(30 \cdot 40 = 1200)$
1554	product

These examples show the underlying structure of multiplication. We can speed up the process considerably by noting some key relationships.

Writing Partial Products

$$
\begin{array}{r}
56 \\
3 \\
\hline
18 \quad (3 \cdot 6 = 18) \\
150 \quad (3 \cdot 50 = 150) \\
\hline
168 \\
\end{array}
$$

Shorter Method

1 ⟵ 1 carried from 18

$$
\begin{array}{r}
56 \\
3 \\
\hline
168 \\
\end{array}
$$

From 18, write 8 below the 3 and carry 10 by writing 1 above the 5. Then multiply $3 \cdot 5$ and add the 1.

The shorter method is the one most people use and is recommended. We illustrate the two methods again with larger numbers.

Writing Partial Products

$$
\begin{array}{r}
63 \\
48 \\
\hline
24 \quad (8 \cdot 3 = 24) \\
480 \quad (8 \cdot 60 = 480) \\
120 \quad (40 \cdot 3 = 120) \\
2400 \quad (40 \cdot 60 = 2400) \\
\hline
3024 \\
\end{array}
$$

Shorter Method

1 carried from 120

2 ⟵ 2 carried from 24

$$
\begin{array}{r}
{}^1 2 \\
63 \\
48 \\
\hline
504 \\
252 \\
\hline
3024 \\
\end{array}
$$

(2 is moved to the tens column because it is from $40 \cdot 3 = 120$.)

Either method is correct and may be used successfully.

Suppose you want to multiply 2000 by 423. Would you write

$$
\begin{array}{r}
2000 \\
423 \\
\hline
6000 \\
4000 \\
8000 \\
\hline
846000 \\
\end{array}
\qquad \text{or} \qquad
\begin{array}{r}
423 \\
2000 \\
\hline
000 \\
000 \\
000 \\
846 \\
\hline
846000 \\
\end{array}
$$

Both answers are correct; neither technique is wrong. However, writing all the 0's is a waste of time, so knowledge about powers of ten is helpful. We can write

$$
\begin{array}{r}
423 \\
2000 \\
\hline
846000
\end{array}
$$

We know $2000 = 2 \cdot 1000$, so we are simply multiplying $423 \cdot 2$, then the result by 1000.

EXAMPLES

1. Multiply $596 \cdot 3000$.

$$
\begin{array}{r}
596 \\
3000 \\
\hline
1,788,000
\end{array}
$$

2. Multiply $265 \cdot 15,000$.

$$
\begin{array}{r}
265 \\
15,000 \\
\hline
1\ 325\ 000 \\
2\ 65 \\
\hline
3,975,000
\end{array}
$$

PRACTICE QUIZ	Find the following products.	ANSWERS
	1. $\begin{array}{r} 18 \\ 24 \\ \hline \end{array}$	**1.** 432
	2. $\begin{array}{r} 300 \\ 500 \\ \hline \end{array}$	**2.** 150,000
	3. $\begin{array}{r} 129 \\ 39 \\ \hline \end{array}$	**3.** 5031

EXERCISES 1.6

Please read the explanations and work through the examples before starting these exercises.

Find the following products by writing in all the partial products.

1. 56	**2.** 27	**3.** 48	**4.** 65	**5.** 43	**6.** 72
4	6	9	5	8	6

7. 91
5

8. 39
2

9. 84
3

10. 95
8

11. 42
56

12. 25
33

13. 15
22

14. 29
41

15. 67
36

16. 54
27

17. 48
20

18. 93
30

19. 83
85

20. 96
62

Multiply each of the following.

21. 17
32

22. 28
91

23. 20
44

24. 16
26

25. 25
15

26. 93
47

27. 24
86

28. 72
65

29. 12
13

30. 81
36

31. 126
41

32. 232
76

33. 114
25

34. 72
106

35. 207
143

36. 420
104

37. 200
49

38. 849
205

39. 673
186

40. 192
467

Multiply, using your knowledge of powers of ten.

41. 52
600

42. 72
930

43. 76
5000

44. 500
8000

45. 68
7300

46. 320
4700

47. 41
5300

48. 157
6000

49. 48
5200

50. 39
23,000

51. Find the sum of eighty-three and two hundred seventy-six. Find the difference between ninety-four and seventy-five. Find the product of the sum and the difference.

52. Find the sum of forty-three and sixty-six. Find the sum of one hundred and two hundred. Find the product of the sums.

53. If your salary is $1300 per month and you are to get a raise once a year of $130 per month and you love your work, what will you earn over a two-year period? Over a five-year period?

54. If you rent an apartment with 3 bedrooms for $650 per month and you know the rent will increase once a year by $30 per month, what will you pay in rent over a three-year period? Over a five-year period?

55.   If you drive at 55 miles per hour for 5 hours, how far will you drive? If you drive at 50 miles per hour in a new four-door car for 4 hours, how far will you drive?

56.   Your company bought 18 new cars at a price of $9750 per car. What did the company pay for the new cars? Each one had an air conditioner.

1.7   DIVISION WITH WHOLE NUMBERS

We know that $6 \cdot 10 = 60$ and that 6 and 10 are **factors** of 60. They are also called **divisors.** The process of **division** can be thought of as the reverse of multiplication. We say that 60 **divided by** 6 is 10 (or $60 \div 6 = 10$) because $6 \cdot 10 = 60$. Also, $60 \div 10 = 6$ because $10 \cdot 6 = 60$.

Division does not always involve factors (or exact divisors). In these cases, division can be thought of as repeated subtraction. (Remember that multiplication can be thought of as repeated addition, Section 1.4.) For example, if we want to divide 185 by 7, we are asking, "How many 7's are in 185?" and we can subtract 7 again and again until there is a number left that is less than 7.

$7\overline{)185}$ is the same as $185 \div 7$.

$$
\begin{array}{r}
7\overline{)185} \\
-140 \leftarrow \text{Subtract 20 sevens} \\
\hline
45 \\
-42 \leftarrow \text{Subtract } 6 \text{ sevens} \\
\hline
3 \quad\quad 26 \text{ sevens total} \\
\uparrow \quad\quad\quad \uparrow \\
\text{remainder} \quad \text{quotient}
\end{array}
\qquad \text{or} \qquad
\begin{array}{r}
26 \leftarrow \text{quotient} \\
\overline{6} \\
20 \\
\text{divisor} \rightarrow 7\overline{)185} \leftarrow \text{dividend} \\
-140 \\
\hline
45 \\
-42 \\
\hline
3 \leftarrow \text{remainder}
\end{array}
$$

185 is called the **dividend.** 7 is called the **divisor.** 26 is called the **quotient.** 3 is called the **remainder.**

Check:

$$
\begin{array}{cc}
26 & 182 \\
\times 7 & +3 \\
\hline
182 & 185
\end{array}
$$

Division is checked by multiplying the quotient and divisor and then adding the remainder. The result should be the dividend.

EXAMPLE

Find 257 ÷ 6 using repeated subtraction and check your work.

$$
\begin{array}{r}
6)\overline{257} \\
-180 \quad \text{Subtract 30 sixes.} \\
\hline
77 \\
-60 \quad \text{Subtract 10 sixes.} \\
\hline
17 \\
-12 \quad \text{Subtract } \underline{2} \text{ sixes.} \\
\hline
5 \qquad 42
\end{array}
$$

[NOTE: You can subtract any number of sixes less than the quotient.]

Check:

$$
\begin{array}{cc}
42 & 252 \\
\times 6 & +5 \\
\hline
252 & 257
\end{array}
$$

The method of repeated subtraction provides a basis for understanding a much shorter method of division called the **division algorithm.***
This process is explained in the following examples.

EXAMPLES

1. Find 2076 ÷ 8.

STEP 1.
$$
\begin{array}{r}
2 \\
8)\overline{2076} \\
-1600 \\
\hline
476
\end{array}
$$
Write 2 in hundreds position.
200 eights (200 · 8 = 1600)

STEP 2.
$$
\begin{array}{r}
25 \\
8)\overline{2076} \\
-1600 \\
\hline
476 \\
-400 \\
\hline
76
\end{array}
$$
Write 5 in tens position.

50 eights (50 · 8 = 400)

STEP 3.
$$
\begin{array}{r}
259 \\
8)\overline{2076} \\
-1600 \\
\hline
476 \\
-400 \\
\hline
76 \\
-72 \\
\hline
4
\end{array}
$$
Write 9 in units position.

9 eights (9 · 8 = 72)

*An algorithm is a process or pattern of steps to be followed in working with numbers.

Summary: **The process can be shortened by not writing all the 0's and "bringing down" only one digit at a time.**

$$
\begin{array}{r}
259 \text{ R}4 \\
8\overline{)2076} \\
16 \\
\hline
47 \\
40 \\
\hline
76 \\
72 \\
\hline
4
\end{array}
$$

"Bring down" the 7 only, then divide 8 into 47.

2. 746 ÷ 32

STEP 1.
$$
\begin{array}{r}
2 \\
32\overline{)746} \\
64 \\
\hline
10
\end{array}
$$

Trial divide 30 into 70 or 3 into 7, giving 2 in the tens position. Note that 10 is less than 32.

STEP 2.
$$
\begin{array}{r}
23 \text{ R}10 \\
32\overline{)746} \\
64 \\
\hline
106 \\
96 \\
\hline
10
\end{array}
$$

Trial divide 30 into 100 or 3 into 10, giving 3 in the units position.

Check:

$$
\begin{array}{r}
23 \\
32 \\
\hline
46 \\
69 \\
\hline
736
\end{array}
\qquad
\begin{array}{r}
736 \\
+10 \\
\hline
746
\end{array}
$$

3. 9325 ÷ 45

STEP 1.
$$
\begin{array}{r}
2 \\
45\overline{)9325} \\
90 \\
\hline
3
\end{array}
$$

Trial divide 40 into 90 or 4 into 9, giving 2 in the hundreds position.

STEP 2.
$$
\begin{array}{r}
20 \\
45\overline{)9325} \\
90 \\
\hline
32 \\
0
\end{array}
$$

45 will not divide into 32, so write 0 in the tens column and multiply $0 \cdot 45 = 0$.

$$\begin{array}{r} 208 \\ \hline 45{\overline{\smash{\big)}\,9325}} \\ 90 \\ \hline 32 \\ 0 \\ \hline 325 \\ 340 \\ \hline \end{array}$$

STEP 3.

Trial divide 45 into 325 or 4 into 32. But the trial quotient 8 is too large, since $8 \cdot 45 = 340$ and 340 is larger than 325.

$$\begin{array}{r} 207 \text{ R}10 \\ \hline 45{\overline{\smash{\big)}\,9325}} \\ 90 \\ \hline 32 \\ 0 \\ \hline 325 \\ 315 \\ \hline 10 \end{array}$$

Now the trial divisor is 7. Since $7 \cdot 45 = 315$ and 315 is smaller than 325, 7 is the desired number.

Check:

$$\begin{array}{r} 207 \\ \times 45 \\ \hline 1035 \\ 828 \\ \hline 9315 \end{array} \qquad \begin{array}{r} 9315 \\ +10 \\ \hline 9325 \end{array}$$

[SPECIAL NOTE: In Step 2 of Example 3, we wrote 0 in the quotient because 45 did not divide into 32. Be sure to write 0 in the quotient whenever the divisor does not divide into any of the partial remainders.]

PRACTICE QUIZ	Find the quotient and remainder for each of the following problems.	ANSWERS
	1. $325 \div 7$	1. 46 R3
	2. $16{\overline{\smash{\big)}\,324}}$	2. 20 R4
	3. $41{\overline{\smash{\big)}\,24682}}$	3. 602 R0

EXERCISES 1.7

Please read the text and study the examples before working these exercises.

Find the quotient and remainder for each of the following problems by using the method of repeated subtraction.

1. $210 \div 7$	2. $140 \div 14$	3. $168 \div 8$	4. $70 \div 5$
5. $132 \div 11$	6. $120 \div 4$	7. $75 \div 15$	8. $51 \div 3$

9. $52 \div 8$ **10.** $44 \div 6$ **11.** $600 \div 25$ **12.** $413 \div 20$

13. $161 \div 15$ **14.** $182 \div 13$ **15.** $150 \div 13$ **16.** $500 \div 14$

17. $205 \div 5$ **18.** $321 \div 7$

19. $1042 \div 22$ **20.** $1461 \div 12$

Divide and check using the division algorithm.

21. $6\overline{)32}$ **22.** $7\overline{)17}$ **23.** $4\overline{)25}$ **24.** $5\overline{)35}$

25. $8\overline{)48}$ **26.** $6\overline{)72}$ **27.** $9\overline{)81}$ **28.** $2\overline{)76}$

29. $3\overline{)98}$ **30.** $14\overline{)52}$ **31.** $12\overline{)108}$ **32.** $11\overline{)424}$

33. $16\overline{)128}$ **34.** $20\overline{)305}$ **35.** $18\overline{)206}$ **36.** $30\overline{)847}$

37. $10\overline{)423}$ **38.** $15\overline{)750}$ **39.** $13\overline{)260}$ **40.** $17\overline{)340}$

41. $12\overline{)360}$ **42.** $19\overline{)7603}$ **43.** $16\overline{)4813}$ **44.** $11\overline{)4406}$

45. $13\overline{)3917}$ **46.** $73\overline{)148}$ **47.** $68\overline{)207}$ **48.** $49\overline{)993}$

49. $50\overline{)3065}$ **50.** $40\overline{)2163}$ **51.** $105\overline{)210}$ **52.** $116\overline{)232}$

53. $213\overline{)4760}$ **54.** $716\overline{)3056}$ **55.** $630\overline{)4768}$ **56.** $414\overline{)83276}$

57. $502\overline{)98762}$ **58.** $317\overline{)70365}$ **59.** $471\overline{)50612}$ **60.** $215\overline{)64930}$

61. Find the quotient if eight hundred twenty-eight is divided by thirty-six. Is the remainder zero? Does this mean that thirty-six is a factor of eight hundred twenty-eight? If so, what is its corresponding factor?

62. Find the quotient if two thousand five hundred forty-two is divided by forty-one. Is the remainder zero? Does this mean that forty-one is a factor of two thousand five hundred forty-two? If so, what is its corresponding factor?

63. If one factor of 810 is 27, what is the corresponding factor?

64. If one factor of 1610 is 35, what is the corresponding factor?

65. What number multiplied by 73 gives a product of 1606? Is 73 a factor of 1606?

66. What number multiplied by 18 gives a product of 3654? Is 18 a factor of 3654?

67. If you bought four textbooks for a price of $16 each, including tax, what did you pay for the four books? The price of a pad of paper was 53 cents.

68. If you bought five plants for your yard and paid $13 for each plant, including tax, what did you pay for the plants? Fertilizer cost $10 per bag.

69. A purchasing agent bought 19 new cars for his company for a total price of $190,665. What price did he pay per car?

70. A restaurant owner bought new chairs for a total cost of $9792. If each chair cost $96, how many chairs did she buy?

1.8 AVERAGE

A topic closely related to addition and division is **average.** Your grade in this course may be based on the average of your exam scores. Newspapers and magazines have information about the Dow Jones stock averages, the average income of American families, the average life expectancy of laboratory rats, and so on. The average of a set of numbers is a kind of "middle number" of the set.* **The average of a set of numbers can be defined as the number found by adding the numbers in the set, then dividing this sum by the number of numbers in the set.†**

EXAMPLE

Find the average of the three numbers 32, 47, 23.

$$
\begin{array}{r}
34 \quad \text{(average)} \\
3\overline{)102} \\
\underline{9} \\
12 \\
\underline{12} \\
0
\end{array}
$$

$$
\begin{array}{r}
32 \\
47 \\
23 \\
\hline
102
\end{array}
$$

The sum, 102, is divided by 3 because there are three numbers being added.

The average of a set of whole numbers need not be a whole number. However, in this section, the problems will be set up so that the averages will be whole numbers. Other cases will be discussed later in the chapters on fractions and decimals (Chapters 3 and 4).

The average of a set of numbers can be very useful, but it can also be misleading. Judging the importance of an average is up to you, the reader of the information. For example, suppose five people had the following incomes for one year: $8,000, $9,000, $10,000, $11,000, $12,000. The average of these numbers is $10,000, as shown on the next page:

---

*Such terms as the **average citizen** or **average voter** are not related to numbers and are not so easily defined as the average of a set of numbers.
†This average is also called the arithmetic average, or mean.

$ 8,000 $10,000
   9,000   5)50,000
  10,000   50,000
  11,000   0
  12,000
$50,000

Now consider these incomes: $1,000, $1,000, $1,000, $1,000, $46,000. Averaging again gives $10,000:

$ 1,000 $10,000
   1,000   5)50,000
   1,000   50,000
   1,000   0
  46,000
$50,000

In the first case, the average of $10,000 serves well as a "middle score" or "representative" of all the incomes. However, in the second case, none of the incomes is even close to $10,000. The one large income completely destroys the "representativeness" of the average. Thus, it is useful to see the numbers or at least know something about the numbers before attaching too much importance to an average.

In statistics, other measures of "representativeness" such as the median (actual middle score) and mode (most frequent score) are discussed. Either the median or the mode gives better information about the second example than the average does.

EXERCISES 1.8

Find the average of each of the following sets of numbers.

1. 102, 113, 97, 100   2. 56, 64, 38, 58

3. 6, 8, 7, 4, 4, 5, 6, 8   4. 5, 4, 5, 6, 5, 8, 9, 6

5. 512, 618, 332, 478   6. 436, 520, 630, 422

7. 1000, 1000, 7000   8. 4000, 5000, 6000

9. 897, 182, 617, 534, 700   10. 648, 930, 556, 852, 544

11. On a math exam, five students' scores were 85, 90, 75, 64, and 96. What was the average of these five scores?

12. On an English exam, six students' scores were 76, 83, 68, 90, 92, and 95. What was their average score?

13. On an exam in history, two students scored 95, six students scored 90, three students scored 80, and one student scored 50. What was the class average?

14. A woman bought ten shares of stock in a company at $35 per share. Later she bought ten shares in another company at $49 per share. What was the average price per share that she paid?

15. Nick bought shares in the stock market from two companies. He paid $450 for nine shares in one company and $690 for eleven shares in a second company. What did he pay as an average price per share for the twenty shares?

16. If you made deposits in your checking account of $640, $830, $1056, $890, and $734 in five months, what was your average deposit?

17. A salesman sold items from his sales list for $972, $834, $1005, $1050, and $799. What was the average price per item?

18. In flying twelve trips in one month (30 days), an airline pilot spent the following amount of hours in preparing and flying for each trip: 6, 8, 9, 6, 7, 7, 7, 5, 6, 6, 6, and 11 hours. What was the average amount of time he spent per flight?

19. Three families, each with two children, had incomes of $8942. Two families, each with four children, had incomes of $10,512. Four families, each with two children, had incomes of $11,111. One family had no children and an income of $12,026. What was the average income per family?

20. During July, Mr. Rodriguez made deposits in his checking account of $400 and $750 and wrote checks totaling $625. During August, his deposits were $632, $322, and $798, and his checks totaled $978. In September, the deposits were $520, $436, $200, and $376; the checks totaled $836. What was the average monthly difference between his deposits and his withdrawals? What was his bank balance at the end of September if he had a balance of $500 on July 1?

SUMMARY: CHAPTER 1

OUR SYSTEM OF REPRESENTING NUMBERS DEPENDS ON THREE THINGS

1. The ten digits $\{0, 1, 2, 3, 4, 5, 6, 7, 8, 9\}$.

2. The placement of the digits.

3. The value of each place.

Numbers being added are called **addends**, and the result of the addition is called the **sum**.

COMMUTATIVE
PROPERTY OF
ADDITION
If a and b are two whole numbers, then $a + b = b + a$.

ASSOCIATIVE
PROPERTY OF
ADDITION
If a, b, and c are whole numbers, then

$$a + b + c = (a + b) + c = a + (b + c)$$

ADDITIVE
IDENTITY
If a is a whole number, then there is a unique whole number 0 with the property that $a + 0 = a$.

Subtraction is a reverse addition, and the missing addend is called the **difference** between the sum and one addend.

The result of multiplying two numbers is called the **product,** and the two numbers are called **factors** of the product.

COMMUTATIVE
PROPERTY OF
MULTIPLICATION
If a and b are whole numbers, then $a \cdot b = b \cdot a$.

ZERO FACTOR
LAW
If a is a whole number, then $a \cdot 0 = 0$

MULTIPLICATIVE
IDENTITY
If a is a whole number, then there is a unique whole number 1 with the property that $a \cdot 1 = a$.

ASSOCIATIVE
PROPERTY OF
MULTIPLICATION
If a, b, and c are whole numbers, then

$$a \cdot b \cdot c = a(b \cdot c) = (a \cdot b)c$$

DISTRIBUTIVE
PROPERTY OF
MULTIPLICATION
OVER ADDITION
If a, b, and c are whole numbers, then

$$a(b + c) = a \cdot b + a \cdot c$$

In division:

The number dividing is called the **divisor.**
The number being divided is called the **dividend.**
The result is called the **quotient.**
The number left over is the **remainder,** and it must be less than the divisor.

An **average** of a set of numbers is the number found by adding the numbers in the set, then dividing this sum by the number of numbers in the set.

REVIEW QUESTIONS: CHAPTER 1

Write the following decimal numerals in expanded notation and in their English word equivalents.

1. 495 **2.** 1975 **3.** 60,308

Write the following numbers as decimal numerals.

4. four thousand eight hundred fifty-six

5. fifteen million, thirty-two thousand, one hundred ninety-seven

6. six hundred seventy-two million, three hundred forty thousand, eighty-three

State which property of addition or multiplication is illustrated.

7. $17 + 32 = 32 + 17$ **8.** $3(22 \cdot 5) = (3 \cdot 22)5$

9. $28 + (6 + 12) = (28 + 6) + 12$ **10.** $72 \cdot 89 = 89 \cdot 72$

11. Show that the following statement is **not true.**

$$32 \div (16 \div 2) = (32 \div 16) \div 2$$

What fact does this illustrate?

Add.

12. 8445
267
1351
478

13. 39
487
966
182

Multiply.

14. $60 \cdot 40$ **15.** $47 \cdot 0$ **16.** $80 \cdot 6000$

Subtract.

17. 647
-139

18. 7036
-4652

19. 5000
-2898

Multiply.

20. 96
62

21. 8973
426

22. 4796
3000

23. Divide using the method of repeated subtraction: $7\overline{)2046}$

Divide.

24. 529)71496   **25.** 38)26721

26. If the product of 17 and 51 is added to the product of 16 and 12, what is the sum?

27. Find the average of 33, 42, 25, and 40.

28. If the quotient of 546 and 6 is subtracted from 100, what is the difference?

29. Two years ago, Ms. Miller bought five shares of stock at $353 per share. One year ago, she bought another ten shares at $290 per share. Yesterday, she sold all her shares at $410 per share. What was her total profit? What was her average profit per share?

30. On a history exam, two students scored 98 points, five students scored 87 points, one student scored 81 points, and six students scored 75 points. What was the average score in the class?

31. State two identity properties, one for addition of whole numbers and one for multiplication of whole numbers.

32. What number should be added to seven hundred forty-three to get a sum of eight hundred thirteen?

33. Fill in the missing numbers in the chart according to the directions.

GIVEN NUMBER	ADD 100	DOUBLE	SUBTRACT 200
3	103	206	?
20	120	?	?
15	?	?	?
?	?	?	16

34. How many times can 35 be repeatedly subtracted from 700?

CHAPTER TEST: CHAPTER 1

1. Write 8952 in expanded notation and in its English word equivalent.

2. The number 1 is called the multiplicative _____.

3. Give an example that illustrates the commutative property of multiplication.

Add.

4. 9586
    345
    2078

5.   37
    486
    493
    162
    557

6. 1,480,900
    2,576,850
    5,200,635
    4,523,276

Subtract.

7.   850
    − 362

8.   5097
    − 3868

9.   6000
    − 293

Multiply.

10. 34
    76

11. 2593
    85

12. 793
    266

Divide.

13. 25)10,075

14. 462)79,852

15. 603)1,209,015

16. Find the average of 82, 96, 49, and 69.

17. If the quotient of 51 and 17 is subtracted from the product of 19 and 3, what is the difference?

18. Robert and his brother were saving money to buy a new TV set for their parents. If Robert saved $23 a week and his brother saved $28 a week, how much did they save in six weeks? What was their average weekly savings? If the set they wanted to buy cost $530 including tax, how much did they still need after the six weeks?

2

PRIME NUMBERS

2.1 EXPONENTS

Repeated addition is shortened using multiplication, as $2 + 2 + 2 = 3 \cdot 2 = 6$. Repeated multiplication can be shortened using **exponents.** Thus, if 2 is used as a factor three times, we can write $2 \cdot 2 \cdot 2 = 2^3 = 8$.

In an expression such as $2^3 = 8$, 2 is called the **base,** 3 is called the **exponent,** and 8 is called the **power.** (Exponents are written slightly to the right and above the base.)

EXAMPLES

Repeated Multiplication	Using Exponents
1. $7 \cdot 7 = 49$	$7^2 = 49$
2. $3 \cdot 3 = 9$	$3^2 = 9$
3. $2 \cdot 2 \cdot 2 \cdot 2 = 16$	$2^4 = 16$
4. $10 \cdot 10 \cdot 10 = 1000$	$10^3 = 1000$

DEFINITION

An **exponent** is a number that tells how many times its base is to be used as a factor.

Expressions with exponent 2 are read "squared," with exponent 3 are read "cubed," and with other exponents are read "to the ＿＿ power." For example, we read

$5^2 = 25$ as "five squared is equal to twenty-five";

$4^3 = 64$ as "four cubed is equal to sixty-four"; and

$3^4 = 81$ as "three to the fourth power is equal to eighty-one."

If there is no exponent, it is understood to be 1. Thus, $8 = 8^1$, $6 = 6^1$, and $942 = 942^1$.

When the exponent 0 is used for any base except 0, the value of the power is defined to be 1:

$$2^0 = 1$$
$$3^0 = 1$$
$$5^0 = 1$$
$$46^0 = 1$$

One of the rules for using exponents (which will be studied in algebra) is related to division and involves subtracting exponents. This will help our understanding of the 0 exponent. To divide $\frac{2^6}{2^2}$ or $\frac{5^4}{5^3}$, we can write

$$\frac{2^6}{2^2} = \frac{\not{2} \cdot \not{2} \cdot 2 \cdot 2 \cdot 2 \cdot 2}{\not{2} \cdot \not{2}} = 2 \cdot 2 \cdot 2 \cdot 2 = 2^4 \quad \text{or} \quad \frac{2^6}{2^2} = 2^{6-2} = 2^4$$

$$\frac{5^4}{5^3} = \frac{\not{5} \cdot \not{5} \cdot \not{5} \cdot 5}{\not{5} \cdot \not{5} \cdot \not{5}} = 5 \quad \text{or} \quad \frac{5^4}{5^3} = 5^{4-3} = 5^1$$

The rule is to **subtract the exponents when dividing numbers with the same base.** So,

$$\frac{3^4}{3^4} = 3^{4-4} = 3^0 \quad \text{and} \quad \frac{5^2}{5^2} = 5^{2-2} = 5^0$$

But,

$$\frac{3^4}{3^4} = \frac{81}{81} = 1 \quad \text{and} \quad \frac{5^2}{5^2} = \frac{25}{25} = 1$$

For the rules of exponents to make sense, we must have $3^0 = 1$ and $5^0 = 1$.

DEFINITION   For any nonzero whole number a, $a^0 = 1$.

EXAMPLES
1. $6^0 = 1$

2. $6^2 = 36$ $(6^2 = 6 \cdot 6)$

3. $5^3 = 125$ $(5^3 = 5 \cdot 5 \cdot 5)$

4. $3^3 = 27$ $(3^3 = 3 \cdot 3 \cdot 3)$

5. $2^5 = 32$ $(2^5 = 2 \cdot 2 \cdot 2 \cdot 2 \cdot 2)$

EXERCISES 2.1

Study the text and examples carefully before working these exercises.

In each of the following expressions, name (a) the exponent and (b) the base. Also, find each power.

1. 2^3	**2.** 2^5	**3.** 5^2	**4.** 6^2	**5.** 7^0
6. 11^2	**7.** 1^4	**8.** 4^3	**9.** 4^0	**10.** 3^6
11. 3^2	**12.** 2^4	**13.** 5^0	**14.** 1^{50}	**15.** 62^1
16. 12^2	**17.** 10^2	**18.** 10^3	**19.** 4^2	**20.** 2^5
21. 10^4	**22.** 5^3	**23.** 6^3	**24.** 10^5	**25.** 19^0

Find a base and exponent form for each of the following powers without using the exponent 1. [Hint: The table inside the back cover may be helpful.]

26. 4	**27.** 25	**28.** 16	**29.** 27	**30.** 32
31. 121	**32.** 49	**33.** 8	**34.** 9	**35.** 36
36. 125	**37.** 81	**38.** 64	**39.** 100	**40.** 1000
41. 10,000	**42.** 216	**43.** 144	**44.** 169	**45.** 243
46. 625	**47.** 225	**48.** 196	**49.** 343	**50.** 100,000

Rewrite the following products using exponents.

51. $6 \cdot 6 \cdot 6 \cdot 6 \cdot 6$ **52.** $7 \cdot 7 \cdot 7 \cdot 7$ **53.** $2 \cdot 2 \cdot 7 \cdot 7$

54. $5 \cdot 5 \cdot 9 \cdot 9 \cdot 9$ **55.** $2 \cdot 2 \cdot 3 \cdot 3 \cdot 3$ **56.** $3 \cdot 3 \cdot 5 \cdot 5 \cdot 5$

57. $7 \cdot 7 \cdot 13$ **58.** $11 \cdot 11 \cdot 11$ **59.** $2 \cdot 3 \cdot 3 \cdot 11 \cdot 11$

60. $5 \cdot 5 \cdot 5 \cdot 11 \cdot 11$

2.2   ORDER OF OPERATIONS

To evaluate the expression $5 \cdot 2 + 14 \div 2$, which of the following procedures would you use?

$$5 \cdot 2 + 14 \div 2 = 10 + 14 \div 2 \qquad 5 \cdot 2 + 14 \div 2 = 10 + 7$$
$$= 24 \div 2 \qquad\qquad\qquad = 17$$
$$= 12$$

Both answers cannot be correct! Some agreement must be made as to the **order of the operations.** Mathematicians have, in fact, agreed on a set of rules for simplifying any numerical expression involving addition, subtraction, multiplication, division, and exponents. Under these rules, the second answer, 17, is correct.

RULES FOR ORDER OF OPERATIONS

1.   First, simplify expressions within parentheses.

2.   Second, find any powers indicated by exponents.

3.   Third, moving from left to right, perform any multiplications or divisions in the order they appear.

4.   Fourth, moving from left to right, perform any additions or subtractions in the order they appear.

The rules are very explicit. Read them carefully. Note that in Rule 3, neither multiplication nor division has priority over the other. Whichever of these operations occurs first, moving left to right, is done first. In Rule 4, addition and subtraction are handled in the same way.

The following examples show how to apply the rules for order of operations.

Use the rules for order of operations to find the value of each of the following expressions.

1. Evaluate $14 \div 7 + 3 \cdot 2 - 5$.

$$\underline{14 \div 7} + \underline{3 \cdot 2} - 5$$

$$= \quad 2 \quad + \quad 6 \quad - 5 = 8 - 5 = 3$$

2. Evaluate $3 \cdot 6 \div 9 - 1 + 4 \cdot 7$.

$$\underline{3 \cdot 6} \div 9 - 1 + \underline{4 \cdot 7}$$

$$= \quad 18 \quad \div 9 - 1 + \quad 28 \quad = 2 - 1 + 28 = 1 + 28 = 29$$

3. Evaluate $(6 + 2) + (8 + 1) \div 9$.

$$\underline{(6 + 2)} + \underline{(8 + 1)} \div 9$$

$$= \quad 8 \quad + \quad 9 \quad \div 9 = 8 + 1 = 9$$

4. Evaluate $30 \div 3 \cdot 2 + 3(6 - 21 \div 7)$.

$$\underline{30 \div 3} \cdot 2 + 3(6 - \underline{21 \div 7})$$

$$= \quad 10 \quad \cdot 2 + 3(6 - \quad 3)$$
$$= \quad 20 \quad + 3(3)$$
$$= \quad 20 \quad + \quad 9 \quad = 29$$

5. Evaluate $2 \cdot 3^2 + 18 \div 3^2$.

$$2 \cdot \underline{3^2} + 18 \div \underline{3^2}$$

$$= \underline{2 \cdot 9} + \underline{18 \div 9}$$

$$= \quad 18 \quad + \quad 2 \quad = 20$$

6. Evaluate $3 \cdot 5^2 \div 15 + 30 - 2^3 \cdot 3$.

$$\underbrace{3 \cdot 5^2}\ \div 15 + 30 - \underbrace{2^3}\ \cdot 3$$

$$= \underbrace{3 \cdot 25}\ \div 15 + 30 - \underbrace{8 \cdot 3}$$

$$= \underbrace{75\ \div 15}\ + 30 - 24$$

$$= \quad 5 \quad + 30 - 24 = 11$$

PRACTICE QUIZ	Find the value for each of the following expressions using the rules for order of operations.	ANSWERS
	1. $15 \div 5 + 10 \cdot 2$	1. 23
	2. $3 \cdot 2^3 - 12 - 3 \cdot 2^2$	2. 0
	3. $4 \div 2^2 + 3 \cdot 2^2$	3. 13
	4. $(5 + 7) \div 3 + 1$	4. 5
	5. $19 - 5(3 - 1)$	5. 9

EXERCISES 2.2

Be sure to study the examples before you begin these exercises.

Find the value of each of the following expressions using the rules for order of operations.

1. $4 \div 2 + 7 - 3 \cdot 2$ 2. $8 \cdot 3 \div 12 + 13$

3. $6 + 3 \cdot 2 - 10 \div 2$ 4. $14 \cdot 3 \div 7 \div 2 + 6$

5. $6 \div 2 \cdot 3 - 1 + 2 \cdot 7$ 6. $5 \cdot 1 \cdot 3 - 4 \div 2 + 6 \cdot 3$

7. $72 \div 4 \div 9 - 2 + 3$ 8. $14 + 63 \div 3 - 35$

9. $(2 + 3 \cdot 4) \div 7 + 3$ 10. $(2 + 3) \cdot 4 \div 5 + 3 \cdot 2$

11. $(7 - 3) + (2 + 5) \div 7$ 12. $16(2 + 4) - 90 - 3 \cdot 2$

13. $35 \div (6 - 1) - 5 + 6 \div 2$ 14. $22 - 11 \cdot 2 + 15 - 5 \cdot 3$

15. $(42 - 2 \div 2 \cdot 3) \div 13$ 16. $18 + 18 \div 2 \div 3 - 3 \cdot 1$

17. $4(7 - 2) \div 10 + 5$ 18. $(33 - 2 \cdot 6) \div 7 + 3 - 6$

19. $72 \div 8 + 3 \cdot 4 - 105 \div 5$ 20. $6(14 - 6 \div 2 - 11)$

21. $48 \div 12 \div 4 - 1 + 6$ 22. $5 - 1 \cdot 2 + 4(6 - 18 \div 3)$

23. $8 - 1 \cdot 5 + 6(13 - 39 \div 3)$ **24.** $(21 \div 7 - 3)42 + 6$

25. $16 - 16 \div 2 - 2 + 7 \cdot 3$ **26.** $(135 \div 3 + 21 \div 7) \div 12 - 4$

27. $(13 - 5) \div 4 + 12 \cdot 4 \div 3 - 72 \div 18 \cdot 2 + 16$

28. $15 \div 3 + 2 - 6 + (3)(2)(18)(0)(5)$

29. $100 \div 10 \div 10 + 1000 \div 10 \div 10 \div 10 - 2$

30. $[(85 + 5) \div 3 \cdot 2 + 15] \div 15$

31. $2 \cdot 5^2 - 4 \div 2 + 3 \cdot 7$ **32.** $16 \div 2^4 - 9 \div 3^2$

33. $(4^2 - 7) \cdot 2^3 - 8 \cdot 5 \div 10$ **34.** $4^2 - 2^4 + 5 \cdot 6^2 - 10^2$

35. $(2^5 + 1) \div 11 - 3 + 7(3^3 - 7)$ **36.** $(6 + 8^2 - 10 \div 2) \div 5 + 5 \cdot 3^2$

37. $(5 + 7) \div 4 + 2$ **38.** $(2^3 + 2) \div 5 + (7^2 \div 7)$

39. $(5^2 + 7) \div 8 - (14 \div 7 \cdot 2)$

40. $(3 \cdot 2^2 - 5 \cdot 2 + 2) - (1 \cdot 2^2 + 5 \cdot 2 - 10)$

41. $2^3 \cdot 3^2 \div 24 - 3 + 6^2 \div 4$

42. $2 \cdot 3^2 + 5 \cdot 3^2 + 15^2 - (21 \cdot 3^2 + 6)$

43. $3 \cdot 2^3 - 2^2 + 4 \cdot 2 - 2^4$

44. $2 \cdot 5^2 - 4(21 \div 3 - 7) + 10^3 - 1000$

45. $(4 + 3)^2 - (2 + 3)^2$

46. $40 \div 2 \cdot 5 + 1 \cdot 3^2 \cdot 2$

47. $20 - 2(3 - 1) + 6^2 \div 2 \cdot 3$

48. $8 \div 2 \cdot 4 + 16 \div 4 \cdot 2 + 3 \cdot 2^2$

49. $50 \cdot 10 \div 2 - 2^2 \cdot 5 + 14 - 2 \cdot 7$

50. $(20 \div 2^2 \cdot 5) + (51 \div 17)^2$

2.3 TESTS FOR DIVISIBILITY $(2, 3, 4, 5, 9, 10)$

In our work with factoring (Section 2.5) and fractions (Chapter 3), we will need to be able to divide quickly by small numbers. We will want to know if a number is **exactly divisible** (remainder 0) by some number **before** we divide. There are simple tests we can use to determine whether a number is divisible by 2, 3, 4, 5, 9, or 10 **without actually dividing.**

For example, can you tell (without dividing) if 6495 is divisible by 2? by 3? The answer is that 6495 is divisible by 3 but not by 2.

$$
\begin{array}{r}
2165 \\
3\overline{)6495} \\
6 \\
\hline
04 \\
3 \\
\hline
19 \\
18 \\
\hline
15 \\
15 \\
\hline
0 \quad \text{Remainder}
\end{array}
\qquad
\begin{array}{r}
3247 \\
2\overline{)6495} \\
6 \\
\hline
04 \\
4 \\
\hline
09 \\
8 \\
\hline
15 \\
14 \\
\hline
1 \quad \text{Remainder (remainder not 0)}
\end{array}
$$

The following list of rules explains how to test for divisibility by 2, 3, 4, 5, 9, and 10. There are other tests for other numbers such as 6, 7, 8, and 15, but the rules given here are sufficient for our purposes.

TESTS FOR DIVISIBILITY BY 2, 3, 4, 5, 9, AND 10

For 2:   If the last digit (units digit) of a whole number is 0, 2, 4, 6, or 8, then the whole number is divisible by 2. In other words, even whole numbers are divisible by 2; odd whole numbers are not divisible by 2.

For 3:   If the sum of the digits of a whole number is divisible by 3, then the number is divisible by 3.

For 4:   If the last two digits of a whole number form a number that is divisible by 4, then the number is divisible by 4. (00 is considered to be divisible by 4.)

For 5:   If the last digit of a whole number is 0 or 5, then the number is divisible by 5.

For 9:   If the sum of the digits of a whole number is divisible by 9, then the number is divisible by 9.

For 10:   If the last digit of a whole number is 0, then the number is divisible by 10.

EXAMPLES

1. 356 is divisible by 2 since the last digit is 6.

2. 6801 is divisible by 3 since $6 + 8 + 0 + 1 = 15$ and 15 is divisible by 3.

3. 9036 is divisible by 4 since 36 (last 2 digits) is divisible by 4.

4. 1365 is divisible by 5 since 5 is the last digit.

5. 9657 is divisible by 9 since $9 + 6 + 5 + 7 = 27$ and 27 is divisible by 9.

6. 3590 is divisible by 10 since 0 is the last digit.

If a number is divisible by 9, then it will also be divisible by 3. In Example 5, $9 + 6 + 5 + 7 = 27$ and 27 is divisible by 9 and by 3, so 9657 is divisible by 9 and by 3.

But, a number divisible by 3 might not be divisible by 9. In Example 2, $6 + 8 + 0 + 1 = 15$ and 15 is *not* divisible by 9, so 6801 is divisible by 3 but not by 9.

Similarly, any number divisible by 10 is also divisible by 5, but a number that is divisible by 5 might not be divisible by 10. The number 2580 is divisible by 10 and also by 5, but 4365 is divisible by 5 and not by 10.

Many numbers are divisible by more than one of the numbers 2, 3, 4, 5, 9, and 10. These numbers will satisfy more than one of the six tests. For example, 4365 is divisible by 5 (last digit is 5) and by 3 $(4 + 3 + 6 + 5 = 18)$ and by 9 $(4 + 3 + 6 + 5 = 18)$.

EXAMPLES

Use all six tests to determine which of the numbers 2, 3, 4, 5, 9, and 10 will divide into the following numbers.

1. 5712

   (a) divisible by 2 (last digit is 2, an even digit)

   (b) divisible by 3 $(5 + 7 + 1 + 2 = 15$ and 15 is divisible by 3)

   (c) divisible by 4 (12 is divisible by 4)

   (d) not divisible by 5 (last digit is not 0 or 5)

   (e) not divisible by 9 $(5 + 7 + 1 + 2 = 15$ and 15 is not divisible by 9)

   (f) not divisible by 10 (last digit is not 0)

2. 2530

   (a) divisible by 2 (last digit is 0, an even digit)

   (b) not divisible by 3 $(2 + 5 + 3 + 0 = 10$ and 10 is not divisible by 3)

   (c) not divisible by 4 (30 is not divisible by 4)

   (d) divisible by 5 (last digit is 0)

   (e) not divisible by 9 $(2 + 5 + 3 + 0 = 10$ and 10 is not divisible by 9)

   (f) divisible by 10 (last digit is 0)

PRACTICE QUIZ	Using the techniques of this section, determine which of the numbers 2, 3, 4, 5, 9, and 10 (if any) will divide exactly into each of the following numbers.	ANSWERS
	1. 842	**1.** 2
	2. 9030	**2.** 2, 3, 5, 10
	3. 4031	**3.** None

EXERCISES 2.3

Be sure to study the text and the examples before working these exercises.

Using the techniques of this section, determine which of the numbers 2, 3, 4, 5, 9, and 10 (if any) will divide exactly into each of the following numbers.

1. 72	**2.** 81	**3.** 105	**4.** 333	**5.** 150
6. 471	**7.** 664	**8.** 154	**9.** 372	**10.** 375
11. 443	**12.** 173	**13.** 567	**14.** 480	**15.** 331
16. 370	**17.** 571	**18.** 466	**19.** 897	**20.** 695
21. 795	**22.** 777	**23.** 45,000	**24.** 885	**25.** 4422
26. 1234	**27.** 4321	**28.** 8765	**29.** 5678	**30.** 402
31. 705	**32.** 732	**33.** 441	**34.** 555	**35.** 666
36. 9000	**37.** 10,000	**38.** 576	**39.** 549	**40.** 792
41. 5700	**42.** 4391	**43.** 5476	**44.** 6930	**45.** 4380
46. 510	**47.** 8805	**48.** 7155	**49.** 8377	**50.** 2222
51. 35,622	**52.** 75,495	**53.** 12,324	**54.** 55,555	
55. 632,448	**56.** 578,400	**57.** 9,737,001	**58.** 17,158,514	
59. 36,762,252	**60.** 20,498,105			

61. If a number is divisible by both 2 and 9, will it be divisible by 18? Give five examples to support your answer.

62. If a number is divisible by both 3 and 9, will it be divisible by 27? Give five examples to support your answer.

2.4 PRIME NUMBERS AND COMPOSITE NUMBERS

We know that $2 \cdot 9 = 18$ and that 2 and 9 are factors (or divisors) of 18. Are 2 and 9 the *only* factors of 18? Of course not; because $1 \cdot 18 = 18$, $2 \cdot 9 = 18$, and $3 \cdot 6 = 18$, the factors of 18 are 1, 2, 3, 6, 9, and 18. Some numbers have only two factors. For example, $1 \cdot 41 = 41$ and 1 and 41 are the only factors of 41. Also, 1 and 23 are the only factors of 23. These numbers are particularly important and are called **prime numbers.**

For this discussion, we will use only nonzero whole numbers. The nonzero whole numbers 1, 2, 3, 4, 5, 6, ... are called **counting numbers** or **natural numbers.**

Every counting number has 1 and the number itself as factors, as the following list shows.

COUNTING NUMBER	FACTORS
18	1, 2, 3, 6, 9, 18
26	1, 2, 13, 26
12	1, 2, 3, 4, 6, 12
17	1, 17
3	1, 3
29	1, 29

The last three numbers in the list (17, 3, 29) have exactly two factors. These numbers are **prime numbers.**

DEFINITION
A **prime number** is a counting number with exactly two different factors (or divisors).

DEFINITION
A **composite number** is a counting number with more than two different factors (or divisors).

Thus, in the list discussed, 18, 26, and 12 are composite numbers, and 17, 3, and 29 are prime numbers.

[NOTE: Since $1 \cdot 1 = 1$ and 1 is the only factor of 1, the number 1 does not have two different factors and it does not have more than two different factors. Therefore, 1 is neither a prime number nor a composite number.]

EXAMPLES

Prime Numbers

1. 7   7 has exactly two different factors, 1 and 7.

2. 11   11 has exactly two different factors, 1 and 11.

3. 23   23 has exactly two different factors, 1 and 23.

Composite Numbers

4. 14 $14 = 1 \cdot 14$ and $14 = 2 \cdot 7$ so 14 has more than two different factors.

5. 15 $15 = 1 \cdot 15$ and $15 = 3 \cdot 5$ so 15 has more than two different factors.

6. 21 $21 = 1 \cdot 21$ and $21 = 3 \cdot 7$ so 21 has more than two different factors.

There is no formula to help us find all the prime numbers. However, there is a technique developed by a Greek mathematician named Eratosthenes. He used the concept of **multiples.** To find the multiples of any counting number, multiply each of the counting numbers by that number.

counting numbers	1, 2, 3, 4, 5, 6, 7, 8, . . .
multiples of 8	8, 16, 24, 32, 40, 48, 56, 64, . . .
multiples of 2	2, 4, 6, 8, 10, 12, 14, 16, . . .
multiples of 3	3, 6, 9, 12, 15, 18, 21, 24, . . .
multiples of 10	10, 20, 30, 40, 50, 60, 70, 80, . . .

None of the multiples of a number, except possibly the number itself, can be prime since they all have that number as a factor. To sift out the prime numbers according to the **Sieve of Eratosthenes,** we proceed by eliminating multiples as the following steps describe.

1. To find the prime numbers from 1 to 50, list all the counting numbers from 1 to 50 in rows of ten.

1	2	3	4	5	6	7	8	9	10
11	12	13	14	15	16	17	18	19	20
21	22	23	24	25	26	27	28	29	30
31	32	33	34	35	36	37	38	39	40
41	42	43	44	45	46	47	48	49	50

2. Start by crossing out 1 (since 1 is not a prime number). Next, circle 2 and cross out all the other multiples of 2; that is, cross out every second number.

1̸	②	3	4̸	5	6̸	7	8̸	9	1̸0̸
11	1̸2̸	13	1̸4̸	15	1̸6̸	17	1̸8̸	19	20
21	2̸2̸	23	2̸4̸	25	2̸6̸	27	2̸8̸	29	3̸0̸
31	3̸2̸	33	3̸4̸	35	3̸6̸	37	3̸8̸	39	4̸0̸
41	4̸2̸	43	4̸4̸	45	4̸6̸	47	4̸8̸	49	5̸0̸

3.   The first number after 2 not crossed out is 3. Circle 3 and cross out all multiples of 3 that are not already crossed out; that is, after 3, every third number should be crossed out.

1̸	②	③	4̸	5	6̸	7	8̸	9̸	1̸0̸
11	1̸2̸	13	1̸4̸	1̸5̸	1̸6̸	17	1̸8̸	19	2̸0̸
2̸1̸	2̸2̸	23	2̸4̸	25	2̸6̸	2̸7̸	28	29	3̸0̸
31	3̸2̸	3̸3̸	3̸4̸	35	3̸6̸	37	38	3̸9̸	4̸0̸
41	4̸2̸	43	4̸4̸	4̸5̸	4̸6̸	47	48	49	5̸0̸

4.   The next number not crossed out is 5. Circle 5 and cross out all multiples of 5 that are not already crossed out. If we proceed this way, we will have the prime numbers circled and the composite numbers crossed out. The final table is as follows:

1̸	②	③	4̸	⑤	6̸	⑦	8̸	9̸	1̸0̸
⑪	1̸2̸	⑬	1̸4̸	1̸5̸	1̸6̸	⑰	1̸8̸	⑲	2̸0̸
2̸1̸	2̸2̸	㉓	2̸4̸	2̸5̸	2̸6̸	2̸7̸	28	㉙	3̸0̸
㉛	3̸2̸	3̸3̸	3̸4̸	3̸5̸	3̸6̸	㊲	38	3̸9̸	4̸0̸
㊀	4̸2̸	㊃	4̸4̸	4̸5̸	4̸6̸	㊼	48	4̸9̸	5̸0̸

Looking at the final table (the Sieve of Eratosthenes), we see that the prime numbers less than 50 are 2, 3, 5, 7, 11, 13, 17, 19, 23, 29, 31, 37, 41, 43, 47. You should also notice that (a) the only even prime number is 2; and (b) all other prime numbers are odd, but not all odd numbers are prime.

Computers can be used to help determine whether very large numbers are prime. To decide whether a relatively small number such as 103 or 221 is prime or not, we can use our tests for divisibility from Section 2.3 for 2, 3, and 5, then continue to divide by progressively larger prime numbers until

1.   we find a 0 remainder (in which case the number is composite); or

2.   we find a quotient less than the divisor (in which case the number is prime).

EXAMPLES   1.   Is 103 prime?

Tests for 2, 3, and 5 fail. (103 is not even; $1 + 0 + 3 = 4$ and 4 is not divisible by 3; the last digit is not 0 or 5.)

Divide by 7:

$$
\begin{array}{r}
14 \quad \text{quotient greater than divisor} \\
7\overline{)103} \\
\underline{7} \\
33 \\
\underline{28} \\
5 \quad \text{remainder not 0}
\end{array}
$$

Divide by 11:

$$
\begin{array}{r}
9 \quad \text{quotient is less than divisor} \\
11\overline{)103} \\
\underline{99} \\
4
\end{array}
$$

So, 103 is prime.

2. Is 221 prime?

Tests for 2, 3, and 5 fail.

Divide by 7:

$$
\begin{array}{r}
31 \quad \text{quotient greater than divisor} \\
7\overline{)221} \\
\underline{21} \\
11 \\
\underline{7} \\
4 \quad \text{remainder not 0}
\end{array}
$$

Divide by 11:

$$
\begin{array}{r}
20 \quad \text{quotient greater than divisor} \\
11\overline{)221} \\
\underline{22} \\
01 \\
\underline{0} \\
1 \quad \text{remainder not 0}
\end{array}
$$

Divide by 13:

$$
\begin{array}{r}
17 \\
13\overline{)221} \\
\underline{13} \\
91 \\
\underline{91} \\
0 \quad \text{remainder is 0}
\end{array}
$$

So, 221 is composite and not prime. [Note: $221 = 13 \cdot 17$.]

EXERCISES 2.4

Be sure you know what prime numbers are and what multiples are before you begin these exercises.

List the multiples for each of the following numbers.

1. 5	**2.** 7	**3.** 11	**4.** 13	**5.** 12
6. 9	**7.** 20	**8.** 17	**9.** 16	**10.** 25

11. Construct a Sieve of Eratosthenes for the numbers from 1 to 100. List the prime numbers from 1 to 100.

Decide whether each of the following numbers is prime or composite. If the number is composite, find at least two pairs of factors for the number.

12. 17	**13.** 19	**14.** 28	**15.** 32	**16.** 47
17. 59	**18.** 16	**19.** 63	**20.** 14	**21.** 51
22. 67	**23.** 89	**24.** 73	**25.** 61	**26.** 52
27. 57	**28.** 98	**29.** 86	**30.** 53	**31.** 37

Two numbers are given. Find two factors of the first number whose sum is the second number.

EXAMPLE: 12, 8—Two factors of 12 whose sum is 8 are 6 and 2 since $6 \cdot 2 = 12$ and $6 + 2 = 8$.

32. 24, 10	**33.** 12, 7	**34.** 16, 10	**35.** 12, 13	**36.** 14, 9
37. 50, 27	**38.** 20, 9	**39.** 24, 11	**40.** 48, 19	**41.** 36, 15
42. 7, 8	**43.** 63, 24	**44.** 51, 20	**45.** 25, 10	**46.** 16, 8
47. 60, 17	**48.** 52, 17	**49.** 27, 12	**50.** 72, 22	

51. List the set of prime numbers less than 100 that **are not** odd.

52. List the set of prime numbers less than 100 that **are** odd.

53. List the set of all factors of 52. Divide each factor into 52 and show that the quotient is another factor of 52.

2.5   PRIME FACTORIZATIONS

When working with fractions (Chapter 3), we will want to find all the prime factors of composite numbers. For example, $28 = 4 \cdot 7$ and 4 and 7 are factors of 28. However, 4 is not prime. Since $4 = 2 \cdot 2$, we can write $28 = 2 \cdot 2 \cdot 7$ and this $(2 \cdot 2 \cdot 7)$ is the **prime factorization** of 28.

The **prime factorization** of a number is the product of all the prime factors of that number, including repeated factors.

There are two techniques for finding prime factorizations.

I. To find the prime factorization of a composite number:

(a) Factor the composite number into **any** two factors.

$$60 = 6 \cdot 10 \qquad \text{(We know 10 is a factor.)}$$

(b) If either or both factors are not prime, factor each of these.

$$60 = \underset{\diagup \diagdown}{6} \cdot \underset{\diagup \diagdown}{10}$$
$$= 2 \cdot 3 \cdot 2 \cdot 5$$

(c) Continue this process until all factors are prime. For convenience, we can write the factors in order and use exponents.

$$60 = 6 \cdot 10$$
$$= 2 \cdot 3 \cdot 2 \cdot 5$$
$$= 2 \cdot 2 \cdot 3 \cdot 5$$
$$= 2^2 \cdot 3 \cdot 5$$

EXAMPLES   Find the prime factorization for each number.

1. $70 = 7 \cdot 10$ (10 is a factor since the last digit is 0.)
    $ = 7 \cdot 2 \cdot 5$
    $ = 2 \cdot 5 \cdot 7$

2. $85 = 5 \cdot 17$ (5 is a factor since the last digit is 5.)

3. $72 = 8 \cdot 9$ (9 is a factor since $7 + 2 = 9$.)
    $ = 2 \cdot 4 \cdot 3 \cdot 3$
    $ = 2 \cdot 2 \cdot 2 \cdot 3 \cdot 3$
    $ = 2^3 \cdot 3^2$

4. $245 = 5 \cdot 49$ (5 is a factor since the last digit is 5.)
    $ = 5 \cdot 7 \cdot 7$
    $ = 5 \cdot 7^2$

Another arrangement, called a **factor tree,** may also be used. As these factor trees illustrate, no matter what two factors are used in the first step, the prime factorization is the same.

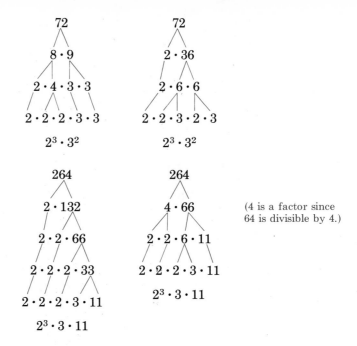

72
$8 \cdot 9$
$2 \cdot 4 \cdot 3 \cdot 3$
$2 \cdot 2 \cdot 2 \cdot 3 \cdot 3$
$2^3 \cdot 3^2$

72
$2 \cdot 36$
$2 \cdot 6 \cdot 6$
$2 \cdot 2 \cdot 3 \cdot 2 \cdot 3$
$2^3 \cdot 3^2$

264
$2 \cdot 132$
$2 \cdot 2 \cdot 66$
$2 \cdot 2 \cdot 2 \cdot 33$
$2 \cdot 2 \cdot 2 \cdot 3 \cdot 11$
$2^3 \cdot 3 \cdot 11$

264
$4 \cdot 66$
$2 \cdot 2 \cdot 6 \cdot 11$
$2 \cdot 2 \cdot 2 \cdot 3 \cdot 11$
$2^3 \cdot 3 \cdot 11$

(4 is a factor since 64 is divisible by 4.)

II. To find the prime factorization of a composite number:

(a) Divide the composite number by any prime number that will divide into it.

$$\begin{array}{r} 30 \\ 2\overline{)60} \end{array}$$

(b) Continue to divide the **quotient** by prime numbers.

$$\begin{array}{r} 30 \\ 2\overline{)60} \end{array} \qquad \begin{array}{r} 15 \\ 2\overline{)30} \end{array} \qquad \begin{array}{r} 5 \\ 3\overline{)15} \end{array}$$

(c) The prime factorization is the product of all the prime divisors and the last prime quotient.

$$60 = 2 \cdot 2 \cdot 3 \cdot 5 = 2^2 \cdot 3 \cdot 5$$

The division may be written in a more compact form as shown:

$$\begin{array}{r} 2\overline{)60} \\ 2\overline{)30} \\ 3\overline{)15} \\ \overline{5} \end{array} \qquad 60 = 2^2 \cdot 3 \cdot 5$$

EXAMPLES

1. $2\overline{)70}$ $70 = 2 \cdot 5 \cdot 7$
    $5\overline{)35}$
    7

2. $5\overline{)85}$ $85 = 5 \cdot 17$
    17

3. $2\overline{)72}$ $72 = 2^3 \cdot 3^2$
    $2\overline{)36}$
    $2\overline{)18}$
    $3\overline{)9}$
    3

4. $5\overline{)245}$ $245 = 5 \cdot 7^2$
    $7\overline{)49}$
    7

Both methods I and II give the same prime factorization for a composite number. Try to learn method I because it is easier to apply in working with fractions.

The **Fundamental Theorem of Arithmetic** says that every composite number has only one prime factorization regardless of what factors you start with or what method you use.

FUNDAMENTAL
THEOREM OF
ARITHMETIC

Every composite number has a unique prime factorization.

The order of the prime factors is not important since the factorization is not changed because of any difference in order. Also, remember that, even though 1 is not a prime number, 1 is always a factor of any whole number.

The only factors (or divisors) of a composite number are (a) 1, (b) the number itself, (c) a prime factor, and (d) products of prime factors. We are to find the products of all possible combinations of prime factors in the prime factorization.

For example, to find the factors of $30 = 2 \cdot 3 \cdot 5$, we have

$$1, 30, 2, 3, 5, 2 \cdot 3, 2 \cdot 5, 3 \cdot 5$$

or $1, 30, 2, 3, 5, 6, 10, 15.$

Since $140 = 2 \cdot 2 \cdot 5 \cdot 7$, the factors of 140 are

$$1, 140, 2, 5, 7, 2 \cdot 2, 2 \cdot 5, 2 \cdot 7, 5 \cdot 7, 2 \cdot 2 \cdot 5, 2 \cdot 2 \cdot 7, 2 \cdot 5 \cdot 7$$

or $1, 140, 2, 5, 7, 4, 10, 14, 35, 20, 28, 70.$

There are no other factors of 140.

PRACTICE QUIZ	Find the prime factorization of each of the following numbers.	ANSWERS
	1.　42	1.　$2 \cdot 3 \cdot 7$
	2.　56	2.　$2^3 \cdot 7$
	3.　230	3.　$2 \cdot 5 \cdot 23$
	4.　Using the prime factorization of 63, find all the factors of 63.	4.　1, 63, 3, 7, 9, 21

EXERCISES 2.5

Study the text carefully before you start these exercises.

Find the prime factorization for each of the following numbers. Use the tests for divisibility given in Section 2.3 whenever you need them to help you get started.

1.　24	2.　28	3.　27	4.　16	5.　36
6.　60	7.　72	8.　90	9.　81	10.　105
11.　125	12.　160	13.　75	14.　150	15.　210
16.　40	17.　250	18.　93	19.　168	20.　360
21.　126	22.　48	23.　17	24.　47	25.　51
26.　144	27.　121	28.　169	29.　225	30.　52
31.　32	32.　98	33.　108	34.　103	35.　101
36.　202	37.　78	38.　500	39.　10,000	40.　100,000

Using the prime factorization of each number, find all the factors (or divisors) of each number.

41.　12	42.　18	43.　28	44.　98	45.　121
46.　45	47.　105	48.　54	49.　97	50.　144

2.6　LEAST COMMON MULTIPLE (LCM)

The techniques discussed in this section are used throughout Chapter 3 on fractions. Study these ideas thoroughly because they will make your work with fractions much easier.

Remember that the multiples of a number are the products of that number with the counting numbers. One multiple of a number is that number, and all other multiples are larger than the number.

counting numbers	1, 2, 3, 4, 5, 6, 7, 8, 9, 10, ...
multiples of 6	6, 12, 18, 24, ⟨30⟩, 36, 42, 48, 54, ⟨60⟩, ...
multiples of 10	10, 20, ⟨30⟩, 40, 50, ⟨60⟩, 70, 80, ⟨90⟩, 100, ...

The common multiples for 6 and 10 are 30, 60, 90, 120, ... The **Least Common Multiple** is 30.

DEFINITION   The **Least Common Multiple (LCM)** of a set of counting numbers is the smallest number common to all the sets of multiples of the given numbers.

Except for the number itself, **factors** (or **divisors**) of the number are smaller than the number, and **multiples** are larger than the number. For example, 6 and 10 are factors of 30, and both are smaller than 30. 30 is a multiple of 6 and 10 and is larger than either 6 or 10.

Listing all the multiples as we did for 6 and 10 and then choosing the least common multiple (LCM) is not very efficient. Either of two other techniques, one involving prime factorizations and the other involving division by prime factors, is easier to apply.

I.   To find the LCM of a set of counting numbers:

(a) Find the prime factorization of each number.

(b) Find the prime factors that appear in **any one** of the prime factorizations.

(c) Form the product of these primes using each prime the most number of times it appears in **any one** of the prime factorizations.

EXAMPLES   1. Find the LCM for 18, 30, 45.

$$18 = 2 \cdot 9 = 2 \cdot 3 \cdot 3 = 2^1 \cdot 3^2$$
$$30 = 6 \cdot 5 = 2 \cdot 3 \cdot 5 = 2^1 \cdot 3^1 \cdot 5^1$$
$$45 = 9 \cdot 5 = 3 \cdot 3 \cdot 5 = 3^2 \cdot 5^1$$

For 2:

$$\left. \begin{cases} 2^1 \text{ is a factor of 18} \\ 2^1 \text{ is a factor of 30} \end{cases} \right\} \quad \text{Use } 2^1 \text{ in the LCM.}$$

For 3:

$$\left.\begin{array}{l} 3^2 \text{ is a factor of 18} \\ 3^1 \text{ is a factor of 30} \\ 3^2 \text{ is a factor of 45} \end{array}\right\} \quad \text{Use } 3^2 \text{ in the LCM.}$$

For 5:

$$\left.\begin{array}{l} 5^1 \text{ is a factor of 30} \\ 5^1 \text{ is a factor of 45} \end{array}\right\} \quad \text{Use } 5^1 \text{ in the LCM.}$$

$$\text{LCM} = 2^1 \cdot 3^2 \cdot 5^1 = 90$$

2. Find the LCM for 36, 24, 48.

$$\left.\begin{array}{l} 36 = 4 \cdot 9 = 2^2 \cdot 3^2 \\ 24 = 8 \cdot 3 = 2^3 \cdot 3 \\ 48 = 16 \cdot 3 = 2^4 \cdot 3 \end{array}\right\} \quad \begin{array}{l} \text{LCM} = 2^4 \cdot 3^2 \\ \phantom{\text{LCM}} = 144 \end{array}$$

3. Find the LCM for 27, 30, 42.

$$\left.\begin{array}{l} 27 = 3 \cdot 9 = 3^3 \\ 30 = 6 \cdot 5 = 2 \cdot 3 \cdot 5 \\ 42 = 2 \cdot 21 = 2 \cdot 3 \cdot 7 \end{array}\right\} \quad \begin{array}{l} \text{LCM} = 2 \cdot 3^3 \cdot 5 \cdot 7 \\ \phantom{\text{LCM}} = 1890 \end{array}$$

The second technique for finding the LCM involves division.

II.   To find the LCM of a set of counting numbers:

   (a) Write the numbers horizontally and find a prime number that will divide into more than one number, if possible.

   (b) Divide by that prime and write the quotients beneath the dividends. Rewrite any numbers not divided beneath themselves.

   (c) Continue the process until no two numbers have a common prime divisor.

   (d) The LCM is the product of all the prime divisors and the last set of quotients.

EXAMPLE

Find the LCM for 20, 25, 18, 6.

$$\left.\begin{array}{l} 5\,\overline{)\,20 \quad 25 \quad 18 \quad 6\,} \\ 2\,\overline{)\,4 \quad 5 \quad 18 \quad 6\,} \\ 3\,\overline{)\,2 \quad 5 \quad 9 \quad 3\,} \\ 2 \quad 5 \quad 3 \quad 1 \end{array}\right\} \quad \begin{array}{l} \text{LCM} = 5 \cdot 2 \cdot 3 \cdot 2 \cdot 5 \cdot 3 \cdot 1 \\ \phantom{\text{LCM}} = 2^2 \cdot 3^2 \cdot 5^2 \\ \phantom{\text{LCM}} = 900 \end{array}$$

Each of the numbers in the set is a factor (or divisor) of the LCM. To find out how many times each number divides into the LCM, we can investigate the prime factorizations. We do not have to do any dividing. All the prime factors of each number are present in the LCM.

EXAMPLES

1. Find the LCM for 12, 18, 20. How many times does each number divide into the LCM?

$$\left.\begin{array}{l} 12 = 2^2 \cdot 3 \\ 18 = 2 \cdot 3^2 \\ 20 = 2^2 \cdot 5 \end{array}\right\} \quad \text{LCM} = 2^2 \cdot 3^2 \cdot 5 = 180$$

$$180 = 2 \cdot 2 \cdot 3 \cdot 3 \cdot 5 = (2 \cdot 2 \cdot 3) \cdot (3 \cdot 5) = 12 \cdot 15$$
$$180 = 2 \cdot 2 \cdot 3 \cdot 3 \cdot 5 = (2 \cdot 3 \cdot 3) \cdot (2 \cdot 5) = 18 \cdot 10$$
$$180 = 2 \cdot 2 \cdot 3 \cdot 3 \cdot 5 = (2 \cdot 2 \cdot 5) \cdot (3 \cdot 3) = 20 \cdot 9$$

Thus,

12 divides 15 times into 180;

18 divides 10 times into 180; and

20 divides 9 times into 180.

2. Find the LCM for 27, 30, 42 and tell how many times each number divides into the LCM.

$$\left.\begin{array}{l} 27 = 3^3 \\ 30 = 2 \cdot 3 \cdot 5 \\ 42 = 2 \cdot 3 \cdot 7 \end{array}\right\} \quad \text{LCM} = 2 \cdot 3^3 \cdot 5 \cdot 7 = 1890$$

$$1890 = 2 \cdot 3 \cdot 3 \cdot 3 \cdot 5 \cdot 7 = (3 \cdot 3 \cdot 3) \cdot (2 \cdot 5 \cdot 7) = 27 \cdot 70$$
$$1890 = 2 \cdot 3 \cdot 3 \cdot 3 \cdot 5 \cdot 7 = (2 \cdot 3 \cdot 5) \cdot (3 \cdot 3 \cdot 7) = 30 \cdot 63$$
$$1890 = 2 \cdot 3 \cdot 3 \cdot 3 \cdot 5 \cdot 7 = (2 \cdot 3 \cdot 7) \cdot (3 \cdot 3 \cdot 5) = 42 \cdot 45$$

We will use this technique for finding how many times a number divides into the LCM throughout Chapter 3 in our work with fractions.

PRACTICE QUIZ	Find the LCM for each of the following sets of numbers.	ANSWERS
	1. 30, 40, 50	**1.** LCM = 600
	2. 28, 70	**2.** LCM = 140
	3. 168, 140	**3.** LCM = 840

EXERCISES 2.6

Study the procedures and examples in this section carefully before you begin these exercises.

Find the LCM for each of the following sets of numbers.

1.	8, 12	**2.**	3, 5, 7	**3.**	4, 6, 9	**4.**	3, 5, 9
5.	2, 5, 11	**6.**	4, 14, 18	**7.**	6, 15, 12	**8.**	6, 8, 27
9.	25, 40	**10.**	40, 75	**11.**	28, 98	**12.**	30, 75
13.	30, 80	**14.**	16, 28	**15.**	25, 100	**16.**	20, 50
17.	35, 100	**18.**	144, 216	**19.**	36, 42	**20.**	40, 100
21.	2, 4, 8	**22.**	10, 15, 35	**23.**	8, 13, 15	**24.**	25, 35, 49
25.	6, 12, 15	**26.**	8, 10, 120	**27.**	6, 15, 80	**28.**	13, 26, 169
29.	45, 125, 150		**30.**	34, 51, 54		**31.**	33, 66, 121
32.	36, 54, 72		**33.**	45, 145, 290		**34.**	54, 81, 108
35.	45, 75, 135		**36.**	35, 40, 72		**37.**	10, 20, 30, 40
38.	15, 25, 30, 40		**39.**	24, 40, 48, 56		**40.**	169, 637, 845

Find the LCM and tell how many times each number divides into the LCM.

41.	8, 10, 15	**42.**	6, 15, 30	**43.**	10, 15, 24
44.	8, 10, 120	**45.**	6, 18, 27, 45	**46.**	12, 95, 228
47.	45, 63, 98	**48.**	40, 56, 196	**49.**	99, 143, 363
50.	125, 135, 225				

51. Two long-distance joggers are running on the same course. They meet and say "Hi." One jogger goes around the course in 10 minutes and the other goes around the course in 14 minutes. They continue to jog until they meet again. In how many minutes will they meet? How many times will each jog around the course?

52. Three night watchmen walk around inspecting buildings at a shopping center. The watchmen take 9, 12, and 14 minutes, respectively, for the inspection trip. If they start at the same time, in how many minutes will they meet? How many inspection trips will each watchman have made?

53. Two astronauts miss connections at their first rendezvous in space. If one astronaut circles the earth every 12 hours and the other every 16 hours, in how many hours will they rendezvous again? How many orbits will each astronaut make before the second rendezvous?

54. Three truck drivers lunch together whenever all three are at the routing station at the same time. The route for the first driver takes 5 days, for the second driver 15 days, and for the third driver 6 days. How often do the three drivers lunch together? If the first driver's route was changed to 6 days, how often would they lunch together?

55. Four book salespersons leave the home office the same day. They take 10 days, 12 days, 15 days, and 18 days, respectively, to travel their own sales regions. In how many days will they all meet again at the home office? How many sales trips will each have made?

SUMMARY: CHAPTER 2

DEFINITION   An **exponent** is a number that tells how many times its base is to be used as a factor.

DEFINITION   For any nonzero whole number a, $a^0 = 1$.

RULES FOR ORDER OF OPERATIONS

1. First, simplify expressions within parentheses.

2. Second, find any powers indicated by exponents.

3. Third, moving from left to right, perform any multiplications or divisions in the order they appear.

4. Fourth, moving from left to right, perform any additions or subtractions in the order they appear.

TESTS FOR DIVISIBILITY BY 2, 3, 4, 5, 9, AND 10

For 2:   If the last digit (units digit) of a whole number is 0, 2, 4, 6, or 8, then the whole number is divisible by 2. In other words, even whole numbers are divisible by 2; odd whole numbers are not divisible by 2.

For 3:   If the sum of the digits of a whole number is divisible by 3, then the number is divisible by 3.

For 4:   If the last two digits of a whole number form a number that is divisible by 4, then the number is divisible by 4. (00 is considered to be divisible by 4.)

For 5:   If the last digit of a whole number is 0 or 5, then the number is divisible by 5.

For 9:   If the sum of the digits of a whole number is divisible by 9, then the number is divisible by 9.

For 10:   If the last digit of a whole number is 0, then the number is divisible by 10.

DEFINITION   A **prime number** is a counting number with exactly two different factors (or divisors).

DEFINITION   A **composite number** is a counting number with more than two different factors (or divisors).

FUNDAMENTAL
THEOREM OF
ARITHMETIC   Every composite number has a unique prime factorization.

DEFINITION   The **Least Common Multiple (LCM)** of a set of natural numbers is the smallest number common to all the sets of multiples of the given numbers.

TO FIND THE LCM OF A SET OF NATURAL NUMBERS

I. (a) Find the prime factorization of each number.

    (b) Find the prime factors that appear in **any one** of the prime factorizations.

    (c) Form the product of these primes using each prime the most number of times it appears in **any one** of the prime factorizations.

II. (a) Write the numbers horizontally and find a prime number that will divide into more than one number, if possible.

    (b) Divide by that prime and write the quotients beneath the dividends. Rewrite any numbers not divided beneath themselves.

    (c) Continue the process until no two numbers have a common prime divisor.

    (d) The LCM is the product of all the prime divisors and the last set of quotients.

REVIEW QUESTIONS: CHAPTER 2

1. In the expression $3^4 = 81$, 3 is called the _____, 4 is called the _____, and 81 is called the _____.

2. A prime number is a counting number with exactly two _____ _____.

3. Every composite number has a unique _____ factorization.

4. Since 36 has more than two different factors, 36 is a _____ number.

Find a base and exponent form for each of the following powers without using the exponent 1.

5. 128

6. 169

Evaluate each of the following expressions.

7. $7 + 3 \cdot 2 - 1 + 9 \div 3$

8. $3 \cdot 2^5 - 2 \cdot 5^2$

9. $14 \div 2 + 2 \cdot 8 + 30 \div 5 \cdot 2$

10. $(16 \div 2^2 + 6) \div 2 + 8$

11. $(75 - 3 \cdot 5) \div 10 - 4$

12. $(7^2 \cdot 2 + 2) \div 10 \div (2 + 3)$

Determine which of the numbers 2, 3, 4, 5, 9, and 10 (if any) will divide exactly into each of the following numbers.

13. 45

14. 72

15. 479

16. 5040

17. 8836

18. 575,493

19. List the multiples of 3. Are any of these multiples prime numbers?

20. Is 223 a prime number? Show your work.

21. List the prime numbers less than 60.

22. Find two factors of 24 whose sum is 10.

23. Find two factors of 60 whose sum is 17.

Find the prime factorization for each of the following numbers.

24. 150

25. 65

26. 84

27. 92

Find the LCM for each of the following sets of numbers.

28. 8, 14, 24   **29.** 8, 12, 25, 36   **30.** 27, 54, 135

31. Find the LCM for the numbers 18, 39, and 63 and tell how many times each number divides into the LCM.

32. One racing car goes around the track every 30 seconds and the other every 35 seconds. If both cars start a race from the same point, in how many seconds will the first car be exactly one lap ahead of the second car? How many laps will each car have made when the first car is two laps ahead of the second?

CHAPTER TEST: CHAPTER 2

1. A number that tells how many times another number is used as a factor is called a/an _____.

2. Write 125 in an exponent-base form without using the exponent 1.

3. Explain why 35 is not a prime number.

4. Without dividing, tell why 4 will divide into 216 but 5 will not.

5. List the multiples of 6. Are any of these multiples prime numbers?

Evaluate each of the following expressions.

6. $2 \cdot 5^2 + 21 \div 3$ 　　　　　　　7. $12 \div 3 \cdot 2 - 5 + 2^3 \div 4$

8. $27 - 2(8 + 10 \div 2)$

Find the prime factorization for each of the following numbers.

9. 300 　　　　10. 52 　　　　11. 64 　　　　12. 180

Find the LCM for each of the following sets of numbers and tell how many times each number divides into the LCM.

13. 15, 25, 75 　　　　　　　14. 42, 105, 147

15. Two girls are exercising on bicycles. One cycles around a track in 60 seconds and the other takes 75 seconds. In how many seconds will the first girl overtake the second girl? How many laps will each have cycled if they stop for a rest when the first girl overtakes the second?

3

RATIONAL NUMBERS

3.1 MULTIPLYING RATIONAL NUMBERS

Rational number is the technical name for the common term **fraction.** There are fractions that are not rational numbers, but we will not study these in any detail in this text.

Rational numbers can be used to indicate

1. equal parts of a whole,

2. division.

(In Chapter 5 we will also use rational numbers to solve algebraic equations such as $5x = 3$.)

DEFINITION

A **rational number** is a number that can be written in the form $\frac{a}{b}$, where a is a whole number and b is a nonzero whole number.

$\dfrac{a}{b}$ numerator
denominator

[NOTE: The numerator a can be 0, but the denominator b cannot be 0.]

Examples of rational numbers are $\frac{1}{2}$, $\frac{3}{4}$, $\frac{9}{10}$, and $\frac{17}{3}$.

Figure 3.1 shows how a whole may be separated into equal parts. We see that the rational numbers $\frac{1}{2}$, $\frac{2}{4}$, $\frac{4}{8}$, and $\frac{6}{12}$ all represent the same amount of the whole. These numbers are **equivalent,** or **equal.**

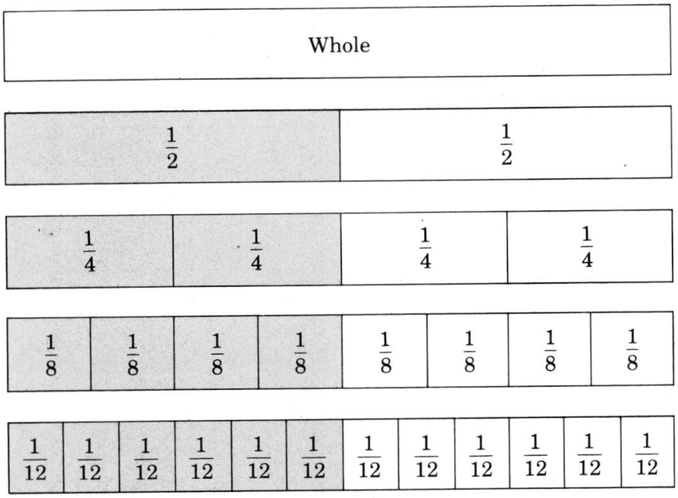

Figure 3.1

$$\frac{1}{2} = \frac{2}{4} = \frac{3}{6} = \frac{4}{8} = \frac{5}{10} = \frac{6}{12} \text{ and so on}$$

Also,

$$\frac{2}{3} = \frac{4}{6} = \frac{6}{9} = \frac{8}{12} = \frac{10}{15} = \frac{12}{18} \text{ and so on}$$

And, with whole numbers,

$$0 = \frac{0}{1} = \frac{0}{2} = \frac{0}{3} = \frac{0}{4} = \frac{0}{5} \text{ and so on}$$

$$1 = \frac{1}{1} = \frac{2}{2} = \frac{3}{3} = \frac{4}{4} = \frac{5}{5} \text{ and so on}$$

$$2 = \frac{2}{1} = \frac{4}{2} = \frac{6}{3} = \frac{8}{4} = \frac{10}{5} \text{ and so on}$$

Rational numbers such as $\frac{6}{3}$, $\frac{15}{7}$, and $\frac{18}{5}$, in which the numerator is larger than the denominator, are sometimes called **improper fractions.** This term is misleading because there is nothing "improper" about such fractions. In fact, in algebra and other courses in mathematics, improper fractions are preferred over mixed numbers such as $2\frac{1}{7}$ and $3\frac{3}{5}$. We will refer to the **fraction form** of a rational number regardless of whether the numerator or the denominator is larger. Mixed numbers will be discussed later.

Under the division concept of rational numbers, $\frac{15}{5} = 3$ because $15 = 3 \cdot 5$. Similarly, $\frac{0}{2} = 0$ since $0 = 0 \cdot 2$. In general, $\frac{0}{b} = 0$ if $b \neq 0$ ($\neq$ means **is not equal to**).

DIVISION BY 0 IS UNDEFINED. NO DENOMINATOR CAN BE 0.

Consider $\frac{5}{0} = \square$. Then $5 = 0 \cdot \square = 0$. But this is impossible; $5 \neq 0$. Next consider $\frac{0}{0} = \square$. Then $0 = 0 \cdot \square = 0$ for any value of $\square$. This means that $\frac{0}{0}$ could be any number. But in arithmetic an operation such as division cannot give more than one answer. Thus, $\frac{a}{0}$ is **undefined for any value of** a.

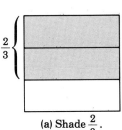 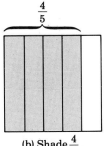

(a) Shade $\frac{2}{3}$.   (b) Shade $\frac{4}{5}$.   (c) The overlapped region is
$$\frac{2}{3} \cdot \frac{4}{5} = \frac{8}{15}.$$

Figure 3.2

Multiplication of $\frac{2}{3}$ by $\frac{4}{5}$ is illustrated in Figure 3.2. Draw a square and (a) separate it into thirds in one direction (shade in 2 thirds), (b) separate it into fifths in the other direction (shade in 4 fifths), then (c) the two squares overlap in fifteenths, and the common shaded portion represents the product $\frac{2}{3} \cdot \frac{4}{5} = \frac{8}{15}$.

The product $\frac{1}{3} \cdot \frac{2}{5}$ is illustrated in Figure 3.3.

The diagrams in Figures 3.2 and 3.3 indicate that to multiply two fractions, multiply their numerators and multiply their denominators.

DEFINITION   The **product** of two rational numbers $\frac{a}{b}$ and $\frac{c}{d}$ is the rational number whose numerator is the product of the numerators $(a \cdot c)$ and whose denominator is the product of the denominators $(b \cdot d)$. That is,

$$\frac{a}{b} \cdot \frac{c}{d} = \frac{a \cdot c}{b \cdot d}$$

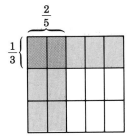

The square is separated into thirds in one direction and into fifths in the other direction. There are 15 equal regions. The overlapping shaded region represents

$$\frac{1}{3} \cdot \frac{2}{5} = \frac{2}{15}$$

Figure 3.3

| EXAMPLES | 1. $\dfrac{2}{3} \cdot \dfrac{5}{7} = \dfrac{2 \cdot 5}{3 \cdot 7} = \dfrac{10}{21}$ |

2. $\dfrac{1}{4} \cdot \dfrac{3}{5} = \dfrac{1 \cdot 3}{4 \cdot 5} = \dfrac{3}{20}$

3. $\dfrac{7}{5} \cdot \dfrac{2}{1} = \dfrac{7 \cdot 2}{5 \cdot 1} = \dfrac{14}{5}$

The definition can be extended to multiplication with more than two fractions.

4. $\dfrac{1}{4} \cdot \dfrac{3}{5} \cdot \dfrac{7}{2} = \dfrac{1 \cdot 3 \cdot 7}{4 \cdot 5 \cdot 2} = \dfrac{21}{40}$

5. $\dfrac{2}{3} \cdot \dfrac{11}{15} \cdot \dfrac{1}{7} = \dfrac{2 \cdot 11 \cdot 1}{3 \cdot 15 \cdot 7} = \dfrac{22}{315}$

As with whole numbers, multiplication with rational numbers is both **commutative** and **associative.**

COMMUTATIVE PROPERTY OF MULTIPLICATION

If $\dfrac{a}{b}$ and $\dfrac{c}{d}$ are rational numbers, then

$$\frac{a}{b} \cdot \frac{c}{d} = \frac{c}{d} \cdot \frac{a}{b}$$

ASSOCIATIVE PROPERTY OF MULTIPLICATION

If $\dfrac{a}{b}, \dfrac{c}{d}$, and $\dfrac{e}{f}$ are rational numbers, then

$$\frac{a}{b} \cdot \frac{c}{d} \cdot \frac{e}{f} = \left(\frac{a}{b} \cdot \frac{c}{d}\right) \cdot \frac{e}{f} = \frac{a}{b} \cdot \left(\frac{c}{d} \cdot \frac{e}{f}\right)$$

The properties are stated here for general knowledge and completeness without any related exercises.

PRACTICE QUIZ	Find the products.	ANSWERS
	1. $\dfrac{3}{4} \cdot \dfrac{5}{7}$	1. $\dfrac{15}{28}$
	2. $\dfrac{1}{3} \cdot \dfrac{1}{4}$	2. $\dfrac{1}{12}$
	3. $\dfrac{1}{5} \cdot \dfrac{3}{7} \cdot \dfrac{11}{2}$	3. $\dfrac{33}{70}$

EXERCISES 3.1

Study the text and examples carefully before you work these exercises.

1. Each of the following products has the same value. What is that value?

(a) $\dfrac{0}{4} \cdot \dfrac{5}{6}$ (b) $\dfrac{3}{10} \cdot \dfrac{0}{8}$ (c) $\dfrac{1}{10} \cdot \dfrac{0}{10}$ (d) $\dfrac{0}{6} \cdot \dfrac{0}{52}$

2. What is the value, if any, of each of the following numbers?

(a) $\dfrac{7}{0}$ (b) $\dfrac{16}{0}$ (c) $\dfrac{75}{0} \cdot \dfrac{2}{0}$ (d) $\dfrac{1}{0} \cdot \dfrac{0}{2}$

Write a rational number that represents the shaded parts in each of the following diagrams.

3. 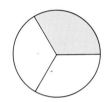   4. 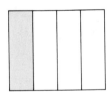   5.

6. 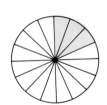   7. 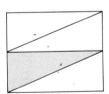   8.

Draw diagrams similar to Figure 3.3 that illustrate each of the following products.

9. $\dfrac{1}{5} \cdot \dfrac{3}{4}$ 10. $\dfrac{5}{6} \cdot \dfrac{5}{6}$ 11. $\dfrac{7}{8} \cdot \dfrac{1}{4}$ 12. $\dfrac{2}{3} \cdot \dfrac{4}{7}$ 13. $\dfrac{2}{9} \cdot \dfrac{2}{3}$

Find the following products.

14. $\dfrac{1}{5} \cdot \dfrac{3}{5}$ 15. $\dfrac{1}{4} \cdot \dfrac{1}{4}$ 16. $\dfrac{1}{3} \cdot \dfrac{2}{3}$ 17. $\dfrac{6}{7} \cdot \dfrac{2}{5}$

18. $\dfrac{4}{7} \cdot \dfrac{3}{5}$ 19. $\dfrac{1}{3} \cdot \dfrac{1}{3}$ 20. $\dfrac{1}{2} \cdot \dfrac{1}{2}$ 21. $\dfrac{3}{16} \cdot \dfrac{1}{2}$

22. $\dfrac{2}{5} \cdot \dfrac{2}{5}$ 23. $\dfrac{3}{7} \cdot \dfrac{3}{7}$ 24. $\dfrac{1}{2} \cdot \dfrac{3}{4}$ 25. $\dfrac{5}{8} \cdot \dfrac{3}{4}$

26. $\dfrac{1}{9} \cdot \dfrac{4}{9}$ 27. $\dfrac{0}{3} \cdot \dfrac{5}{7}$ 28. $\dfrac{0}{4} \cdot \dfrac{7}{6}$ 29. $\dfrac{7}{6} \cdot \dfrac{5}{2}$

30. $\dfrac{4}{1} \cdot \dfrac{3}{1}$ **31.** $\dfrac{2}{1} \cdot \dfrac{5}{1}$ **32.** $\dfrac{14}{1} \cdot \dfrac{0}{2}$ **33.** $\dfrac{15}{1} \cdot \dfrac{3}{2}$

34. $\dfrac{6}{5} \cdot \dfrac{7}{1}$ **35.** $\dfrac{8}{5} \cdot \dfrac{4}{3}$ **36.** $\dfrac{5}{6} \cdot \dfrac{11}{3}$ **37.** $\dfrac{9}{4} \cdot \dfrac{11}{5}$

38. $\dfrac{1}{5} \cdot \dfrac{2}{7} \cdot \dfrac{3}{11}$ **39.** $\dfrac{4}{13} \cdot \dfrac{2}{5} \cdot \dfrac{6}{7}$ **40.** $\dfrac{7}{8} \cdot \dfrac{7}{9} \cdot \dfrac{7}{3}$ **41.** $\dfrac{1}{6} \cdot \dfrac{1}{10} \cdot \dfrac{1}{6}$

42. $\dfrac{9}{100} \cdot 1 \cdot 3$ **43.** $\dfrac{5}{1} \cdot \dfrac{12}{1} \cdot \dfrac{14}{1}$ **44.** $\dfrac{1}{3} \cdot \dfrac{1}{10} \cdot \dfrac{7}{3}$

45. $\dfrac{3}{11} \cdot \dfrac{0}{8} \cdot \dfrac{6}{7}$ **46.** $\dfrac{0}{4} \cdot \dfrac{3}{8} \cdot \dfrac{1}{5}$ **47.** $\dfrac{5}{6} \cdot \dfrac{5}{11} \cdot \dfrac{5}{7} \cdot \dfrac{0}{3}$

48. $\left(\dfrac{3}{4}\right)\left(\dfrac{5}{8}\right)\left(\dfrac{11}{13}\right)$ **49.** $\left(\dfrac{7}{5}\right)\left(\dfrac{8}{3}\right)\left(\dfrac{13}{3}\right)$ **50.** $\left(\dfrac{1}{10}\right)\left(\dfrac{3}{5}\right)\left(\dfrac{3}{10}\right)$

3.2   CHANGING TERMS AND REDUCING PRODUCTS

We know that 1 is the **multiplicative identity** for whole numbers; that is, $a \cdot 1 = a$. The number 1 is also the **multiplicative identity** for rational numbers since

$$\frac{a}{b} \cdot 1 = \frac{a}{b} \cdot \frac{1}{1} = \frac{a \cdot 1}{b \cdot 1} = \frac{a}{b}$$

MULTIPLICATIVE
IDENTITY   If $\dfrac{a}{b}$ is a rational number, then

$$\frac{a}{b} \cdot 1 = \frac{a}{b}$$

In general,

$$1 = \frac{2}{2} = \frac{3}{3} = \frac{4}{4} = \cdots = \frac{k}{k} = \cdots$$

where k is any counting number.
Therefore,

$$\frac{a}{b} = \frac{a}{b} \cdot 1 = \frac{a}{b} \cdot \frac{k}{k} = \frac{a \cdot k}{b \cdot k} \quad \text{where } k \neq 0$$

We can use this idea two ways: (1) **to raise to higher terms** (find an equivalent fraction with a larger denominator), and (2) **to reduce** (find an equivalent fraction with a smaller denominator). **To raise to higher terms,** multiply the numerator and denominator by the same counting number.

EXAMPLES

1. Find a rational number with denominator 28 equivalent to $\frac{3}{4}$. $\left(\text{That is,}\right.$ $\left.\frac{3}{4} = \frac{?}{28}.\right)$

$$\frac{3}{4} = \frac{3}{4} \cdot \frac{7}{7} = \frac{21}{28} \qquad \text{(Here } k = 7 \text{ since } 4 \cdot 7 = 28.)$$

2. $\frac{9}{10} = \frac{?}{30}$

$$\frac{9}{10} = \frac{9}{10} \cdot \frac{3}{3} = \frac{27}{30} \qquad \text{(Here } k = 3 \text{ since } 10 \cdot 3 = 30.)$$

3. $\frac{9}{10} = \frac{?}{40}$

$$\frac{9}{10} = \frac{9}{10} \cdot \frac{4}{4} = \frac{36}{40} \qquad \text{(Here } k = 4 \text{ since } 10 \cdot 4 = 40.)$$

4. $\frac{11}{8} = \frac{?}{40}$

$$\frac{11}{8} = \frac{11}{8} \cdot \frac{5}{5} = \frac{55}{40} \qquad \text{(Here } k = 5 \text{ since } 8 \cdot 5 = 40.)$$

A rational number is in **lowest terms** if the numerator and denominator have no common factor other than 1. **To reduce to lowest terms,** factor the numerator and denominator into prime factors and use the fact that $\frac{k}{k} = 1$.

EXAMPLES

Reduce each fraction to lowest terms.

1. $\frac{14}{21}$

$$\frac{14}{21} = \frac{2 \cdot 7}{3 \cdot 7} = \frac{2}{3} \cdot \frac{7}{7} = \frac{2}{3} \cdot 1 = \frac{2}{3}$$

2. $\dfrac{15}{20}$

$$\frac{15}{20} = \frac{3 \cdot 5}{2 \cdot 2 \cdot 5} = \frac{3}{4} \cdot \frac{5}{5} = \frac{3}{4} \cdot 1 = \frac{3}{4}$$

Usually we just cross out common factors (prime or not) with the understanding that $\dfrac{k}{k} = 1$.

3. $\dfrac{44}{20}$

$$\frac{44}{20} = \frac{\cancel{4} \cdot 11}{\cancel{4} \cdot 5} = \frac{11}{5} \qquad \left(\text{Note that } \frac{11}{5} \text{ is a reduced improper fraction.} \right)$$

If the numerator is a factor of the denominator, or vice versa, 1 must be used as a factor.

4. $\dfrac{5}{35}$

$$\frac{5}{35} = \frac{1 \cdot \cancel{5}}{7 \cdot \cancel{5}} = \frac{1}{7} \qquad \text{(1 must be used as a factor.)}$$

5. $\dfrac{8}{72}$

$$\frac{8}{72} = \frac{2 \cdot 2 \cdot 2}{6 \cdot 12} = \frac{\cancel{2} \cdot \cancel{2} \cdot \cancel{2} \cdot 1}{\cancel{2} \cdot 3 \cdot \cancel{2} \cdot \cancel{2} \cdot 3} = \frac{1}{9}$$

or $\qquad \dfrac{8}{72} = \dfrac{\cancel{8} \cdot 1}{\cancel{8} \cdot 9} = \dfrac{1}{9}$

If you see that 8 is a common factor, fine. Otherwise, use the prime factors.

EXAMPLES

1. $\dfrac{15}{28} \cdot \dfrac{7}{9} = \dfrac{15 \cdot 7}{28 \cdot 9} = \dfrac{\cancel{3} \cdot 5 \cdot \cancel{7}}{4 \cdot \cancel{7} \cdot \cancel{3} \cdot 3} = \dfrac{5}{4 \cdot 3} = \dfrac{5}{12}$

2. $\dfrac{9}{10} \cdot \dfrac{25}{32} \cdot \dfrac{44}{33} = \dfrac{9 \cdot 25 \cdot 44}{10 \cdot 32 \cdot 33} = \dfrac{3 \cdot \cancel{3} \cdot 5 \cdot \cancel{5} \cdot \cancel{4} \cdot \cancel{11}}{2 \cdot \cancel{5} \cdot \cancel{4} \cdot 8 \cdot \cancel{3} \cdot \cancel{11}} = \dfrac{3 \cdot 5}{2 \cdot 8} = \dfrac{15}{16}$

3. $\dfrac{36}{49} \cdot \dfrac{14}{75} \cdot \dfrac{15}{18} = \dfrac{36 \cdot 14 \cdot 15}{49 \cdot 75 \cdot 18}$

$$= \frac{\cancel{2} \cdot 2 \cdot \cancel{3} \cdot \cancel{3} \cdot 2 \cdot \cancel{7} \cdot \cancel{3} \cdot \cancel{5}}{\cancel{7} \cdot 7 \cdot \cancel{3} \cdot 5 \cdot \cancel{5} \cdot \cancel{2} \cdot \cancel{3} \cdot \cancel{3}} = \frac{2 \cdot 2}{7 \cdot 5} = \frac{4}{35}$$

4. $\dfrac{55}{26} \cdot \dfrac{8}{44} \cdot \dfrac{91}{35} = \dfrac{55 \cdot 8 \cdot 91}{26 \cdot 44 \cdot 35}$

$$= \frac{\cancel{5} \cdot \cancel{11} \cdot \cancel{2} \cdot \cancel{2} \cdot \cancel{2} \cdot \cancel{7} \cdot \cancel{13}}{\cancel{2} \cdot \cancel{13} \cdot \cancel{2} \cdot \cancel{2} \cdot \cancel{11} \cdot \cancel{7} \cdot \cancel{5}} = \frac{1}{1} = 1$$

The use of prime factors in multiplying and reducing rational numbers is a sound technique and particularly thorough for students who have difficulty with fractions. It also provides a consistent approach to multiplying with rational numbers.

When common factors are easily seen, even though they are not prime numbers, they can be divided into both the numerator and denominator, and the quotients can be written. This technique is commonly used, but you must be careful and organized. The last four examples are shown again.

EXAMPLES

1'. $\dfrac{\cancel{15}}{\cancel{28}} \cdot \dfrac{\cancel{7}}{\cancel{9}} = \dfrac{5}{12}$

3 is divided into both 15 and 9.
7 is divided into both 7 and 28.

2'. $\dfrac{\cancel{9}}{\cancel{10}} \cdot \dfrac{\cancel{25}}{\cancel{32}} \cdot \dfrac{\cancel{44}}{\cancel{33}} = \dfrac{15}{16}$

11 is divided into both 44 and 33.
5 is divided into both 25 and 10.
4 is divided into both 4 and 32.
3 is divided into both 3 and 9.

3'. $\dfrac{\cancel{36}}{\cancel{49}} \cdot \dfrac{\cancel{14}}{\cancel{75}} \cdot \dfrac{\cancel{15}}{\cancel{18}} = \dfrac{4}{35}$

18 is divided into both 18 and 36.
7 is divided into both 14 and 49.
15 is divided into both 15 and 75.

4'. $\dfrac{\cancel{55}}{\cancel{26}} \cdot \dfrac{\cancel{8}}{\cancel{44}} \cdot \dfrac{\cancel{91}}{\cancel{35}} = \dfrac{1}{1}$

11 is divided into both 55 and 44.
13 is divided into both 26 and 91.
5 is divided into both 5 and 35.
7 is divided into both 7 and 7.
2 is divided into both 2 and 8.
4 is divided into both 4 and 4.

Choose the technique that best suits your own needs and abilities.

3

Raise to higher terms as indicated.

ANSWERS

1. $\dfrac{3}{5} = \dfrac{60}{100}$

$2/3$

1. $\dfrac{60}{100}$

2. $\dfrac{7}{2} = \dfrac{35}{10}$

2. $\dfrac{35}{10}$

Reduce to lowest terms.

3. $\dfrac{25}{55}$ $\dfrac{5}{11}$

3. $\dfrac{5}{11}$

4. $\dfrac{34}{51}$

4. $\dfrac{2}{3}$

Multiply. $\left[\text{HINT: Write 6 as } \dfrac{6}{1} \text{ and}\right.$

factor before multiplying. $\Big]$

16

5. $\dfrac{17}{100} \cdot \dfrac{27}{34} \cdot \dfrac{25}{9} \cdot \dfrac{6}{1}$

5. $\dfrac{9}{4}$

EXERCISES 3.2

Review the text material before you start these exercises.

Raise to higher terms or reduce to lower terms as indicated.

1. $\dfrac{7}{8} = \dfrac{21}{24}$ $\dfrac{7}{8}$ 2. $\dfrac{1}{16} = \dfrac{4}{64} \dfrac{1}{16}$ 3. $\dfrac{2}{5} = \dfrac{10}{25} \dfrac{1}{5}$ 4. $\dfrac{6}{7} = \dfrac{42}{49}$

5. $\dfrac{1}{9} = \dfrac{5}{45} \dfrac{1}{9}$ 6. $\dfrac{3}{4} = \dfrac{15}{20} \dfrac{3}{4}$ 7. $\dfrac{5}{8} = \dfrac{10}{16} \dfrac{5}{8}$ 8. $\dfrac{6}{5} = \dfrac{54}{45}$

9. $\dfrac{14}{3} = \dfrac{}{9}$ 10. $\dfrac{5}{8} = \dfrac{}{96}$ 11. $\dfrac{9}{16} = \dfrac{}{96}$ 12. $\dfrac{7}{2} = \dfrac{}{20}$

13. $\dfrac{10}{11} = \dfrac{}{44}$ 14. $\dfrac{3}{16} = \dfrac{}{80}$ 15. $\dfrac{11}{12} = \dfrac{}{48}$ 16. $\dfrac{3}{7} = \dfrac{}{105}$

17. $\dfrac{5}{21} = \dfrac{}{42}$ 18. $\dfrac{2}{3} = \dfrac{}{48}$ 19. $\dfrac{5}{12} = \dfrac{}{108}$ 20. $\dfrac{1}{13} = \dfrac{}{39}$

21. $\dfrac{12}{14} = \dfrac{}{7}$ 22. $\dfrac{18}{40} = \dfrac{}{20}$ 23. $\dfrac{90}{100} = \dfrac{}{10}$ 24. $\dfrac{80}{100} = \dfrac{}{10}$

25. $\dfrac{60}{100} = \dfrac{}{10}$ 26. $\dfrac{42}{70} = \dfrac{}{10}$ 27. $\dfrac{10}{100} = \dfrac{}{10}$ 28. $\dfrac{66}{60} = \dfrac{}{10}$

29. $\dfrac{88}{80} = \dfrac{}{10}$ 30. $\dfrac{64}{48} = \dfrac{}{6}$ 31. $\dfrac{3}{100} = \dfrac{}{1000}$ 32. $\dfrac{9}{10} = \dfrac{}{1000}$

Reduce the following rational numbers to lowest terms. Just rewrite the fraction if it is already reduced.

33. $\frac{3}{9}$ $\frac{1}{3}$ 34. $\frac{16}{24}$ $\frac{2}{3}$ 35. $\frac{9}{12}$ $\frac{3}{5}$ 36. $\frac{6}{20}$ 37. $\frac{16}{40}$

38. $\frac{24}{30}$ 39. $\frac{14}{36}$ 40. $\frac{5}{11}$ 41. $\frac{0}{25}$ 42. $\frac{75}{100}$

43. $\frac{22}{55}$ 44. $\frac{60}{75}$ 45. $\frac{30}{36}$ 46. $\frac{7}{28}$ 47. $\frac{26}{39}$

48. $\frac{27}{56}$ 49. $\frac{34}{51}$ 50. $\frac{36}{48}$ 51. $\frac{24}{100}$ 52. $\frac{16}{32}$

53. $\frac{30}{45}$ 54. $\frac{28}{42}$ 55. $\frac{12}{35}$ 56. $\frac{66}{84}$ 57. $\frac{14}{63}$

58. $\frac{30}{70}$ 59. $\frac{25}{76}$ 60. $\frac{70}{84}$ 61. $\frac{50}{100}$ 62. $\frac{48}{12}$

63. $\frac{54}{9}$ 64. $\frac{51}{6}$ 65. $\frac{6}{51}$ 66. $\frac{27}{72}$ 67. $\frac{18}{40}$

68. $\frac{144}{156}$ 69. $\frac{150}{135}$ 70. $\frac{121}{165}$ 71. $\frac{140}{112}$ 72. $\frac{96}{108}$

Multiply. [HINT: Factor before multiplying.]

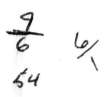

73. $\frac{23}{36} \cdot \frac{20}{46}$ 74. $\frac{1}{8} \cdot \frac{4}{21}$ $\frac{1}{6}$ 75. $\frac{5}{15} \cdot \frac{18}{24}$

76. $\frac{20}{32} \cdot \frac{9}{13} \cdot \frac{26}{7}$ 77. $\frac{69}{15} \cdot \frac{30}{8} \cdot \frac{14}{46}$ 78. $\frac{42}{52} \cdot \frac{27}{22} \cdot \frac{33}{9}$

79. $\frac{3}{4} \cdot 18 \cdot \frac{7}{2} \cdot \frac{22}{54}$ 80. $\frac{9}{10} \cdot \frac{35}{40} \cdot \frac{65}{15}$ 81. $\frac{66}{84} \cdot \frac{12}{5} \cdot \frac{28}{33}$

82. $\frac{24}{100} \cdot \frac{36}{48} \cdot \frac{15}{9}$ 83. $\frac{17}{10} \cdot \frac{5}{42} \cdot \frac{18}{51} \cdot \frac{4}{1}$ 84. $\frac{75}{8} \cdot \frac{16}{36} \cdot \frac{9}{1} \cdot \frac{7}{25}$

3.3 ADDING RATIONAL NUMBERS

Consider the two rational numbers $\frac{3}{7}$ and $\frac{1}{7}$ illustrated by the shaded portions in the diagrams in Figure 3.4.

$\frac{3}{7}$ $\frac{1}{7}$

(a)   (b)

Figure 3.4

If these two shaded portions are combined, that is, put into one diagram, the result is the **sum** of the two numbers $\frac{3}{7} + \frac{1}{7}$, as in Figure 3.5.

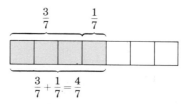

Figure 3.5

Similar diagrams can be used to illustrate the following procedure:

> To add two fractions with the same denominator, add the numerators and use the common denominator.

We state this as a definition.

DEFINITION

The **sum** of two rational numbers $\frac{a}{b}$ and $\frac{c}{b}$ is the rational number whose numerator is the sum of the numerators $(a + c)$ and whose denominator is the common denominator (b). That is,

$$\frac{a}{b} + \frac{c}{b} = \frac{a + c}{b} \qquad (b \neq 0)$$

For the remainder of this chapter, we will follow the customary procedure of reducing all answers to lowest terms, whether adding, subtracting, multiplying, or dividing with rational numbers.

EXAMPLES

1. $\dfrac{1}{3} + \dfrac{1}{3} = \dfrac{1+1}{3} = \dfrac{2}{3}$

2. $\dfrac{4}{15} + \dfrac{6}{15} = \dfrac{4+6}{15} = \dfrac{10}{15} = \dfrac{2 \cdot \cancel{5}}{3 \cdot \cancel{5}} = \dfrac{2}{3}$

Addition can be extended to more than two numbers by adding all the numerators and using the same denominator.

3. $\dfrac{1}{4} + \dfrac{2}{4} + \dfrac{3}{4} = \dfrac{1+2+3}{4} = \dfrac{6}{4} = \dfrac{\cancel{2} \cdot 3}{\cancel{2} \cdot 2} = \dfrac{3}{2}$

4. $\dfrac{2}{7} + \dfrac{3}{7} + \dfrac{1}{7} + \dfrac{6}{7} = \dfrac{2+3+1+6}{7} = \dfrac{12}{7}$

Of course, numbers to be added will not always have the same denominator. In such a case, find numbers with the same denominator that are equivalent to the numbers to be added. For example, to add $\dfrac{1}{6} + \dfrac{3}{10}$, we can write

$$\frac{1}{6} = \frac{1}{6} \cdot \frac{5}{5} = \frac{5}{30} \quad \text{and} \quad \frac{3}{10} = \frac{3}{10} \cdot \frac{3}{3} = \frac{9}{30}$$

so

$$\frac{1}{6} + \frac{3}{10} = \frac{5}{30} + \frac{9}{30}$$

Why was 30 the common denominator? Another common denominator would be 60, but 30 is smaller. We want to find the **smallest common denominator (SCD).**

To find the smallest common denominator (SCD), find the least common multiple (LCM) of the denominator. Then change each number to an equivalent number with the SCD and add these numbers.

EXAMPLES

1. $\dfrac{1}{4} + \dfrac{3}{8} + \dfrac{7}{10}$

Find the LCM of 4, 8, 10.

$$\left.\begin{array}{l} 4 = 2 \cdot 2 \\ 8 = 2 \cdot 2 \cdot 2 \\ 10 = 2 \cdot 5 \end{array}\right\} \quad \begin{array}{l} \text{LCM} = 2 \cdot 2 \cdot 2 \cdot 5 = 40 \\ \phantom{\text{LCM}} = 4 \cdot 10 \\ \phantom{\text{LCM}} = 8 \cdot 5 \\ \phantom{\text{LCM}} = 10 \cdot 4 \end{array}$$

$$\frac{1}{4} + \frac{3}{8} + \frac{7}{10} = \frac{1}{4} \cdot \frac{10}{10} + \frac{3}{8} \cdot \frac{5}{5} + \frac{7}{10} \cdot \frac{4}{4} = \frac{10}{40} + \frac{15}{40} + \frac{28}{40} = \frac{53}{40}$$

$\dfrac{10}{10}$ is used since $4 \cdot 10 = 40$ $\dfrac{5}{5}$ is used since $8 \cdot 5 = 40$ $\dfrac{4}{4}$ is used since $10 \cdot 4 = 40$ all fractions have SCD = 40

2. $\dfrac{5}{21} + \dfrac{5}{28}$

$$\left.\begin{array}{l} 21 = 3 \cdot 7 \\ 28 = 2 \cdot 2 \cdot 7 \end{array}\right\} \quad LCM = 2 \cdot 2 \cdot 3 \cdot 7 = 84$$
$$= 21 \cdot 4$$
$$= 28 \cdot 3$$

$$\dfrac{5}{21} + \dfrac{5}{28} = \dfrac{5}{21} \cdot \dfrac{4}{4} + \dfrac{5}{28} \cdot \dfrac{3}{3} = \dfrac{20}{84} + \dfrac{15}{84} = \dfrac{35}{84} = \dfrac{7 \cdot 5}{2 \cdot 2 \cdot 3 \cdot 7} = \dfrac{5}{12}$$

3. $\dfrac{2}{3} + \dfrac{1}{6} + \dfrac{5}{12}$

The LCM for 3, 6, 12 is obviously 12, and no formal work is necessary.

$$\dfrac{2}{3} + \dfrac{1}{6} + \dfrac{5}{12} = \dfrac{2}{3} \cdot \dfrac{4}{4} + \dfrac{1}{6} \cdot \dfrac{2}{2} + \dfrac{5}{12} = \dfrac{8}{12} + \dfrac{2}{12} + \dfrac{5}{12}$$

$$= \dfrac{15}{12} = \dfrac{3 \cdot 5}{2 \cdot 2 \cdot 3} = \dfrac{5}{4}$$

The numbers can be written vertically:

$$\dfrac{2}{3} = \dfrac{2}{3} \cdot \dfrac{4}{4} = \dfrac{8}{12}$$

$$\dfrac{1}{6} = \dfrac{1}{6} \cdot \dfrac{2}{2} = \dfrac{2}{12}$$

$$\dfrac{5}{12} = \dfrac{5}{12} = \dfrac{5}{12}$$

$$\overline{}$$

$$\dfrac{15}{12} = \dfrac{3 \cdot 5}{2 \cdot 2 \cdot 3} = \dfrac{5}{4}$$

4. $\dfrac{1}{4} + \dfrac{1}{16}$

The LCM of 4, 16 is 16.

$$\dfrac{1}{4} + \dfrac{1}{16} = \dfrac{1}{4} \cdot \dfrac{4}{4} + \dfrac{1}{16} = \dfrac{4}{16} + \dfrac{1}{16} = \dfrac{5}{16}$$

An alternative method is illustrated as

$$\dfrac{a}{b} \;\;\lower2pt{\Large\times}\;\; \dfrac{c}{d} = \dfrac{ad + bc}{bd}$$

This method can be used only for adding **two** fractions and will not generally give the SCD.

$$\frac{1}{4} + \frac{1}{16} = \frac{1 \cdot 16 + 4 \cdot 1}{4 \cdot 16} = \frac{16 + 4}{64} = \frac{20}{64} = \frac{\cancel{2} \cdot \cancel{2} \cdot 5}{\cancel{2} \cdot \cancel{2} \cdot 2 \cdot 2 \cdot 2 \cdot 2} = \frac{5}{16}$$

This second method is particularly convenient when a calculator is available and large numbers are not a problem. But remember, it only works with two fractions.

For completeness and reference purposes, we state the commutative and associative properties for addition with rational numbers. There are no related exercises.

COMMUTATIVE
PROPERTY OF
ADDITION

If $\frac{a}{b}$ and $\frac{c}{d}$ are rational numbers, then

$$\frac{a}{b} + \frac{c}{d} = \frac{c}{d} + \frac{a}{b}$$

ASSOCIATIVE
PROPERTY OF
ADDITION

If $\frac{a}{b}, \frac{c}{d},$ and $\frac{e}{f}$ are rational numbers, then

$$\frac{a}{b} + \frac{c}{d} + \frac{e}{f} = \left(\frac{a}{b} + \frac{c}{d}\right) + \frac{e}{f} = \frac{a}{b} + \left(\frac{c}{d} + \frac{e}{f}\right)$$

Adding 0 to a rational number gives the same rational number. Thus, 0 is the **additive identity** for rational numbers.

ADDITIVE
IDENTITY

If $\frac{a}{b}$ is a rational number, then

$$\frac{a}{b} + 0 = \frac{a}{b}$$

EXAMPLE

$$\frac{9}{16} + 0 = \frac{9}{16} + \frac{0}{16} = \frac{9 + 0}{16} = \frac{9}{16}$$

$$\left[\text{NOTE: If } b \neq 0, \frac{0}{b} = 0.\right]$$

PRACTICE QUIZ	Find the following sums. Reduce all answers.	ANSWERS
	1. $\dfrac{1}{8} + \dfrac{3}{8} + \dfrac{2}{8}$	1. $\dfrac{3}{4}$
	2. $\dfrac{}{3} + \dfrac{}{8} + \dfrac{}{6}$	2. $\dfrac{35}{24}$
	3. $\dfrac{7}{10} + \dfrac{10}{100} + \dfrac{5}{1000}$	3. $\dfrac{715}{1000} = \dfrac{143}{200}$

EXERCISES 3.3

Work through the examples in the text before you begin these exercises.

Draw diagrams similar to Figure 3.4 and Figure 3.5 illustrating the following sums.

1. $\dfrac{2}{5} + \dfrac{2}{5}$ 2. $\dfrac{3}{10} + \dfrac{7}{10}$ 3. $\dfrac{3}{7} + \dfrac{6}{7}$

4. Show that $\dfrac{5}{16} = \dfrac{1}{16} \cdot 5$ and $\dfrac{7}{16} = \dfrac{1}{16} \cdot 7$.

5. Use Exercise 4 and the distributive property (Section 1.6) to show that $\dfrac{5}{16} + \dfrac{7}{16} = \dfrac{12}{16}$.

Add the following rational numbers and reduce all answers.

6. $\dfrac{6}{10} + \dfrac{4}{10}$ 7. $\dfrac{3}{14} + \dfrac{2}{14}$ 8. $\dfrac{1}{20} + \dfrac{3}{20}$ 9. $\dfrac{3}{4} + \dfrac{3}{4}$

10. $\dfrac{5}{6} + \dfrac{4}{6}$ 11. $\dfrac{7}{5} + \dfrac{3}{5}$ 12. $\dfrac{11}{15} + \dfrac{7}{15}$ 13. $\dfrac{7}{9} + \dfrac{8}{9}$

14. $\dfrac{3}{25} + \dfrac{12}{25}$ 15. $\dfrac{7}{90} + \dfrac{37}{90} + \dfrac{21}{90}$ 16. $\dfrac{11}{75} + \dfrac{12}{75} + \dfrac{62}{75}$

17. $\dfrac{14}{32} + \dfrac{7}{32} + \dfrac{1}{32}$ 18. $\dfrac{4}{100} + \dfrac{35}{100} + \dfrac{76}{100}$ 19. $\dfrac{21}{95} + \dfrac{33}{95} + \dfrac{3}{95}$

20. $\dfrac{1}{200} + \dfrac{17}{200} + \dfrac{25}{200}$ 21. $\dfrac{1}{12} + \dfrac{2}{3} + \dfrac{1}{4}$ 22. $\dfrac{3}{8} + \dfrac{5}{16}$

23. $\dfrac{2}{5} + \dfrac{6}{10} + \dfrac{3}{20}$ 24. $\dfrac{28}{4} + \dfrac{7}{16} + \dfrac{6}{32}$ 25. $\dfrac{2}{7} + \dfrac{4}{21} + \dfrac{1}{3}$

26. $\dfrac{1}{6} + \dfrac{1}{4} + \dfrac{1}{3}$ 27. $\dfrac{2}{39} + \dfrac{1}{3} + \dfrac{4}{13}$ 28. $\dfrac{1}{2} + \dfrac{3}{10} + \dfrac{4}{5}$

29. $\dfrac{1}{27} + \dfrac{4}{18} + \dfrac{1}{6}$ 30. $\dfrac{2}{7} + \dfrac{3}{20} + \dfrac{9}{14}$ 31. $\dfrac{1}{8} + \dfrac{1}{12} + \dfrac{1}{9}$

32. $\dfrac{2}{5} + \dfrac{4}{7} + \dfrac{3}{8}$ **33.** $\dfrac{2}{3} + \dfrac{3}{4} + \dfrac{5}{6}$ **34.** $\dfrac{1}{5} + \dfrac{7}{30} + \dfrac{1}{6}$

35. $\dfrac{1}{5} + \dfrac{2}{15} + \dfrac{1}{6}$ **36.** $\dfrac{1}{5} + \dfrac{1}{10} + \dfrac{1}{4}$ **37.** $\dfrac{1}{5} + \dfrac{7}{40} + \dfrac{1}{4}$

38. $\dfrac{1}{3} + \dfrac{5}{12} + \dfrac{1}{15}$ **39.** $\dfrac{1}{4} + \dfrac{1}{20} + \dfrac{8}{15}$ **40.** $\dfrac{7}{10} + \dfrac{3}{25} + \dfrac{3}{4}$

41. $\dfrac{5}{8} + \dfrac{4}{27} + \dfrac{1}{48}$ **42.** $\dfrac{3}{16} + \dfrac{5}{48} + \dfrac{1}{32}$

43. $\dfrac{72}{105} + \dfrac{2}{45} + \dfrac{15}{21}$ **44.** $\dfrac{1}{63} + \dfrac{2}{27} + \dfrac{1}{45}$

45. $\dfrac{0}{27} + \dfrac{0}{16} + \dfrac{1}{5}$ **46.** $\dfrac{5}{6} + \dfrac{0}{100} + \dfrac{0}{70} + \dfrac{1}{3}$

47. $\dfrac{3}{10} + \dfrac{1}{100} + \dfrac{7}{1000}$ **48.** $\dfrac{11}{100} + \dfrac{15}{10} + \dfrac{1}{10}$

49. $\dfrac{17}{1000} + \dfrac{1}{100} + \dfrac{1}{10,000}$ **50.** $6 + \dfrac{1}{100} + \dfrac{3}{10}$

51. $8 + \dfrac{1}{10} + \dfrac{9}{100} + \dfrac{1}{1000}$ **52.** $\dfrac{1}{10} + \dfrac{3}{10} + \dfrac{9}{100}$

53. $\dfrac{7}{10} + \dfrac{5}{100} + \dfrac{3}{1000}$ **54.** $\dfrac{1}{2} + \dfrac{3}{4} + \dfrac{1}{100}$

55. $\dfrac{1}{4} + \dfrac{1}{8} + \dfrac{7}{100}$ **56.** $\dfrac{9}{1000} + \dfrac{7}{1000} + \dfrac{21}{10,000}$

57. $\dfrac{11}{100} + \dfrac{1}{2} + \dfrac{3}{1000}$ **58.** $\dfrac{3}{4} + \dfrac{17}{1000} + \dfrac{13}{10,000} + 2$

59. $5 + \dfrac{1}{10} + \dfrac{3}{100} + \dfrac{4}{1000}$ **60.** $\dfrac{13}{10,000} + \dfrac{1}{100,000} + \dfrac{21}{1,000,000}$

3.4 SUBTRACTING RATIONAL NUMBERS

Diagrams can be used to illustrate the difference between two rational numbers, such as $\dfrac{5}{8} - \dfrac{3}{8}$, as shown in Figure 3.6.

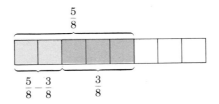

Figure 3.6

The relative size of the numbers is important because we want to subtract the smaller from the larger. Which is larger, $\frac{2}{3}$ or $\frac{5}{7}$? First find equivalent fractions with a common denominator, then compare numerators.

$$\frac{2}{3} = \frac{2}{3} \cdot \frac{7}{7} = \frac{14}{21} \quad \text{and} \quad \frac{5}{7} = \frac{5}{7} \cdot \frac{3}{3} = \frac{15}{21}$$

Since 15 is larger than 14,

$$\frac{15}{21} \text{ is larger than } \frac{14}{21}$$

and $\frac{5}{7}$ is larger than $\frac{2}{3}$

EXAMPLES

1. Which is larger, $\frac{5}{6}$ or $\frac{7}{8}$? How much larger?

$$\frac{5}{6} = \frac{5}{6} \cdot \frac{4}{4} = \frac{20}{24} \quad \text{and} \quad \frac{7}{8} = \frac{7}{8} \cdot \frac{3}{3} = \frac{21}{24}$$

$\frac{7}{8}$ is larger since $\frac{21}{24}$ is larger than $\frac{20}{24}$.

$$\frac{7}{8} - \frac{5}{6} = \frac{21}{24} - \frac{20}{24} = \frac{1}{24}$$

So, $\frac{7}{8}$ is $\frac{1}{24}$ larger than $\frac{5}{6}$.

2. Arrange $\frac{2}{3}$, $\frac{7}{10}$, and $\frac{9}{15}$ in order, smallest to largest.

(The SCD, smallest common denominator, is 30.)

$$\frac{2}{3} = \frac{2}{3} \cdot \frac{10}{10} = \frac{20}{30}; \quad \frac{7}{10} = \frac{7}{10} \cdot \frac{3}{3} = \frac{21}{30}; \quad \frac{9}{15} = \frac{9}{15} \cdot \frac{2}{2} = \frac{18}{30}$$

In order, smallest to largest, $\frac{9}{15}$, $\frac{2}{3}$, $\frac{7}{10}$.

We now give a formal definition of subtraction with two rational numbers.

DEFINITION   The **difference** of two rational numbers $\frac{a}{b}$ and $\frac{c}{b}$ with a greater than or equal to c is a rational number whose numerator is the difference of the numerators and whose denominator is the common denominator b. In symbols,

$$\frac{a}{b} - \frac{c}{b} = \frac{a-c}{b} \qquad (b \neq 0)$$

EXAMPLES   1. $\dfrac{5}{6} - \dfrac{1}{6} = \dfrac{5-1}{6} = \dfrac{4}{6} = \dfrac{\cancel{2} \cdot 2}{\cancel{2} \cdot 3} = \dfrac{2}{3}$

2. $\dfrac{9}{10} - \dfrac{7}{10} = \dfrac{9-7}{10} = \dfrac{2}{10} = \dfrac{\cancel{2} \cdot 1}{\cancel{2} \cdot 5} = \dfrac{1}{5}$

If the numbers do not have the same denominator, find equivalent numbers with the SCD, then subtract.

3. $\dfrac{9}{10} - \dfrac{2}{15} = \dfrac{9}{10} \cdot \dfrac{3}{3} - \dfrac{2}{15} \cdot \dfrac{2}{2} = \dfrac{27}{30} - \dfrac{4}{30} = \dfrac{27-4}{30} = \dfrac{23}{30}$

4. $5 - \dfrac{1}{3} = \dfrac{5}{1} \cdot \dfrac{3}{3} - \dfrac{1}{3} = \dfrac{15}{3} - \dfrac{1}{3} = \dfrac{15-1}{3} = \dfrac{14}{3}$

PRACTICE QUIZ	Find the following differences. Reduce all answers.	ANSWERS
	1. $\dfrac{5}{9} - \dfrac{1}{9}$	1. $\dfrac{4}{9}$
	2. $\dfrac{7}{6} - \dfrac{4}{6}$	2. $\dfrac{1}{2}$
	3. $\dfrac{21}{10} - \dfrac{14}{15}$	3. $\dfrac{7}{30}$

EXERCISES 3.4

1. Using figures similar to that in Figure 3.6, illustrate the differences.

(a) $\dfrac{2}{3} - \dfrac{1}{3}$ (b) $\dfrac{3}{4} - \dfrac{5}{12}$ (c) $2 - \dfrac{5}{4}$

Find the larger number of each of the following pairs of numbers and tell how much larger it is. [HINT: Find the SCD, smallest common denominator.]

2. $\dfrac{2}{3}, \dfrac{3}{4}$ **3.** $\dfrac{5}{6}, \dfrac{7}{8}$ **4.** $\dfrac{4}{5}, \dfrac{17}{20}$ **5.** $\dfrac{4}{10}, \dfrac{3}{8}$

6. $\dfrac{13}{20}, \dfrac{5}{8}$ **7.** $\dfrac{13}{16}, \dfrac{21}{25}$ **8.** $\dfrac{14}{35}, \dfrac{12}{30}$ **9.** $\dfrac{10}{36}, \dfrac{7}{24}$

10. $\dfrac{17}{80}, \dfrac{11}{48}$ **11.** $\dfrac{37}{100}, \dfrac{24}{75}$

Arrange the following numbers in order, smallest to largest.

12. $\dfrac{2}{3}, \dfrac{3}{5}, \dfrac{7}{10}$ **13.** $\dfrac{8}{9}, \dfrac{9}{10}, \dfrac{11}{12}$ **14.** $\dfrac{7}{6}, \dfrac{11}{12}, \dfrac{19}{20}$

15. $\dfrac{17}{12}, \dfrac{40}{36}, \dfrac{31}{24}$ **16.** $\dfrac{1}{3}, \dfrac{5}{42}, \dfrac{3}{7}$ **17.** $\dfrac{7}{8}, \dfrac{31}{36}, \dfrac{13}{18}$

18. $\dfrac{1}{100}, \dfrac{3}{1000}, \dfrac{20}{10,000}$ **19.** $\dfrac{32}{100}, \dfrac{298}{1000}, \dfrac{3333}{10,000}, \dfrac{3}{10}$

20. $\dfrac{72}{120}, \dfrac{80}{150}, \dfrac{35}{60}, \dfrac{15}{24}$

Find the difference in each of the following exercises. Reduce all answers.

21. $\dfrac{4}{7} - \dfrac{1}{7}$ **22.** $\dfrac{9}{10} - \dfrac{3}{10}$ **23.** $\dfrac{5}{8} - \dfrac{1}{8}$ **24.** $\dfrac{11}{12} - \dfrac{7}{12}$

25. $\dfrac{13}{15} - \dfrac{4}{15}$ **26.** $\dfrac{5}{6} - \dfrac{2}{3}$ **27.** $\dfrac{22}{15} - \dfrac{8}{10}$ **28.** $\dfrac{3}{4} - \dfrac{2}{3}$

29. $\dfrac{15}{16} - \dfrac{21}{32}$ **30.** $\dfrac{5}{8} - \dfrac{3}{5}$ **31.** $\dfrac{14}{27} - \dfrac{7}{18}$ **32.** $\dfrac{8}{45} - \dfrac{11}{72}$

33. $\dfrac{46}{55} - \dfrac{10}{33}$ **34.** $\dfrac{5}{36} - \dfrac{1}{30}$ **35.** $\dfrac{5}{36} - \dfrac{1}{32}$ **36.** $5 - \dfrac{3}{4}$

37. $4 - \dfrac{5}{8}$ **38.** $2 - \dfrac{9}{16}$ **39.** $1 - \dfrac{13}{16}$ **40.** $6 - \dfrac{2}{3}$

41. $\dfrac{9}{10} - \dfrac{3}{100}$ **42.** $\dfrac{159}{1000} - \dfrac{1}{10}$

43. $\dfrac{76}{100} - \dfrac{82}{10,000}$ **44.** $\dfrac{999}{1000} - \dfrac{99}{100}$

45. Find the sum of $\dfrac{1}{4}, \dfrac{1}{8}$, and $\dfrac{3}{16}$. Subtract $\dfrac{1}{3}$ from the sum. What is the difference?

46. Find the difference between $\dfrac{2}{3}$ and $\dfrac{5}{9}$. Add $\dfrac{5}{6}$ to the difference. What is the sum?

47. Find the sum of $\dfrac{7}{16}$ and $\dfrac{5}{32}$. Multiply the sum by $\dfrac{3}{19}$. What is the product?

48. Find the product of $\dfrac{9}{10}$ and $\dfrac{9}{10}$. Find the difference between $\dfrac{3}{4}$ and $\dfrac{1}{25}$. Find the sum of the product and the difference.

3.5 MIXED NUMBERS

The **sum** of a whole number and a rational number less than 1 can be indicated by writing the numbers side by side. This form of a rational number is called a **mixed number.**

$$5 + \frac{2}{3} = 5\frac{2}{3} \qquad \text{(Mixed number read ``five and two-thirds'')}$$

$$2 + \frac{3}{4} = 2\frac{3}{4} \qquad \text{(Mixed number read ``two and three-fourths'')}$$

The rational number that is not the whole number is called the **fraction part of the mixed number.**

Remember that a mixed number indicates addition of the whole number and the fraction part. Using this fact, we can change mixed numbers to the form of improper fractions (or just fraction form).

EXAMPLES

1. Change $5\dfrac{2}{3}$ to fraction form.

$$5\frac{2}{3} = 5 + \frac{2}{3} = \frac{5}{1} \cdot \frac{3}{3} + \frac{2}{3} = \frac{15}{3} + \frac{2}{3} = \frac{17}{3}$$

2. Change $6\dfrac{2}{9}$ to fraction form.

$$6\frac{2}{9} = 6 + \frac{2}{9} = \frac{6}{1} \cdot \frac{9}{9} + \frac{2}{9} = \frac{54}{9} + \frac{2}{9} = \frac{56}{9}$$

3. Change $10\dfrac{1}{3}$ to fraction form.

$$10\frac{1}{3} = 10 + \frac{1}{3} = \frac{10}{1} \cdot \frac{3}{3} + \frac{1}{3} = \frac{30}{3} + \frac{1}{3} = \frac{31}{3}$$

There is a very convenient shortcut for changing mixed numbers to fraction form. In all three examples, the numerator of the improper fraction is found by multiplying the whole number by the denominator, then adding the numerator of the fraction part. The denominator is the same as the denominator of the fraction part. You can do most of the steps of the shortcut in your head without writing anything down except the answer.

Change the following mixed numbers to fraction form using the shortcut just described.

1. $5\dfrac{2}{3}$

Multiply $3 \cdot 5 = 15$ and add 2, getting $15 + 2 = 17$. Thus,

$$5\frac{2}{3} = \frac{17}{3}$$

2. $3\dfrac{5}{8}$

Multiply $8 \cdot 3 = 24$ and add 5, getting $24 + 5 = 29$. Thus,

$$3\frac{5}{8} = \frac{29}{8}$$

3. $4\dfrac{1}{2} = \dfrac{9}{2}$

To change an improper fraction to a mixed number, we use the fact that a fraction can indicate division. Divide the numerator by the denominator. If there is a remainder, write it as the numerator of the fraction part of the mixed number. The divisor is the denominator of the fraction part.

1. Change $\dfrac{7}{3}$ to a mixed number: $3\overline{)7}$ ... $\dfrac{7}{3} = 2 + \dfrac{1}{3} = 2\dfrac{1}{3}$

$$\begin{array}{r} 2 \\ 3\overline{)7} \\ 6 \\ \hline 1 \end{array}$$

2. Change $\dfrac{29}{4}$ to a mixed number: $4\overline{)29}$ ... $\dfrac{29}{4} = 7 + \dfrac{1}{4} = 7\dfrac{1}{4}$

$$\begin{array}{r} 7 \\ 4\overline{)29} \\ 28 \\ \hline 1 \end{array}$$

3. Change $\dfrac{59}{3}$ to a mixed number:

$$3\overline{)59} \quad \begin{array}{r} 19 \\ \hline 59 \\ 3 \\ \hline 29 \\ 27 \\ \hline 2 \end{array}$$

$$\dfrac{59}{3} = 19 + \dfrac{2}{3} = 19\dfrac{2}{3}$$

TO CHANGE RATIONAL NUMBERS TO MIXED NUMBERS

1. Reduce first, then change to mixed numbers; or

2. Change to mixed numbers first, then reduce the fraction part.

EXAMPLES

Change $\dfrac{14}{10}$ to a mixed number and reduce the answer.

1. $\dfrac{14}{10} = \dfrac{\cancel{2} \cdot 7}{\cancel{2} \cdot 5} = \dfrac{7}{5}$ $5\overline{)7} \quad \begin{array}{r} 1 \\ \hline 7 \\ 5 \\ \hline 2 \end{array}$ $\dfrac{7}{5} = 1 + \dfrac{2}{5} = 1\dfrac{2}{5}$

2. $\dfrac{14}{10}$ $10\overline{)14} \quad \begin{array}{r} 1 \\ \hline 14 \\ 10 \\ \hline 4 \end{array}$ $\dfrac{14}{10} = 1 + \dfrac{4}{10} = 1\dfrac{4}{10}$

$\dfrac{4}{10} = \dfrac{\cancel{2} \cdot 2}{\cancel{2} \cdot 5} = \dfrac{2}{5}$ $1\dfrac{4}{10} = 1\dfrac{2}{5}$

Changing a rational number to a mixed number is not reducing it. **Reducing** involves finding common factors in the numerator and denominator, while **changing to a mixed number** means dividing and writing the whole number and fraction part. Common factors are not involved.

PRACTICE QUIZ

1. Reduce $\dfrac{18}{16}$ to lowest terms.

2. Change $\dfrac{51}{34}$ to a mixed number.

3. Change $6\dfrac{2}{3}$ to fraction form.

ANSWERS

1. $\dfrac{9}{8}$

2. $1\dfrac{1}{2}$

3. $\dfrac{20}{3}$

EXERCISES 3.5

Reduce the following rational numbers to lowest terms.

1. $\dfrac{24}{18}$ 2. $\dfrac{25}{10}$ 3. $\dfrac{16}{12}$ 4. $\dfrac{10}{8}$ 5. $\dfrac{39}{26}$

6. $\dfrac{48}{32}$ 7. $\dfrac{35}{25}$ 8. $\dfrac{18}{16}$ 9. $\dfrac{80}{64}$ 10. $\dfrac{75}{60}$

Change the following rational numbers to mixed numbers.

11. $\dfrac{100}{24}$ 12. $\dfrac{25}{10}$ 13. $\dfrac{16}{12}$ 14. $\dfrac{10}{8}$ 15. $\dfrac{39}{26}$

16. $\dfrac{42}{8}$ 17. $\dfrac{43}{7}$ 18. $\dfrac{34}{16}$ 19. $\dfrac{45}{6}$ 20. $\dfrac{75}{12}$

21. $\dfrac{56}{18}$ 22. $\dfrac{31}{15}$ 23. $\dfrac{36}{12}$ 24. $\dfrac{48}{16}$ 25. $\dfrac{72}{16}$

26. $\dfrac{70}{34}$ 27. $\dfrac{45}{15}$ 28. $\dfrac{60}{36}$ 29. $\dfrac{35}{20}$ 30. $\dfrac{185}{100}$

Change the following mixed numbers to fraction form.

31. $4\dfrac{5}{8}$ 32. $3\dfrac{3}{4}$ 33. $5\dfrac{1}{15}$ 34. $1\dfrac{3}{5}$ 35. $4\dfrac{2}{11}$

36. $2\dfrac{11}{44}$ 37. $2\dfrac{9}{27}$ 38. $4\dfrac{6}{7}$ 39. $10\dfrac{8}{12}$ 40. $11\dfrac{3}{8}$

41. $6\dfrac{8}{10}$ 42. $14\dfrac{1}{5}$ 43. $16\dfrac{2}{3}$ 44. $12\dfrac{4}{8}$ 45. $20\dfrac{3}{15}$

46. $9\dfrac{4}{10}$ 47. $13\dfrac{1}{7}$ 48. $49\dfrac{0}{12}$ 49. $17\dfrac{0}{3}$ 50. $3\dfrac{1}{50}$

3.6   ADDING AND SUBTRACTING MIXED NUMBERS

Since a mixed number itself represents addition, two or more mixed numbers can be added by adding the whole numbers and the fraction parts separately. The following examples illustrate the procedure.

EXAMPLES

1. $4\dfrac{2}{7} + 6\dfrac{3}{7} = 4 + \dfrac{2}{7} + 6 + \dfrac{3}{4}$

$$= (4 + 6) + \left(\dfrac{2}{7} + \dfrac{3}{7}\right)$$

$$= 10 + \dfrac{5}{7}$$

$$= 10\dfrac{5}{7}$$

Vertically,

$$4\frac{2}{7}$$
$$6\frac{3}{7}$$
$$\overline{10\frac{5}{7}}$$

2. $13\dfrac{1}{6} + 2\dfrac{5}{18} = 13 + \dfrac{1}{6} + 2 + \dfrac{5}{18}$

$\qquad\qquad\quad = (13 + 2) + \left(\dfrac{1}{6} + \dfrac{5}{18}\right)$

$\qquad\qquad\quad = 15 + \left(\dfrac{1}{6}\cdot\dfrac{3}{3} + \dfrac{5}{18}\right)$

$\qquad\qquad\quad = 15 + \left(\dfrac{3}{18} + \dfrac{5}{18}\right)$

$\qquad\qquad\quad = 15\dfrac{8}{18} = 15\dfrac{4}{9}$

Vertically,

$$13\,\frac{1}{6} = 13\frac{1}{6}\cdot\frac{3}{3} = 13\frac{3}{18}$$
$$2\frac{5}{18} = \;\; 2\frac{5}{18} \;\; = \;\; 2\frac{5}{18}$$
$$\overline{\phantom{2\frac{5}{18}}}\qquad\overline{\phantom{2\frac{5}{18}}}\qquad\overline{\phantom{2\frac{5}{18}}}$$
$$15\frac{8}{18} = 15\frac{4}{9} \qquad \text{(Reduce the fraction part.)}$$

3. $7\dfrac{2}{3} + 9\dfrac{4}{5} = 7 + \dfrac{2}{3} + 9 + \dfrac{4}{5}$

$\qquad\qquad\quad = (7 + 9) + \left(\dfrac{2}{3}\cdot\dfrac{5}{5} + \dfrac{4}{5}\cdot\dfrac{3}{3}\right)$

$\qquad\qquad\quad = 16 + \left(\dfrac{10}{15} + \dfrac{12}{15}\right)$

$\qquad\qquad\quad = 16 + \dfrac{22}{15} = 16 + 1\dfrac{7}{15}\qquad$ (Change fraction part to mixed number.)

$\qquad\qquad\quad = 16 + 1 + \dfrac{7}{15}$

$\qquad\qquad\quad = 17\dfrac{7}{15}$

Vertically,

$$7\frac{2}{3} = 7\frac{2}{3} \cdot \frac{5}{5} = 7\frac{10}{15}$$

$$\underline{9\frac{4}{5} = 9\frac{4}{5} \cdot \frac{3}{3} = 9\frac{12}{15}}$$

$$16\frac{22}{15} = 16 + 1\frac{7}{15} = 17\frac{7}{15}$$

(If the fraction part is greater than 1, change it to a mixed number and add it to the whole number.)

Subtraction with mixed numbers also involves working with the whole numbers and fraction parts separately. Subtract the fraction parts, then subtract the whole numbers, and write the answer as a mixed number. The following examples illustrate the procedure.

EXAMPLES

1. $4\frac{3}{5} - 1\frac{2}{5} = (4 - 1) + \left(\frac{3}{5} - \frac{2}{5}\right) = 3 + \frac{1}{5} = 3\frac{1}{5}$

2. $5\frac{6}{7} - 3\frac{2}{7} = (5 - 3) + \left(\frac{6}{7} - \frac{2}{7}\right) = 2 + \frac{4}{7} + 2\frac{4}{7}$

3. $10\frac{3}{5} - 6\frac{3}{20} = (10 - 6) + \left(\frac{3}{5} - \frac{3}{20}\right) = 4 + \left(\frac{12}{20} - \frac{3}{20}\right)$

$$= 4 + \frac{9}{20} = 4\frac{9}{20}$$

Or, writing the numbers one under the other,

$$\begin{array}{c} 4\frac{3}{5} \\ -1\frac{2}{5} \\ \hline 3\frac{1}{5} \end{array} \qquad \begin{array}{c} 5\frac{6}{7} \\ -3\frac{2}{7} \\ \hline 2\frac{4}{7} \end{array} \qquad \begin{array}{c} 10\frac{3}{5} = 10\frac{12}{20} \\ -\;6\frac{3}{20} = 6\frac{3}{20} \\ \hline 4\frac{9}{20} \end{array}$$

Sometimes the fraction part of the number being subtracted is larger than the fraction part of the first number. In this case, in the first number, "borrow" 1 from the whole number by adding 1 to the fraction part. The first number will be written as a whole number plus an improper fraction. Now subtract.

<u>EXAMPLES</u>   1. $4\frac{1}{5} - 2\frac{3}{5}$

$$4\frac{1}{5} = 3 + 1 + \frac{1}{5} \qquad \left(\frac{3}{5} \text{ is larger than } \frac{1}{5}, \text{ so "borrow" 1 from 4.}\right)$$

$$= 3 + \frac{5}{5} + \frac{1}{5} = 3\frac{6}{5}$$

$$4\frac{1}{5} - 2\frac{3}{5} = 3\frac{6}{5} - 2\frac{3}{5}$$

$$= (3 - 2) + \left(\frac{6}{5} - \frac{3}{5}\right)$$

$$= 1 + \frac{3}{5}$$

$$= 1\frac{3}{5}$$

Vertically,

$$4\frac{1}{5} = 3\frac{6}{5}$$
$$-2\frac{3}{5} = 2\frac{3}{5}$$
$$\overline{\qquad 1\frac{3}{5}}$$

2. $19\frac{2}{3} - 5\frac{3}{4}$

$$19\frac{2}{3} = 19\frac{8}{12} = 18\frac{20}{12} \qquad \left(1 + \frac{8}{12} = \frac{12}{12} + \frac{8}{12} = \frac{20}{12}\right)$$
$$-\;5\frac{3}{4} = \;5\frac{9}{12} = \;5\frac{9}{12}$$
$$\overline{\qquad\qquad\qquad 13\frac{11}{12}}$$

PRACTICE QUIZ	Add or subtract as indicated. Reduce all answers.	ANSWERS
	1. $2\frac{1}{2} + 3\frac{1}{3}$ $\frac{1}{2} + \frac{1}{3}$ $\acute{c}$	1. $5\frac{5}{6}$
	2. $4\frac{5}{8} + 10\frac{7}{8}$ $\frac{12}{8}$ $\frac{1}{4}$	2. $15\frac{1}{2}$
	3. $9\frac{7}{10} - 3\frac{1}{5}$	3. $6\frac{1}{2}$
	4. $12\frac{1}{9} - 8\frac{1}{3}$	4. $3\frac{7}{9}$

EXERCISES 3.6

Find the indicated sums. The problems may be written vertically if you prefer.

1. $4\frac{1}{2} + 3\frac{1}{6}$ 2. $3\frac{1}{4} + 7\frac{1}{8}$ 3. $25\frac{1}{10} + 17\frac{1}{4}$

4. $5\frac{1}{7} + 3\frac{1}{3}$ 5. $6\frac{5}{12} + 4\frac{1}{3}$ 6. $5\frac{3}{10} + 2\frac{1}{14}$

7. $8\frac{2}{9} + 4\frac{1}{27}$ 8. $11\frac{3}{4} + 2\frac{5}{16}$ 9. $6\frac{4}{9} + 12\frac{1}{15}$

10. $4\frac{1}{6} + 13\frac{9}{10}$ 11. $21\frac{3}{4} + 6\frac{3}{4}$ 12. $3\frac{5}{8} + 3\frac{5}{8}$

13. $7\frac{3}{5} + 2\frac{1}{8}$ 14. $9\frac{1}{8} + 3\frac{7}{12}$ 15. $3\frac{1}{3} + 4\frac{1}{4} + 5\frac{1}{5}$

16. $\frac{3}{7} + 2\frac{1}{14} + 2\frac{1}{6}$ 17. $20\frac{5}{8} + 42\frac{5}{6}$ 18. $25\frac{2}{3} + 1\frac{1}{16}$

19. $32\frac{1}{64} + 4\frac{1}{24} + 17\frac{3}{8}$ 20. $3\frac{1}{20} + 7\frac{1}{15} + 2\frac{3}{10}$

Find the indicated differences. The problems may be written vertically if you prefer.

21. $5\frac{3}{4} - 2\frac{1}{4}$ 22. $7\frac{9}{10} - 3\frac{3}{10}$ 23. $4\frac{7}{8} - 1\frac{1}{4}$

24. $9\frac{5}{6} - 2\frac{1}{4}$ 25. $15\frac{5}{8} - 11\frac{3}{4}$ 26. $14\frac{6}{10} - 3\frac{4}{5}$

27. $8\frac{3}{32} - 4\frac{3}{16}$ 28. $12\frac{3}{4} - 7\frac{1}{6}$ 29. $8\frac{11}{12} - 5\frac{9}{10}$

30. $4\frac{7}{16} - 3$ **31.** $5\frac{9}{10} - 2$ **32.** $7 - 6\frac{2}{3}$

33. $12 - 4\frac{1}{5}$ **34.** $2 - 1\frac{3}{8}$ **35.** $75 - 17\frac{5}{6}$

36. $4\frac{9}{16} - 2\frac{7}{8}$ **37.** $3\frac{7}{10} - 2\frac{5}{6}$ **38.** $15\frac{11}{16} - 13\frac{7}{8}$

39. $20\frac{3}{6} - 3\frac{4}{8}$ **40.** $17\frac{3}{12} - 12\frac{2}{8}$

41. A bus trip consists of three parts. The first part takes $2\frac{1}{3}$ hours, the second part takes $2\frac{1}{2}$ hours, and the third part takes $3\frac{3}{4}$ hours. How many hours does the trip take?

42. A construction company built three sections of highway. One section was $20\frac{7}{10}$ kilometers, a second section was $3\frac{4}{10}$ kilometers, and a third section was $11\frac{6}{10}$ kilometers. What was the total length built?

43. Sara can paint a room in $3\frac{3}{5}$ hours, and Emily can paint the same size room in $4\frac{1}{5}$ hours. How many hours are saved by having Sara paint the room? How many minutes are saved?

44. A teacher graded two sets of test papers. The first set took $3\frac{3}{4}$ hours to grade, and the second set took $2\frac{3}{5}$ hours. How much faster did the teacher grade the second set?

45. Mike takes $1\frac{1}{2}$ hours to clean a pool, and Tom takes $2\frac{1}{3}$ hours to clean the same pool. How much longer does Tom take?

46. During each week of six weeks of dieting, Mr. Johnson, who originally weighed 240 pounds, lost $5\frac{1}{2}$ pounds, $2\frac{3}{4}$ pounds, $4\frac{5}{16}$ pounds, $1\frac{3}{4}$ pounds, $2\frac{5}{8}$ pounds, and $3\frac{1}{4}$ pounds. What did he weigh at the end of the six weeks if he was 35 years old?

47. A triangle (a three-sided figure) has sides of $42\frac{3}{4}$ feet, $23\frac{1}{2}$ feet, and $22\frac{7}{8}$ feet. What is the perimeter (total distance around) of the triangle?

48. A quadrilateral (a four-sided figure) has sides of $3\frac{1}{2}$ inches, $2\frac{1}{4}$ inches, $3\frac{5}{8}$ inches, and $2\frac{3}{4}$ inches. What is the perimeter (total distance around) of the quadrilateral?

49. A long-distance runner was in training. He ran ten miles in $50\frac{3}{10}$ minutes. Three months later, he ran the same ten miles in $43\frac{7}{10}$ minutes. By how much did his time improve?

50. Certain shares of stock were selling for $43\frac{7}{8}$ dollars per share. One month later, the same stock was selling for $48\frac{1}{2}$ dollars per share. By how much did the stock increase in price?

3.7   MULTIPLYING MIXED NUMBERS

Probably the simplest way to multiply mixed numbers is to change them to fraction form and then multiply. The answers may be changed back to mixed numbers for convenience, or they may be left in fraction form. As the following examples illustrate, numerators and denominators should be factored and numbers reduced before multiplication.

EXAMPLES

1. $\left(1\frac{2}{3}\right)\left(2\frac{3}{4}\right) = \left(\frac{5}{3}\right)\left(\frac{11}{4}\right) = \frac{5 \cdot 11}{3 \cdot 4} = \frac{55}{12}$ or $4\frac{7}{12}$

2. $\left(5\frac{2}{3}\right)\left(2\frac{1}{4}\right) = \left(\frac{17}{3}\right)\left(\frac{9}{4}\right) = \frac{17 \cdot 3 \cdot \cancel{3}}{\cancel{3} \cdot 4} = \frac{51}{4}$ or $12\frac{3}{4}$

3. $\left(2\frac{1}{3}\right)\left(4\frac{1}{5}\right) = \left(\frac{7}{3}\right)\left(\frac{21}{5}\right) = \frac{7 \cdot \cancel{3} \cdot 7}{\cancel{3} \cdot 5} = \frac{49}{5}$ or $9\frac{4}{5}$

4. $\frac{5}{6} \cdot 3\frac{3}{10} = \left(\frac{5}{6}\right)\left(\frac{33}{10}\right) = \frac{5 \cdot 33}{6 \cdot 10} = \frac{\cancel{5} \cdot \cancel{3} \cdot 11}{2 \cdot \cancel{3} \cdot 2 \cdot \cancel{5}} = \frac{11}{4}$ or $2\frac{3}{4}$

5. $4\frac{1}{2} \cdot 1\frac{1}{6} \cdot 3\frac{1}{3} = \frac{9}{2} \cdot \frac{7}{6} \cdot \frac{10}{3} = \frac{9 \cdot 7 \cdot 10}{2 \cdot 6 \cdot 3} = \frac{\cancel{3} \cdot \cancel{3} \cdot 7 \cdot \cancel{2} \cdot 5}{2 \cdot \cancel{2} \cdot \cancel{3} \cdot \cancel{3}} = \frac{35}{2}$ or $17\frac{1}{2}$

Large mixed numbers can be multiplied the same way. The numbers will be large, but the procedure is simple. For example,

$$24\frac{3}{8} \cdot 45\frac{1}{4} = \frac{195}{8} \cdot \frac{181}{4} = \frac{35,295}{32} = 1102\frac{31}{32}$$

The same product can be found by writing the numbers one under the other, but the procedure involves four products and addition of fractions as outlined below. The procedure is not recommended; it is shown for comparison and for your information.

$$24\frac{3}{8}$$

$$45\frac{1}{4}$$

$$\frac{3}{32} \quad \longleftarrow \quad \frac{1}{4} \cdot \frac{3}{8} = \frac{3}{32}$$

$$6 \quad \longleftarrow \quad \frac{1}{4} \cdot 24 = 6$$

$$16\frac{7}{8} \quad \longleftarrow \quad 45 \cdot \frac{3}{8} = \frac{135}{8} = 16\frac{7}{8}$$

$$\left.\begin{array}{c}120\\960\end{array}\right\} \quad \longleftarrow \quad 45 \cdot 24$$

$$1102\frac{31}{32} \qquad \left(\frac{3}{32} + \frac{7}{8} = \frac{3}{32} + \frac{28}{32} = \frac{31}{32}\right)$$

The use of the word **of** with fractions deserves special clarification. When we write a phrase such as "$\frac{3}{4}$ of 40," we mean to multiply $\frac{3}{4} \cdot 40$. In effect, **of** means multiply when used with fractions. (This is also true of percents and will be discussed again in Chapter 5.)

<hr>

EXAMPLES   1.   Find $\frac{3}{4}$ of 40.

$$\frac{3}{4} \cdot 40 = \frac{3}{\cancel{4}} \cdot \frac{\cancel{40}^{10}}{1} = 30$$

2.   Find $\frac{3}{4}$ of 80.

$$\frac{3}{4} \cdot 80 = \frac{3}{\cancel{4}} \cdot \frac{\cancel{80}^{20}}{1} = 60$$

3. Find $\frac{2}{3}$ of $\frac{9}{10}$.

$$\frac{2}{3} \cdot \frac{9}{10} = \frac{2 \cdot 3 \cdot 3}{3 \cdot 2 \cdot 5} = \frac{3}{5} \quad ,$$

PRACTICE QUIZ

Find the indicated products.

1. $4\frac{1}{3} \cdot \frac{2}{13} =$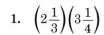

2. $5\frac{1}{2} \cdot 3\frac{2}{3}$

3. Find $\frac{9}{10}$ of 70.

ANSWERS

1. $\frac{2}{3}$

2. $\frac{121}{6}$ or $20\frac{1}{6}$

3. 63

EXERCISES 3.7

Find the indicated products.

1. $\left(2\frac{1}{3}\right)\left(3\frac{1}{4}\right)$
2. $\left(1\frac{1}{5}\right)\left(1\frac{1}{7}\right)$
3. $4\frac{1}{2}\left(2\frac{1}{3}\right)$

4. $3\frac{1}{3}\left(2\frac{1}{5}\right)$
5. $6\frac{1}{4}\left(3\frac{3}{5}\right)$
6. $5\frac{1}{3}\left(2\frac{1}{4}\right)$

7. $\left(8\frac{1}{2}\right)\left(3\frac{2}{3}\right)$
8. $\left(9\frac{1}{3}\right)2\frac{1}{7}$
9. $\left(6\frac{2}{7}\right)1\frac{3}{11}$

10. $\left(11\frac{1}{4}\right)1\frac{1}{15}$
11. $6\frac{2}{3} \cdot 4\frac{1}{2}$
12. $4\frac{3}{8} \cdot 2\frac{2}{7}$

13. $9\frac{3}{4} \cdot 2\frac{6}{26}$
14. $7\frac{1}{2} \cdot \frac{2}{15}$
15. $\frac{3}{4} \cdot 1\frac{1}{3}$

16. $3\frac{4}{5} \cdot 2\frac{1}{7}$
17. $12\frac{1}{2} \cdot 2\frac{1}{5}$
18. $9\frac{3}{5} \cdot 1\frac{1}{16}$

19. $6\frac{1}{8} \cdot 3\frac{1}{7}$
20. $5\frac{1}{4} \cdot 11\frac{1}{3}$
21. $\frac{1}{4} \cdot \frac{2}{3} \cdot \frac{6}{7}$

22. $\frac{7}{8} \cdot \frac{24}{25} \cdot \frac{5}{21}$
23. $\frac{3}{16} \cdot \frac{8}{9} \cdot \frac{3}{5}$
24. $\frac{2}{5} \cdot \frac{1}{5} \cdot \frac{4}{7}$

25. $\frac{6}{7} \cdot \frac{2}{11} \cdot \frac{3}{5}$
26. $\left(3\frac{1}{2}\right)\left(2\frac{1}{7}\right)\left(5\frac{1}{4}\right)$
27. $\left(4\frac{3}{8}\right)\left(2\frac{1}{5}\right)\left(1\frac{1}{7}\right)$

28. $\left(6\frac{3}{16}\right)\left(2\frac{1}{11}\right)\left(5\frac{3}{5}\right)$
29. $7\frac{1}{3} \cdot 5\frac{1}{4} \cdot 6\frac{2}{7}$
30. $2\frac{5}{8} \cdot 3\frac{2}{5} \cdot 1\frac{3}{4}$

31. $2\dfrac{1}{16} \cdot 4\dfrac{1}{3} \cdot 1\dfrac{3}{11}$ **32.** $5\dfrac{1}{10} \cdot 3\dfrac{1}{7} \cdot 2\dfrac{1}{17}$ **33.** $2\dfrac{1}{4} \cdot 6\dfrac{3}{8} \cdot 1\dfrac{5}{27}$

34. $1\dfrac{3}{32} \cdot 1\dfrac{1}{7} \cdot 1\dfrac{1}{25}$ **35.** $1\dfrac{5}{16} \cdot 1\dfrac{1}{3} \cdot 1\dfrac{1}{5}$ **36.** $24\dfrac{1}{5} \cdot 35\dfrac{1}{6}$

37. $72\dfrac{3}{5} \cdot 25\dfrac{1}{6}$ **38.** $42\dfrac{5}{6} \cdot 30\dfrac{1}{7}$ **39.** $75\dfrac{1}{3} \cdot 40\dfrac{1}{25}$

40. $36\dfrac{3}{4} \cdot 17\dfrac{5}{12}$

41. Find $\dfrac{2}{3}$ of 60.   **42.** Find $\dfrac{1}{4}$ of 80.   **43.** Find $\dfrac{1}{5}$ of 100.

44. Find $\dfrac{3}{5}$ of 100.   **45.** Find $\dfrac{3}{4}$ of 100.   **46.** Find $\dfrac{5}{6}$ of 120.

47. Find $\dfrac{1}{2}$ of $\dfrac{5}{8}$.   **48.** Find $\dfrac{1}{6}$ of $\dfrac{3}{4}$.   **49.** Find $\dfrac{9}{10}$ of $\dfrac{15}{21}$.

50. Find $\dfrac{7}{8}$ of $\dfrac{8}{10}$.

51. A telephone pole is 32 feet long. If $\dfrac{5}{16}$ of the pole must be underground and $\dfrac{11}{16}$ of the pole aboveground, how much of the pole is underground? How much is aboveground?

52. The total distance around a square (its perimeter) is found by multiplying the length of one side by 4. Find the perimeter of a square if the length of one side is $5\dfrac{1}{16}$ inches.

53. A man driving to work drives $17\dfrac{7}{10}$ miles one way five days a week. How many miles does he drive each week going to and from work?

54. A length of pipe is $27\dfrac{3}{4}$ feet. What would be the total length if $36\dfrac{1}{2}$ of these pipes were laid end to end?

55. A woman reads $\dfrac{1}{6}$ of a book in 3 hours. If the book contains 540 pages, how many pages will she read in 3 hours? How long will she take to read the entire book?

56. Three towns are located on the same highway (assume a very straight section of highway). Two towns are 53 kilometers apart. If one of the towns is $45\dfrac{9}{10}$ kilometers from the third town, how far is the other town from the third town? (A sketch might be helpful.)

3.8   DIVIDING RATIONAL NUMBERS

If the product of two rational numbers is 1, then the numbers are called **reciprocals** of each other. For example,

$$\text{Since } \frac{5}{6} \cdot \frac{6}{5} = 1, \quad \frac{5}{6} \text{ and } \frac{6}{5} \text{ are reciprocals.}$$

$$\text{Since } \frac{10}{17} \cdot \frac{17}{10} = 1, \quad \frac{10}{17} \text{ and } \frac{17}{10} \text{ are reciprocals}$$

DEFINITION   The **reciprocal** of a rational number $\frac{a}{b}$ where $a \neq 0$ and $b \neq 0$ is $\frac{b}{a}$ because

$$\frac{a}{b} \cdot \frac{b}{a} = 1$$

[NOTE: If $a = 0$, then $\frac{0}{b}$ has no reciprocal since $\frac{b}{0}$ is undefined.]

In division with whole numbers, $6 \div 3 = 2$ because $6 = 2 \cdot 3$. The same is true for rational numbers. That is, if

$$\frac{3}{4} \div \frac{1}{5} = \square \quad \text{then} \quad \frac{3}{4} = \square \cdot \frac{1}{5}$$

What goes in the box? Try $\frac{3}{4} \cdot \frac{5}{1}$.

$$\text{Does } \frac{3}{4} = \boxed{\frac{3}{4} \cdot \frac{5}{1}} \cdot \frac{1}{5} ? \quad \boxed{\frac{3}{4} \cdot \frac{5}{1}} \cdot \frac{1}{5} = \frac{3}{4} \cdot \left(\frac{5}{1} \cdot \frac{1}{5} \right) = \frac{3}{4} \cdot 1$$
$$= \frac{3}{4}$$

Therefore, $\frac{3}{4} \cdot \frac{5}{1}$ is the answer. That is,

$$\frac{3}{4} \div \frac{1}{5} = \boxed{\frac{3}{4} \cdot \frac{5}{1}}$$

This procedure works for all nonzero rational numbers. Since a fraction indicates division, we can write

$$\frac{a}{b} \div \frac{c}{d} = \frac{\frac{a}{b}}{\frac{c}{d}} \qquad \text{and} \qquad \frac{\frac{d}{c}}{\frac{d}{c}} = 1$$

So,

$$\frac{a}{b} \div \frac{c}{d} = \frac{\frac{a}{b}}{\frac{c}{d}} = \frac{\frac{a}{b}}{\frac{c}{d}} \cdot 1 = \frac{\frac{a}{b}}{\frac{c}{d}} \cdot \frac{\frac{d}{c}}{\frac{d}{c}} = \frac{\frac{a}{b} \cdot \frac{d}{c}}{\frac{c}{d} \cdot \frac{d}{c}} = \frac{\frac{a}{b} \cdot \frac{d}{c}}{1} = \frac{a}{b} \cdot \frac{d}{c}$$

Thus,

$$\frac{a}{b} \div \frac{c}{d} = \frac{a}{b} \cdot \frac{d}{c} \quad \text{where } b, c, d \neq 0$$

In words: **To divide by any number (except 0), multiply by its reciprocal.**

EXAMPLES

1. $\dfrac{3}{4} \div \dfrac{2}{3} = \dfrac{3}{4} \cdot \dfrac{3}{2} = \dfrac{3 \cdot 3}{4 \cdot 2} = \dfrac{9}{8}$ or $1\dfrac{1}{8}$

Note that the divisor is $\dfrac{2}{3}$, and we multiply by its reciprocal, $\dfrac{3}{2}$.

2. $\dfrac{7}{16} \div 7 = \dfrac{7}{16} \cdot \dfrac{1}{7} = \dfrac{7 \cdot 1}{16 \cdot 7} = \dfrac{7}{7} \cdot \dfrac{1}{16} = \dfrac{1}{16}$

Note that the divisor is 7, and we multiply by its reciprocal, $\dfrac{1}{7}$.

3. $\dfrac{16}{27} \div \dfrac{4}{9} = \dfrac{16}{27} \cdot \dfrac{9}{4} = \dfrac{4 \cdot \cancel{4} \cdot \cancel{9}}{\cancel{9} \cdot 3 \cdot \cancel{4}}$

$= \dfrac{4}{3}$ or $1\dfrac{1}{3}$

Note that the divisor is $\dfrac{4}{9}$, and we multiply by its reciprocal, $\dfrac{9}{4}$.

4. $3\dfrac{1}{4} \div 7\dfrac{\cancel{4}}{\cancel{4}} = \dfrac{\cancel{13}}{\cancel{4}} \div \dfrac{39}{5} = \dfrac{\cancel{13}}{4} \cdot \dfrac{5}{\cancel{39}} = \dfrac{\cancel{13} \cdot 5}{4 \cdot \cancel{13} \cdot 3}^{\frac{5}{12}}$

$= \dfrac{5}{12}$

Note that the divisor is $\dfrac{39}{5}$, and we multiply by its reciprocal, $\dfrac{5}{39}$.

5. $\dfrac{2\dfrac{1}{4}}{4\dfrac{1}{2}} = \dfrac{\dfrac{9}{4}}{\dfrac{9}{2}} = \dfrac{\cancel{9}}{\cancel{4}} \cdot \dfrac{\cancel{2}}{\cancel{9}} = \dfrac{1}{2}$

With whole numbers, if we know a product and one factor, we can find the other factor by dividing. For example, if the product is 24 and one factor is 8, the other factor is 3 because

$$\begin{array}{r} 3 \\ 8\overline{)24} \\ \underline{24} \\ 0 \end{array} \qquad \text{or we know} \qquad 8 \cdot 3 = 24$$

The same procedure applies to rational numbers. If we know the product and one multiplier, we can find the other multiplier by division. Divide the product by the known multiplier. (The word **multiplier** is used here because the word factor implies whole numbers.)

EXAMPLE

If the product of $\frac{3}{4}$ with another number is $\frac{8}{15}$, what is the other number?

$$\frac{3}{4} \cdot ? = \frac{8}{15}$$

So, divide $\frac{8}{15}$ by $\frac{3}{4}$.

$$\frac{8}{15} \div \frac{3}{4} = \frac{8}{15} \cdot \frac{4}{3} = \frac{32}{45}$$

The other number is $\frac{32}{45}$.

PRACTICE QUIZ	Find the quotients as indicated.	ANSWERS
	1. $\frac{5}{2} \div \frac{2}{5}$	1. $\frac{25}{4}$ or $6\frac{1}{4}$
	2. $\frac{12}{27} \div \frac{16}{18}$	2. $\frac{1}{2}$
	3. Find the average of $\frac{7}{8}$, $\frac{9}{16}$, and $\frac{7}{16}$. [HINT: Add, then divide by 3.]	3. $\frac{5}{8}$

EXERCISES 3.8

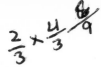

Find the quotients as indicated.

1. $\dfrac{2}{3} \div \dfrac{3}{4}$

2. $\dfrac{1}{5} \div \dfrac{3}{4}$

3. $\dfrac{3}{7} \div \dfrac{3}{5}$

4. $\dfrac{2}{11} \div \dfrac{2}{3}$

5. $\dfrac{3}{5} \div \dfrac{3}{7}$

6. $\dfrac{2}{3} \div \dfrac{2}{11}$

7. $\dfrac{5}{16} \div \dfrac{15}{16}$

8. $\dfrac{7}{18} \div \dfrac{3}{9}$

9. $\dfrac{3}{14} \div \dfrac{2}{7}$

10. $\dfrac{13}{40} \div \dfrac{26}{35}$

11. $\dfrac{5}{12} \div \dfrac{15}{16}$

12. $\dfrac{12}{27} \div \dfrac{10}{18}$

13. $\dfrac{17}{48} \div \dfrac{51}{90}$

14. $\dfrac{3}{5} \div \dfrac{7}{8}$

15. $\dfrac{13}{16} \div \dfrac{2}{3}$

16. $\dfrac{5}{6} \div \dfrac{3}{4}$

17. $\dfrac{3}{4} \div \dfrac{5}{6}$

18. $\dfrac{14}{15} \div \dfrac{21}{25}$

19. $\dfrac{3}{7} \div \dfrac{3}{7}$

20. $\dfrac{6}{13} \div \dfrac{6}{13}$

21. $\dfrac{16}{27} \div \dfrac{7}{18}$

22. $\dfrac{20}{21} \div \dfrac{15}{42}$

23. $\dfrac{25}{36} \div \dfrac{5}{24}$

24. $\dfrac{17}{20} \div \dfrac{3}{14}$

25. $\dfrac{26}{35} \div \dfrac{39}{40}$

26. $\dfrac{5}{6} \div 3\dfrac{1}{4}$

27. $\dfrac{7}{8} \div 7\dfrac{1}{2}$

28. $\dfrac{29}{50} \div 3\dfrac{1}{10}$

29. $4\dfrac{1}{5} \times 3\dfrac{3}{7}$

30. $2\dfrac{1}{17} \div 1\dfrac{4}{4}$

31. $5\dfrac{1}{6} \div 3\dfrac{41}{4}$

32. $2\dfrac{2}{49} \div 3\dfrac{1}{14}$

33. $6\dfrac{5}{6} \div 2$

34. $4\dfrac{1}{5} \div 3$

35. $6\dfrac{5}{6} \div \dfrac{1}{2}$

36. $4\dfrac{5}{8} \div 4$

37. $4\dfrac{5}{8} \div \dfrac{1}{4}$

38. $1\dfrac{1}{32} \div 3\dfrac{2}{3}$

39. $7\dfrac{5}{11} \div 4\dfrac{1}{10}$

40. $13\dfrac{1}{7} \div 4\dfrac{2}{11}$

Evaluate the following expressions, first using the rules of multiplication and division from left to right, then using addition and subtraction from left to right.

41. $\dfrac{1}{2} \div \dfrac{7}{8} + \dfrac{1}{7} \cdot \dfrac{2}{3}$

42. $\dfrac{3}{5} \cdot \dfrac{1}{6} + \dfrac{1}{5} \div 2$

43. $\dfrac{1}{2} \div \dfrac{1}{2} + 1 - \dfrac{2}{3} \cdot 3$

44. $\dfrac{3}{4} + 4\dfrac{1}{2} \cdot \dfrac{1}{3} \div \dfrac{5}{6}$

45. $\dfrac{2}{15} \cdot \dfrac{1}{4} \div \dfrac{3}{5} + \dfrac{1}{25}$

46. $3\dfrac{1}{2} \cdot 5\dfrac{1}{3} + \dfrac{5}{12} \div \dfrac{15}{16}$

47. $2\dfrac{1}{4} + 1\dfrac{1}{5} + 2 \div \dfrac{20}{21}$

48. $\dfrac{5}{8} - \dfrac{1}{3} \cdot \dfrac{2}{5} + 6\dfrac{1}{10}$

49. $1\dfrac{1}{6} \cdot 1\dfrac{2}{19} \div \dfrac{7}{8} + \dfrac{1}{38}$

50. $\dfrac{3}{10} + \dfrac{5}{6} \div \dfrac{1}{4} \cdot \dfrac{1}{8} - \dfrac{7}{60}$

51. Find the average of the numbers $\frac{5}{6}, \frac{7}{15}, \frac{8}{21}$.

52. Find the average of the numbers $\frac{7}{8}, \frac{9}{10}, 2\frac{1}{2}, 1\frac{3}{4}$.

53. The product of $\frac{9}{10}$ with another number is $\frac{5}{3}$. What is the other number?

54. The result of multiplying two numbers is $10\frac{1}{3}$. If one of the numbers is $7\frac{1}{6}$, what is the other one?

55. The product of $7\frac{2}{3}$ with some other rational number is $4\frac{1}{2}$. What is the other number?

56. An airplane is carrying 150 passengers. This is only $\frac{5}{7}$ of its capacity. What is the capacity?

57. The sale price of a coat is \$36. This is $\frac{3}{4}$ of the original price. What was the original price?

58. An estate of \$180,000 was to be shared by two nephews and a son. Each nephew received $\frac{1}{8}$ of the estate. How much did the son receive?

3.9 COMPLEX FRACTIONS

If the numerator of a fraction or the denominator of a fraction or both contain rational numbers that are not whole numbers, then the fraction is called a **complex fraction.** Since addition, subtraction, multiplication, and division (nonzero division) with rational numbers always give rational numbers, a complex fraction is a rational number. The problem discussed in this section is how to write a complex fraction in the form $\frac{a}{b}$ where a and b are whole numbers.

Generally speaking, there are two ways to approach such a problem.

TECHNIQUES FOR SIMPLIFYING COMPLEX FRACTIONS

1. Treat the numerator and denominator as separate problems and simplify each of them first, then perform the division of the numerator by the denominator.

2. Find the least common multiple (LCM) of all the denominators of the fractions that are in the numerator and denominator, then multiply the numerator and denominator by this LCM using the distributive principle.

EXAMPLES

1. Simplify $\dfrac{\dfrac{3}{4} + \dfrac{1}{2}}{1 - \dfrac{1}{3}}$ using the first technique.

Simplifying the numerator, we have $\dfrac{3}{4} + \dfrac{1}{2} = \dfrac{3}{4} + \dfrac{2}{4} = \dfrac{5}{4}$.

Simplifying the denominator, we have $1 - \dfrac{1}{3} = \dfrac{3}{3} - \dfrac{1}{3} = \dfrac{2}{3}$.

Therefore,

$$\frac{\dfrac{3}{4} + \dfrac{1}{2}}{1 - \dfrac{1}{3}} = \frac{\dfrac{5}{4}}{\dfrac{2}{3}} = \frac{5}{4} \cdot \frac{3}{2} = \frac{15}{8}.$$

2. Simplify $\dfrac{\dfrac{3}{4} + \dfrac{1}{2}}{1 - \dfrac{1}{3}}$ using the second technique.

The denominators are 4, 2, and 3. The LCM of 4, 2, 3 is 12. Therefore,

$$\frac{\dfrac{3}{4} + \dfrac{1}{2}}{1 - \dfrac{1}{3}} = \frac{\left(\dfrac{3}{4} + \dfrac{1}{2}\right)}{\left(1 - \dfrac{1}{3}\right)} \cdot \frac{12}{12} = \frac{\left(\dfrac{3}{4} + \dfrac{1}{2}\right)12}{\left(1 - \dfrac{1}{3}\right)12}$$

$$= \frac{\dfrac{3}{4} \cdot 12 + \dfrac{1}{2} \cdot 12}{1 \cdot 12 - \dfrac{1}{3} \cdot 12} = \frac{9 + 6}{12 - 4} = \frac{15}{8}$$

By using the distributive principle and multiplying by the LCM of the denominators, we are assured that the new equivalent fraction will have only whole numbers in its numerator and denominator. This technique should be used only if you are confident in using the distributive principle of multiplication.

EXAMPLE　　Simplify $\dfrac{2\frac{1}{3}}{\frac{1}{4}+\frac{1}{3}}$

Using the first technique:

$$2\frac{1}{3}=\frac{7}{3}\quad\text{and}\quad\frac{1}{4}+\frac{1}{3}=\frac{1}{4}\cdot\frac{3}{3}+\frac{1}{3}\cdot\frac{4}{4}=\frac{3}{12}+\frac{4}{12}=\frac{7}{12}$$

[NOTE: The mixed number is changed to fraction form.]

$$\frac{2\frac{1}{3}}{\frac{1}{4}+\frac{1}{3}}=\frac{\frac{7}{3}}{\frac{7}{12}}=\frac{7}{3}\cdot\frac{12}{7}=\frac{\cancel{7}\cdot\cancel{3}\cdot 4}{\cancel{3}\cdot\cancel{7}\cdot 1}=\frac{4}{1}=4$$

Using the second technique:

The LCM of 4, 3 is 12. Apply the distributive principle and multiply the numerator and denominator by 12.

$$\frac{2\frac{1}{3}}{\frac{1}{4}+\frac{1}{3}}=\frac{\left(\frac{7}{3}\right)}{\left(\frac{1}{4}+\frac{1}{3}\right)}\cdot\frac{12}{12}=\frac{\frac{7}{3}\cdot 12}{\frac{1}{4}\cdot 12+\frac{1}{3}\cdot 12}=\frac{28}{3+4}$$

$$=\frac{28}{7}=4$$

EXERCISES 3.9

Rewrite each complex fraction in the form $\dfrac{a}{b}$ where a and b are whole numbers.

1. $\dfrac{1+\frac{3}{7}}{\frac{2}{3}}$ 　 2. $\dfrac{2+\frac{1}{5}}{1+\frac{1}{4}}$ 　 3. $\dfrac{\frac{1}{5}+\frac{1}{6}}{2\frac{1}{3}}$ 　 4. $\dfrac{\frac{2}{3}+\frac{1}{5}}{4\frac{1}{2}}$

5. $\dfrac{\dfrac{5}{6}-\dfrac{1}{3}}{\dfrac{1}{2}+\dfrac{1}{5}}$

6. $\dfrac{5\dfrac{1}{7}}{2+1}$

7. $\dfrac{3\dfrac{4}{5}}{11+8}$

8. $\dfrac{4+\dfrac{1}{3}}{6+\dfrac{1}{4}}$

9. $\dfrac{7+\dfrac{2}{5}}{2+\dfrac{1}{15}}$

10. $\dfrac{2-\dfrac{1}{3}}{1-\dfrac{1}{3}}$

11. $\dfrac{\dfrac{2}{3}-\dfrac{1}{4}}{\dfrac{3}{5}-\dfrac{1}{4}}$

12. $\dfrac{\dfrac{5}{6}-\dfrac{2}{3}}{\dfrac{5}{8}-\dfrac{1}{16}}$

13. $\dfrac{\dfrac{7}{8}-\dfrac{3}{16}}{\dfrac{1}{3}-\dfrac{1}{4}}$

14. $\dfrac{\dfrac{3}{5}+\dfrac{4}{7}}{\dfrac{3}{8}+\dfrac{1}{10}}$

15. $\dfrac{\dfrac{4}{15}+\dfrac{6}{25}}{\dfrac{3}{5}+\dfrac{3}{10}}$

16. $\dfrac{3\dfrac{1}{4}+2\dfrac{1}{2}}{5\dfrac{1}{8}+1\dfrac{5}{8}}$

17. $\dfrac{7\dfrac{1}{3}+2\dfrac{1}{5}}{6\dfrac{1}{9}+2}$

18. $\dfrac{5\dfrac{2}{3}-1\dfrac{1}{6}}{3\dfrac{1}{2}+3\dfrac{1}{6}}$

19. $\dfrac{2\dfrac{4}{9}+1\dfrac{1}{18}}{1\dfrac{2}{9}-\dfrac{1}{6}}$

20. $\dfrac{7\dfrac{1}{2}+3\dfrac{5}{11}}{2\dfrac{3}{4}+1\dfrac{7}{8}}$

21. $\dfrac{8\dfrac{1}{10}+\dfrac{3}{10}}{\dfrac{6}{100}}$

22. $\dfrac{6\dfrac{1}{100}+5\dfrac{3}{100}}{2\dfrac{1}{2}+3\dfrac{1}{10}}$

23. $\dfrac{4\dfrac{7}{10}-2\dfrac{9}{10}}{5\dfrac{1}{100}}$

24. $\dfrac{7\dfrac{1}{2}+3\dfrac{1}{1000}}{\dfrac{99}{1000}-\dfrac{9}{100}}$

25. $\dfrac{\dfrac{16}{10,000}+\dfrac{3}{10}}{\dfrac{3}{10}-\dfrac{16}{10,000}}$

26. $\dfrac{17\dfrac{1}{10}-16\dfrac{3}{10}}{\dfrac{4}{5}}$

27. $\dfrac{21\dfrac{1}{2}-21\dfrac{50}{100}}{16\dfrac{1}{10}-5\dfrac{1}{10}}$

28. $\dfrac{5\dfrac{1}{10}+13\dfrac{1}{2}}{2\dfrac{9}{10}-1\dfrac{3}{10}}$

29. $\dfrac{46\dfrac{83}{100}-45\dfrac{7}{10}}{32\dfrac{57}{100}-31\dfrac{11}{25}}$

30. $\dfrac{1\dfrac{1}{100}+1\dfrac{1}{10}}{6\dfrac{1}{100}+6\dfrac{1}{10}}$

SUMMARY: CHAPTER 3

DEFINITION   A **rational number** is a number that can be written in the form $\dfrac{a}{b}$ where a is a whole number and b is a nonzero whole number.

DIVISION BY 0 IS UNDEFINED. NO DENOMINATOR CAN BE 0.

Consider $\dfrac{5}{0} = \square$. Then $5 = 0 \cdot \square = 0$. But this is impossible; $5 \neq 0$.

Next consider $\dfrac{0}{0} = \square$. Then $0 = 0 \cdot \square = 0$ for any value of $\square$. This

means that $\dfrac{0}{0}$ could be any number. But in arithmetic an operation such

as division cannot give more than one answer. Thus, $\dfrac{a}{0}$ **is undefined for**

any value of a.

DEFINITION

The **product** of two rational numbers $\dfrac{a}{b}$ and $\dfrac{c}{d}$ is the rational number whose numerator is the product of the numerators $(a \cdot c)$ and whose denominator is the product of the denominators $(b \cdot d)$. That is,

$$\frac{a}{b} \cdot \frac{c}{d} = \frac{a \cdot c}{b \cdot d}$$

COMMUTATIVE PROPERTY OF MULTIPLICATION

If $\dfrac{a}{b}$ and $\dfrac{c}{d}$ are rational numbers, then

$$\frac{a}{b} \cdot \frac{c}{d} = \frac{c}{d} \cdot \frac{a}{b}$$

ASSOCIATIVE PROPERTY OF MULTIPLICATION

If $\dfrac{a}{b}$, $\dfrac{c}{d}$, and $\dfrac{e}{f}$ are rational numbers, then

$$\frac{a}{b} \cdot \frac{c}{d} \cdot \frac{e}{f} = \left(\frac{a}{b} \cdot \frac{c}{d}\right) \cdot \frac{e}{f} = \frac{a}{b} \cdot \left(\frac{c}{d} \cdot \frac{e}{f}\right)$$

MULTIPLICATIVE IDENTITY

If $\dfrac{a}{b}$ is a rational number, then

$$\frac{a}{b} \cdot 1 = \frac{a}{b}$$

$$\frac{a}{b} = \frac{a}{b} \cdot 1 = \frac{a}{b} \cdot \frac{k}{k} = \frac{a \cdot k}{b \cdot k} \qquad \text{where } k \neq 0$$

A rational number is in **lowest terms** (reduced) if the numerator and denominator have no common factor other than 1.

DEFINITION   The **sum** of two rational numbers $\frac{a}{b}$ and $\frac{c}{b}$ is the rational number whose numerator is the sum of the numerators $(a + c)$ and whose denominator is the common denominator (b). That is,

$$\frac{a}{b} + \frac{c}{b} = \frac{a + c}{b} \qquad (b \neq 0)$$

COMMUTATIVE
PROPERTY OF
ADDITION

If $\frac{a}{b}$ and $\frac{c}{d}$ are rational numbers, then

$$\frac{a}{b} + \frac{c}{d} = \frac{c}{d} + \frac{a}{b}$$

ASSOCIATIVE
PROPERTY OF
ADDITION

If $\frac{a}{b}$, $\frac{c}{d}$, and $\frac{e}{f}$ are rational numbers, then

$$\frac{a}{b} + \frac{c}{d} + \frac{e}{f} = \left(\frac{a}{b} + \frac{c}{d}\right) + \frac{e}{f} = \frac{a}{b} + \left(\frac{c}{d} + \frac{e}{f}\right)$$

ADDITIVE
IDENTITY

If $\frac{a}{b}$ is a rational number, then

$$\frac{a}{b} + 0 = \frac{a}{b}$$

DEFINITION   The **difference** of two rational numbers $\frac{a}{b}$ and $\frac{c}{b}$ with a greater than or equal to c is a rational number whose numerator is the difference of the numerators and whose denominator is the common denominator b. In symbols,

$$\frac{a}{b} - \frac{c}{b} = \frac{a - c}{b} \qquad (b \neq 0)$$

TO CHANGE RATIONAL NUMBERS TO MIXED NUMBERS

1. Reduce first, then change to mixed numbers; or

2. Change to mixed numbers first, then reduce the fraction part.

DEFINITION   The **reciprocal** of a rational number $\dfrac{a}{b}$ where $a \neq 0$ and $b \neq 0$ is $\dfrac{b}{a}$ because

$$\frac{a}{b} \cdot \frac{b}{a} = 1$$

[NOTE: If $a = 0$, then $\dfrac{0}{b}$ has no reciprocal since $\dfrac{b}{0}$ is undefined.]

A **complex fraction** is a fraction in which the numerator or denominator or both contain rational numbers that are not whole numbers.

REVIEW QUESTIONS: CHAPTER 3

1. The denominator of a rational number cannot be _____.

2. $\dfrac{0}{7} = 0$, but $\dfrac{7}{0}$ is _____.

3. The reciprocal of $\dfrac{2}{3}$ is ____, and the reciprocal of $\dfrac{3}{2}$ is ____.

4. Which property of addition is illustrated by the following statement?

$$\frac{1}{3} + \left(\frac{5}{6} + \frac{1}{2}\right) = \left(\frac{1}{3} + \frac{5}{6}\right) + \frac{1}{2}$$

5. Draw a diagram illustrating the product $\dfrac{2}{3} \cdot \dfrac{2}{5}$.

Multiply and reduce all answers.

6. $\dfrac{1}{3} \cdot \dfrac{1}{2} \cdot \dfrac{1}{5}$ **7.** $\dfrac{1}{7} \cdot \dfrac{3}{7}$ **8.** $\dfrac{35}{56} \cdot \dfrac{4}{15} \cdot \dfrac{5}{10}$

Fill in the missing terms so that each equation is true.

9. $\dfrac{1}{6} = \dfrac{}{12}$ **10.** $\dfrac{9}{10} = \dfrac{54}{}$ **11.** $\dfrac{15}{13} = \dfrac{75}{}$

Reduce each fraction to lowest terms.

12. $\dfrac{15}{30}$ **13.** $\dfrac{99}{88}$ **14.** $\dfrac{0}{4}$ **15.** $\dfrac{150}{120}$

Add or subtract as indicated and reduce all answers.

16. $\dfrac{3}{7} + \dfrac{2}{7}$

17. $\dfrac{5}{6} - \dfrac{1}{6}$

18. $\dfrac{5}{8} - \dfrac{3}{8}$

19. $\dfrac{1}{12} + \dfrac{5}{36} + \dfrac{11}{24}$

20. $\dfrac{13}{22} - \dfrac{9}{33}$

21. $\dfrac{5}{27} + \dfrac{5}{18}$

Change to mixed numbers.

22. $\dfrac{47}{6}$

23. $\dfrac{342}{100}$

Change to fraction form.

24. $5\dfrac{1}{10}$

25. $13\dfrac{2}{3}$

Add or subtract as indicated and reduce all answers.

26. $27\dfrac{1}{4} + 3\dfrac{1}{2}$

27. $12\dfrac{5}{6} - 6\dfrac{1}{4}$

28. $15 - \dfrac{9}{10}$

29. $4\dfrac{5}{8} + 2\dfrac{3}{14}$

Multiply or divide as indicated and reduce all answers.

30. $\dfrac{7}{8} \cdot \dfrac{16}{25} \cdot \dfrac{5}{14}$

31. $7\dfrac{1}{11}\left(2\dfrac{3}{4}\right)\left(5\dfrac{1}{3}\right)$

32. $\dfrac{5}{6} \div 3\dfrac{3}{4}$

33. $16\dfrac{2}{3} \div 22\dfrac{2}{9}$

34. (a) Find $\dfrac{3}{4}$ of 60.   (b) Find $\dfrac{2}{3}$ of 96.

35. Find the average of $\dfrac{3}{4}$, $\dfrac{5}{8}$, and $\dfrac{9}{10}$.

36. Which is larger, $\dfrac{2}{3}$ or $\dfrac{4}{5}$? How much larger?

37. Evaluate $\dfrac{5}{8} \cdot \dfrac{3}{10} + \dfrac{1}{14} \div 2$.

38.   Simplify

$$\frac{\dfrac{7}{8} - \dfrac{3}{16}}{\dfrac{1}{3} - \dfrac{1}{4}}$$

39.   If a telephone pole is $42\frac{1}{2}$ feet long and $\frac{1}{4}$ of the pole is below ground level, how many feet of the pole are aboveground?

40.   A triangle has three sides. The lengths are $43\frac{1}{2}$ feet, $52\frac{2}{3}$ feet, and $26\frac{1}{4}$ feet. What is the perimeter (total distance around) of the triangle?

41.   The product of $20\frac{2}{3}$ with some other rational number is $6\frac{1}{2}$. What is the other number?

42.   A refrigerator is on sale for $800. This price is $\frac{4}{5}$ of the original price. What was the original price?

CHAPTER TEST: CHAPTER 3

1. $\frac{9}{0}$ is undefined, but $\frac{0}{9} =$ ——— .

2. The numbers $\frac{5}{8}$ and $\frac{8}{5}$ are ————— of each other.

3. Draw a diagram illustrating the product $\frac{1}{2} \cdot \frac{3}{4}$.

4. Which property of multiplication is illustrated by the following statement?

$$\frac{9}{10} \cdot \frac{3}{10} = \frac{3}{10} \cdot \frac{9}{10}$$

5. What is $\frac{3}{10}$ of $\frac{9}{10}$? What is $\frac{3}{4}$ of 100?

Multiply and reduce all answers.

6. $\frac{1}{5} \cdot \frac{1}{7} \cdot \frac{2}{3}$

7. $\frac{54}{12} \cdot \frac{7}{18} \cdot \frac{10}{35}$

Reduce to lowest terms.

8. $\frac{44}{55}$

9. $\frac{51}{68}$

10. $\frac{216}{264}$

Add or subtract as indicated and reduce all answers.

11. $\frac{3}{90} + \frac{6}{90}$

12. $\frac{5}{12} + \frac{5}{36} + \frac{1}{3}$

13. $\frac{13}{22} - \frac{5}{33}$

14. $\frac{71}{56} - \frac{96}{84}$

Multiply or divide as indicated and reduce all answers.

15. $\frac{4}{13} \div 3\frac{1}{4}$

16. $2\frac{1}{5}\left(3\frac{3}{11}\right)\left(5\frac{5}{12}\right)$

17. $4\frac{3}{8} \div 7\frac{1}{2}$

18. $\frac{9}{10} \cdot \frac{4}{5} \cdot \frac{20}{81} \cdot \frac{3}{8}$

19. Find the average of $\frac{1}{3}$, $\frac{1}{4}$, $\frac{1}{5}$, and $\frac{1}{6}$.

20. Evaluate $\frac{3}{5} \div \frac{1}{2} + \frac{7}{8} \cdot \frac{1}{2}$.

21. Simplify

$$\frac{6\frac{1}{2} + 2\frac{3}{4}}{2 - \frac{1}{8}}$$

22. The number $2\frac{1}{2}$ is the product of $\frac{3}{4}$ with some other number. What is the other number?

23. A bookstand is marked for sale at \$56. The store manager plans to discount the price by $\frac{1}{4}$. What will be the new price?

24. A rectangle* is a geometric figure with four sides in which the opposite sides are equal. If the lengths of two sides are $7\frac{1}{2}$ feet and $10\frac{2}{3}$ feet, what is the perimeter (distance around) of the rectangle?

$10\frac{2}{3}$ ft

$7\frac{1}{2}$ ft

---

*A rectangle also has four right angles (angles which measure 90°). Rectangles will be discussed in more detail in Chapter 7.

4

DECIMAL NUMBERS AND PROPORTIONS

4.1   DECIMAL NUMBERS

The powers of ten are 1, 10, 100, 1000, 10,000, 100,000, 1,000,000, and so on. Rational numbers with powers of ten in the denominator are **decimal numbers.** Thus,

$$\frac{3}{10}, \frac{41}{100}, \frac{53}{1000}, \frac{89}{10}, \frac{7}{1}, \text{ and } \frac{836}{1}$$

are all decimal numbers.

DEFINITION

A **decimal number** is a rational number that has a power of ten as its denominator.

The usual notation for decimal numbers is an extension of the place value system. Whole numbers are still written to the left of the decimal point. The values of each place to the right of the decimal point are the reciprocals of powers of ten: $\frac{1}{10}, \frac{1}{100}, \frac{1}{1000}, \frac{1}{10,000}$, and so on. (See Figure 4.1.)

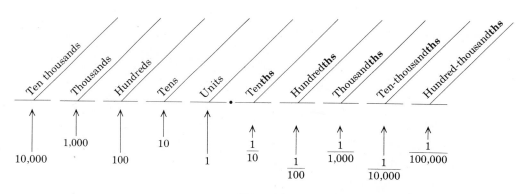

Figure 4.1

Note that **th** is used to indicate the fraction parts.

Reading or Writing a Decimal Number

If the decimal number (or decimal) has a fraction part, the whole number part is read as before; the word **and** indicates the decimal point; the fraction part is read as a whole number with the name of the place of the last digit.

EXAMPLES

1. 37.56 is read "thirty-seven **and** fifty-six hundredths."

   [NOTE: **And** indicates the decimal point, and the digit 6 is in the hundredths position.]

   We can also think of 37.56 as the mixed number $37\frac{56}{100}$.

2. 5.398 is read "five **and** three hundred ninety-eight thousandths."

   5.398 is the same as the mixed number $5\frac{398}{1000}$.

3. Write the decimal number represented by the words "fourteen **and** thirty-six ten-thousandths."

   Answer: 14.0036

   [NOTE: The hyphen (-) in ten-thousandths makes ten-thousandths one word.]

4. Write the decimal number represented by the words "three hundred **and** seventy-eight thousandths."

   Answer: 300.078

SPECIAL NOTES

A. The **th** at the end of a word indicates a fraction part (a part to the right of the decimal point).
      eight hundred = 800
      eight hundred**ths** = 0.08

B. The hyphen (-) indicates one word.
      eight hundred thousand = 800,000
      eight hundred-thousand**ths** = 0.00008

Writing checks points out the need to write numbers in word form. Because of the problem of spelling, poor penmanship, and so on, we are required to write the amount of the check both in words and in numerals as a safety measure (Figure 4.2).

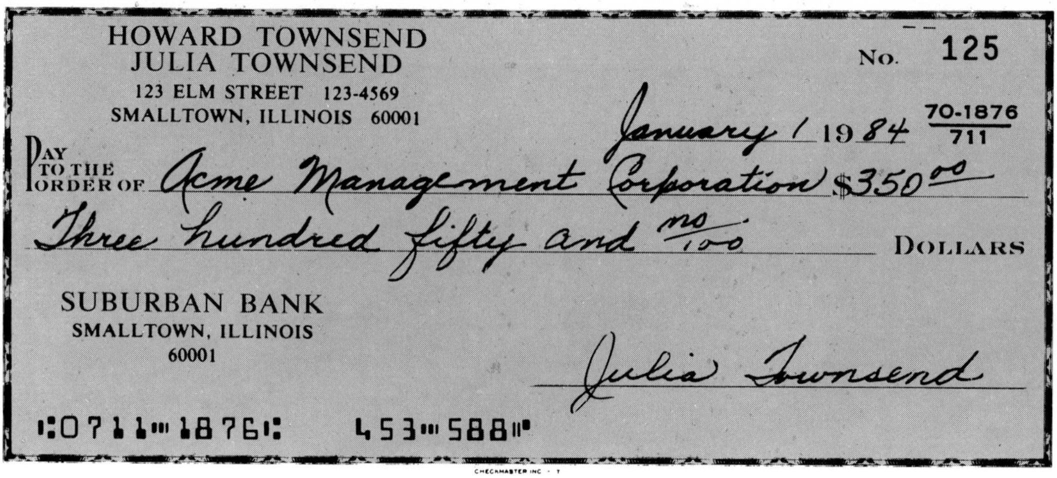

Figure 4.2

PRACTICE QUIZ		ANSWERS
	1. Write five thousandths in decimal notation.	1. 0.005
	2. Write seventy-two and forty-three hundredths in decimal notation.	2. 72.43
	3. Write 56.3 in words.	3. fifty-six and three tenths
	4. Write the mixed number $9\dfrac{32}{1000}$ in decimal notation.	4. 9.032

EXERCISES 4.1

Write the following mixed numbers in decimal notation.

1. $37\dfrac{498}{1000}$ **2.** $18\dfrac{76}{100}$ **3.** $4\dfrac{11}{100}$ **4.** $56\dfrac{3}{100}$

5. $87\dfrac{3}{1000}$ **6.** $95\dfrac{2}{10}$ **7.** $62\dfrac{7}{10}$ **8.** $100\dfrac{25}{100}$

9. $100\dfrac{38}{100}$ **10.** $250\dfrac{623}{1000}$

Write the following decimal numbers in mixed number form.

11. 82.56 **12.** 93.07 **13.** 10.576 **14.** 100.6

15. 65.003

Write the following numbers in decimal notation.

16. three tenths   **17.** fourteen thousandths

18. seventeen hundredths

19. six and twenty-eight hundredths

20. sixty and twenty-eight thousandths

21. seventy-two and three hundred ninety-two thousandths

22. eight hundred fifty and thirty-six ten-thousandths

23. seven hundred and seventy-seven hundredths

24. eight thousand four hundred ninety-two and two hundred sixty-three thousandths

25. six hundred thousand, five hundred and four hundred two thousandths

Write the following decimal numbers in words.

26.	0.5	**27.**	0.93	**28.**	5.06	**29.**	32.58
30.	71.06	**31.**	35.078	**32.**	7.003	**33.**	18.102
34.	50.008	**35.**	607.607	**36.**	593.86	**37.**	593.860
38.	4700.617	**39.**	5000.005	**40.**	603.0065	**41.**	900.4638

42. Duplicate the check form in the text and write samples checks for

    (a) $372.58   (b) $577.50   (c) $2405.37

4.2 ROUNDING OFF

We all **round off** numbers whenever we measure anything. Any measuring device made by humans gives only approximate numbers. One-half cup of sugar, a pencil six inches long, a bolt with diameter three centimeters, a

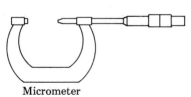

Micrometer

(a) The micrometer is marked to give approximate measures of circular objects.

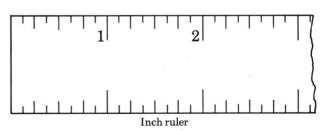

Inch ruler

(b) The ruler is marked to give approximate measures of lengths in fourths and eighths of an inch.

Figure 4.3

room 20 feet wide and $32\frac{1}{2}$ feet long, all represent approximate measuring and rounded off numbers.

The rules for adding, subtracting, multiplying, and dividing with approximate numbers are discussed in shop classes, engineering classes, surveying classes, and so on. These rules will not be discussed here. We will discuss one method of rounding off numbers.

To **round off** a number means to find another number close to the original number. In fact, we want to find the closer of two other numbers. Is 872 closer to 800 or 900? Is 872 closer to 870 or 880?

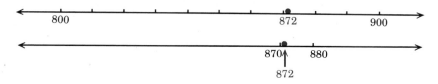

Figure 4.4

As Figure 4.4 illustrates, 872 is closer to 900 than to 800. So, **to the nearest hundred,** 872 rounds off to 900. Also, 872 is closer to 870 than to 880. So, **to the nearest ten,** 872 rounds off to 870.

Using figures as an aid to understanding is fine, but we need to have a specific rule for practical purposes.

RULE FOR ROUNDING OFF DECIMAL NUMBERS

1. Look at the single digit just to the right of the digit that is in the place of desired accuracy.

2. If this digit is 5 or greater, make the digit in the desired place of accuracy one larger and replace all digits to the right with zeros.

3. If this digit is less than 5, leave the digit that is in the place of desired accuracy as it is and replace all digits to the right with zeros.

[This rule is used in many situations, but not all. For example, if the price of an item in a store involves a fraction of a cent, the merchant will always round off to the next highest cent. If three cans of beans cost $1.00, then one can of beans will cost 34¢, not $33\frac{1}{3}$¢ or 33¢.]

EXAMPLES

Round off as indicated.

1.   7283.5   (nearest hundred)

Answer: 7300.0 or 7300

(Look only at the 8. Since 8 is more than 5, change 2 to 3. Replace the remaining digits to the right with zeros.)

2. 5.749 (nearest tenth)

Answer: 5.700 or 5.7

(Look only at the 4. Since 4 is less than 5, the 7 is unchanged. 0's replace the remaining digits to the right. Those replacement 0's to the right of the decimal point may be omitted if you choose. But replacement 0's to the left of the decimal point must remain as shown in Example 1.)

3. 239.53 (nearest unit)

Answer: 240.00 or 240

(Look only at the 5. To make 9 one larger, change 39 to 40. In this case, two digits are affected.)

4. 6.4579 (nearest thousandth)

Answer: 6.4580 or 6.458

(Look only at the 9. Since 9 is more than 5, change 7 to 8. Zero replaces 9.)

If a decimal number has no decimal point, the number is understood to be a whole number. The decimal point is understood to be to the right of the rightmost digit. For example, 325 and 325. are the same.

PRACTICE QUIZ	Round off as indicated.	ANSWERS
	1. 572.3 (nearest ten)	1. 570
	2. 6.749 (nearest tenth)	2. 6.7
	3. 7558 (nearest thousand)	3. 8000
	4. 0.07921 (nearest thousandth)	4. 0.079

EXERCISES 4.2

Round off each of the following decimal numbers as indicated.

To the nearest tenth:

1. 4.763	**2.** 5.031	**3.** 76.349	**4.** 76.352
5. 89.015	**6.** 7.555	**7.** 18.009	**8.** 37.666
9. 14.3338	**10.** 0.036		

To the nearest hundredth:

11. 0.385	**12.** 0.296	**13.** 5.722	**14.** 8.987
15. 6.996	**16.** 13.1346	**17.** 0.0782	**18.** 6.0035
19. 5.7092	**20.** 2.8347		

To the nearest thousandth:

21. 0.0672	**22.** 0.05550	**23.** 0.6338	**24.** 7.6666
25. 32.4785	**26.** 9.4302	**27.** 17.36371	**28.** 4.44449
29. 0.00191	**30.** 20.76962		

To the nearest whole number (or nearest unit):

31. 479.23	**32.** 6.8	**33.** 17.5	**34.** 19.999
35. 382.48	**36.** 649.66	**37.** 439.78	**38.** 701.413
39. 6333.11	**40.** 8122.825		

To the nearest ten:

41. 5163.	**42.** 6475	**43.** 495	**44.** 572.5
45. 998.5	**46.** 378.92	**47.** 5476.2	**48.** 76,523.1
49. 92,540.9	**50.** 7007.7		

To the nearest thousand:

51. 7398	**52.** 62,275	**53.** 47,823.4	**54.** 103,499
55. 217,480.2	**56.** 9872.5	**57.** 379,500	**58.** 4,500,762
59. 7,305,438	**60.** 573,333.3		

61. .0005783   (nearest hundred-thousandth)

62. .5449   (nearest hundredth)

63. 473.8   (nearest ten)

64. 5.00632   (nearest thousandth)

65. 473.8   (nearest hundred)

66. 5750   (nearest thousand)

67. 3.2296   (nearest thousandth)

68. 15.548   (nearest tenth)

69. 78,419   (nearest ten thousand)

70. 78,419   (nearest ten)

4.3 ADDING AND SUBTRACTING DECIMAL NUMBERS

Decimal numbers can be written in an expanded form, such as

$$5.237 = 5 + 2\left(\frac{1}{10}\right) + 3\left(\frac{1}{100}\right) + 7\left(\frac{1}{1000}\right)$$

$$= 5 + \frac{2}{10} + \frac{3}{100} + \frac{7}{1000}$$

Thus, to add $5.237 + 6.15$, we can write

$$5 + \frac{2}{10} + \frac{3}{100} + \frac{7}{1000} + 6 + \frac{1}{10} + \frac{5}{100}$$

$$= (5 + 6) + \left(\frac{2}{10} + \frac{1}{10}\right) + \left(\frac{3}{100} + \frac{5}{100}\right) + \frac{7}{1000}$$

This procedure can be accomplished in a much easier way by writing the decimal numbers one under the other and keeping the decimal points in line. In this way, the whole numbers will be added properly, tenths added to tenths, hundredths to hundredths, and so on. The decimal point in the sum is in line with the other decimal points.

$$
\begin{array}{r}
\text{Add} \quad 5.237 \\
6.150 \\
\hline
11.387
\end{array}
$$

Zeros may be written to the right of the last digit in the fraction part to help keep the digits in the correct line. This will not change the value of any number.

1. Find the sum $5.2 + 6.32 + 13.06$.

$$
\begin{array}{r}
5.20 \\
6.32 \\
13.06 \\
\hline
24.58
\end{array}
$$

[NOTE: Writing 5.2 as 5.20 helps keep the digits lined up.]

2. Find the sum $8 + 3.76 + 47.689 + .2$.

$$
\begin{array}{r}
8.000 \\
3.760 \\
47.689 \\
0.200 \\
\hline
59.649
\end{array}
$$

[NOTE: The zeros are filled in to help keep the digits in line.]

Subtraction with decimal numbers also requires that the decimal points be in a vertical line when the numbers are written one under the other. The reason is so that fractions with the same denominator will be subtracted. Again, the rules are the same as with whole numbers after the decimal points have been lined up. The decimal point in the difference must be in line with the other decimal points.

EXAMPLES

1. Find the difference $5.438 - 2.653$.

$$\begin{array}{r} 5.438 \\ -2.653 \\ \hline 2.785 \end{array}$$

2. Find the difference $17.2 - 3.6954$.

$$\begin{array}{r} 17.2000 \\ -\ 3.6954 \\ \hline 13.5046 \end{array}$$

PRACTICE QUIZ	Find each indicated sum or difference.	ANSWERS
	1. $46.2 + 3.07 + 2.6$	1. 51.87
	2. $9 + 5.6 + 0.58$	2. 15.18
	3. $6.4 - 3.7$	3. 2.7
	4. $18 - 0.4384$	4. 17.5616

EXERCISES 4.3

Find each of the indicated sums.

1. $0.6 + 0.4 + 1.3$ 2. $5 + 6.1 + 0.4$ 3. $0.59 + 6.91 + 0.05$

4. $3.488 + 16.593 + 25.002$ 5. $37.02 + 25 + 6.4 + 3.89$

6. $4.0086 + 0.034 + 0.6 + 0.05$ 7. $43.766 + 9.33 + 17 + 206$

8. $52.3 + 6 + 21.01 + 4.005$ 9. $2.051 + 0.2006 + 5.4 + 37$

10. $5 + 2.37 + 463 + 10.88$

11.	47.3	12.	1.007	13.	4.128	14.	5.0015
	42.03		20.063		0.02		2.443
	29.003		0.49		3.		0.0469

15. 75.2
3.682
14.995

16. 107.39
5.061
23.54
64.9801

17. 34.967
50.6
8.562
9.3

18. 4.156
3.7
25.682
13.405

19. 74.
3.529
52.62
7.001

20. 983.4
47.518
805.411
300.766

Find each of the indicated differences.

21. 5.2 − 3.76

22. 17.83 − 8.9

23. 29.5 − 13.61

24. 1.0057 − 0.03

25. 78.015 − 13.068

26. 22.418
−17.523

27. 4.8
−0.0026

28. 31.009
− 0.534

29. 4.
−1.0566

30. 40.718
− 6.532

31. Mrs. Johnson bought the following items at a department store: dress, $47.25; shoes, $35.75; purse, $12.50. How much did she spend? What was her change if she gave the clerk a $100 bill? (Tax was included in the prices.)

32. The inside radius of a pipe is 2.38 inches, and the outside radius is 2.63 inches. What is the thickness of the pipe?

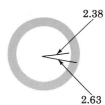

33. Mr. Johnson bought the following items at a department store: slacks, $32.50; shoes, $43.75; shirt, $18.60. How much did he spend? What was his change if he gave the clerk a $100 bill? (Tax was included in the prices.)

34. An architect's scale drawing shows a rectangular lot 2.38 inches on one side and 3.76 inches on the other side. What was the perimeter (distance around) of the rectangle on the drawing?

35. Find the difference between (a) seven and thirty-nine thousandths and (b) four and one hundred six ten-thousandths. Also, find the difference between (a) two hundred and seventeen hundredths and (b) one hundred five and nine hundredths. Find the sum of these two differences.

36. The inside **diameter** of a pipe is 3.85 centimeters, and the outside **diameter** is 3.93 centimeters. What is the thickness of the pipe? (HINT: Draw a sketch.)

4.4   MULTIPLYING DECIMAL NUMBERS

Multiplying tenths by hundredths $\left(\dfrac{1}{10} \cdot \dfrac{1}{100} = \dfrac{1}{1000}\right)$ gives thousandths. Thus,

$$\frac{3}{10} \cdot \frac{5}{100} = \frac{15}{1000} \qquad \text{or} \qquad (.3)(.05) = .015$$

and

$$\frac{6}{10} \cdot \frac{4}{1000} = \frac{24}{100,000} \qquad \text{or} \qquad (.06)(.004) = .00024$$

Vertically,

.05	(total of 3 places to
.3	right of decimal points
.015	in both multipliers)

.004	(total of 5 places to
.06	right of decimal points
.00024	in both multipliers)

We don't want to change to fraction form whenever we multiply two decimal numbers. There is a simple rule suggested by the two examples just shown that makes multiplication with decimal numbers relatively easy. In multiplication, there is no need to keep the decimal points in a line.

RULE FOR MULTIPLYING TWO DECIMAL NUMBERS

1. Multiply the two numbers as if they were whole numbers.

2. Count the total number of places to the right of the decimal points in both multipliers.

3. This sum is the number of places to the right of the decimal point in the product.

EXAMPLES

1.
```
    2.432 ←——3 places
      5.1 ←——1 place
     2432
   12 160
  12.4032 ←——4 places
```

2.
```
     4.35
    12.6
   2 610
   8 70
  43 5
  54.810
```

3.
```
    1.76
      25
    8 80
   35 2
   44.00
```

Multiplication of whole numbers by powers of ten was discussed in Section 1.5. In that section, we simply added zeros to the right of the number.

$$1000(357) = 357,000$$
$$100(46) = 4600$$

Another way of accomplishing the same result is to move the decimal point to the right.

$$1000(357.) = 357,000.$$
$$100(46.) = 4600.$$

The number of places the decimal point is moved is the same as the number of zeros in the power of ten or the exponent of the power of ten. This procedure applies to multiplication with powers of ten and all decimal numbers.

EXAMPLES

1. $10(3.57) = 35.7$ (Move decimal point 1 place to the right.)

2. $100(3.57) = 357.$ (Move decimal point 2 places to the right.)

3. $10^2(4.963) = 496.3$ (Move decimal point 2 places to the right.)

4. $1000(0.9641) = 964.1$ (Move decimal point 3 places to the right.)

PRACTICE QUIZ	Find each of the indicated products.	ANSWERS
	1. (.8)(.2)	1. 0.16
	2. (5.6)(.04)	2. 0.224
	3. $10^4(3.781)$	3. 37810.

EXERCISES 4.4

Find each of the indicated products.

1. (.6)(.7)	2. 3(2.5)	3. 1.4(.2)	4. (3.5)(.6)
5. 6(3.1)	6. .2(.02)	7. .5(.05)	8. .03(.03)
9. 4.1(.06)	10. .7(.1)	11. .06(.01)	12. .23(.12)
13. 4.7(.02)	14. .51(.13)	15. 4.15(2.6)	16. 5.9(.25)
17. 4(.75)	18. 16(.875)	19. 8(.125)	20. .66(.33)

21. $10^2(3.46)$ 22. $10^2(20.57)$ 23. $100(7.82)$

24. $100(16.1)$ 25. $10^1(.435)$ 26. $10^3(4.1782)$

27. $10^0(.38)$ 28. $10^4(51.329)$ 29. $10^4(7.12)$

30. $10^5(2.5148)$

31.	.005	32.	.137	33.	1.06	34.	.01063	35.	71.222
	.009		.06		.14		.087		.111

36. If an architect makes a drawing to the scale that 1 inch represents 6.75 feet, what distance is represented by 5.5 inches?

37. To buy a car, a man can pay $2036.50 cash, or he can put $400 down and make 18 monthly payments of $104.30. How much does he save by paying cash?

38. If an automobile dealer makes $150.70 on each used car he sells and $425.30 on each new car he sells, how much did he make the month that he sold 11 used and 6 new cars?

39. Suppose a tax assessor figures the tax at 0.07 of the assessed value of a home. If the assessed value is figured at a rate of 0.32 of the market value, what taxes are paid on a home with a market value of $136,500?

40. If the sales price of a new refrigerator is $583 and if sales tax is figured at 0.06 times the price, what is the total amount paid for the refrigerator?

41. If you drive south at 57.6 miles per hour for 3 hours, then west at 52.4 miles per hour for 4 hours, how far have you driven in the 7 hours? (Assume you started at least 300 miles east of the Pacific Ocean and 200 miles north of the Gulf of Mexico.)

42. Multiply the numbers 2.456 and 3.16, then round off the product to the nearest tenth. Next, round off each of the factors to the nearest tenth and then multiply and round off this product to the nearest tenth. Did you get the same answer?

4.5   DIVIDING DECIMAL NUMBERS

Division with decimal numbers can be related to division with whole numbers. For example, to divide $4.9\overline{)51.45}$, we can write

$$\frac{51.45}{4.9} = \frac{51.45}{4.9} \times \frac{100}{100} = \frac{5145}{490}$$

Thus, $4.9\overline{)51.45}$ is the same as $490.\overline{)5145}$.

```
        10.
490.)5145.
      490
      245
        0
      245    Remainder
```

But we now have decimal numbers, and we can continue to divide by putting one or more zeros in the dividend:

```
        10.5
490.)5145.0
      490
      245
        0
      2450
      2450
         0
```

This procedure indicates that a whole number for the divisor will be sufficient. That is, in the case of $4.9\overline{)51.45}$, since $(4.9)(10) = 49$, we need only multiply the dividend and divisor by 10. This multiplication by 10 is indicated by arrows showing the new placements of the decimal points.

```
                              1 0.5
4.9 )51.4 5           4.9 )51.4 5
                             49
Multiply both numbers by 10.  2 4
                              0
                             2 4 5
                             2 4 5
                                0
```

The decimal point in the quotient goes directly above the new decimal point in the dividend. Study the following examples carefully.

1. $12.87 \div 4.5$

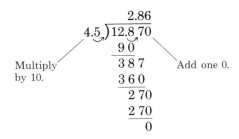

```
                    2.86
        4.5 )12.8 70
               9 0
               3 8 7
               3 6 0
                 2 70
                 2 70
                    0
```

Multiply by 10.   Add one 0.

2.   5.1 ÷ 1.36

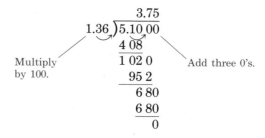

```
                          3.75
                 1.36 )5.10 00
                        4 08
                        1 02 0
                          95 2
                           6 80
                           6 80
                              0
```

Multiply   Add three 0's.
by 100.

3.   6.3252 ÷ 6.3

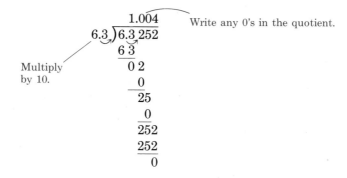

```
                    1.004    Write any 0's in the quotient.
            6.3 )6.3 252
                  6 3
                    0 2
                      0
                     25
                      0
                    252
                    252
                      0
```

Multiply
by 10.

There must be a digit to the right of the decimal point in the quotient above every digit to the right of the decimal point in the dividend.

4.   Find 2 ÷ 3.1 to the nearest thousandth.

This division will not give a 0 remainder no matter how far we divide. We must agree before dividing to round off the quotient to some desired place of accuracy. We divide until we find one digit beyond the desired place of accuracy, then round off the quotient.

approximately

```
         .6451 ≈ 0.645
  3.1 )2.0 0000
       1 8 6
        1 40
        1 24
         160
         155
          50
          31
          19
```

(We divided to the ten-thousandth place before rounding off to the nearest thousandth.)

Multiplying by powers of ten moves the decimal point to the right, giving a larger number. Dividing by powers of ten moves the decimal point to the left, giving a smaller number. For example,

$$100(4.736) = 473.6 \quad \text{and} \quad \frac{4.736}{100} = 0.04736$$

Long division by powers of ten will verify the following results.

> Division by 10 moves the decimal point to the left one place.
>
> Division by 100 moves the decimal point to the left two places.
>
> Division by 1000 moves the decimal point to the left three places, and so on.

EXAMPLES

1. $\dfrac{4.16}{100} = 0.416$

2. $\dfrac{782}{10} = 78.2$

3. $\dfrac{593.3}{1000} = 0.5933$

4. $\dfrac{186.4}{10^2} = 1.864$

PRACTICE QUIZ

Find each of the indicated quotients. ANSWERS

1. $4\overline{)1.83}$ (nearest hundredth) 1. 0.46

2. $.06\overline{)43.721}$ (nearest thousandth) 2. 728.683

3. $\dfrac{42.31}{10^3}$ 3. 0.4231

EXERCISES 4.5

Find the quotients to the nearest thousandth if the remainders are not zero (0).

1. $2\overline{)4.68}$ 2. $3\overline{)1.71}$ 3. $.5\overline{)4.95}$ 4. $.9\overline{)1.62}$

5. $.8\overline{).064}$ 6. $.7\overline{).63}$ 7. $.04\overline{)82.24}$ 8. $.03\overline{)16.02}$

9. $2.4\overline{)48}$ 10. $5.6\overline{)28}$ 11. $1.8\overline{).0036}$ 12. $.14\overline{).042}$

13. $7\overline{)6.6}$ 14. $3.2\overline{).416}$ 15. $9\overline{)7.6}$ 16. $1.6\overline{)9.76}$

17. $6\overline{)1}$ 18. $1.2\overline{)1.56}$ 19. $3.02\overline{)9.1506}$ 20. $4.6\overline{)5}$

21. $7.05\overline{).49773}$ 22. $.37\overline{)4.683}$ 23. $.21\overline{)65.226}$

24. $100\overline{)5.682}$ **25.** $1.62\overline{)34}$ **26.** $4.6 \div .009$

27. $5.2 \div .71$ **28.** $.03 \div .008$ **29.** $.71 \div .025$

30. $29.3 \div 6.9$

Find the exact quotients without rounding off.

31. $100\overline{)78.4}$ **32.** $1000\overline{)16.4963}$ **33.** $.5036 \div 10^3$

34. $45.621 \div 10^2$ **35.** $7.682 \div 10^0$ **36.** $\dfrac{.00167}{10^3}$

37. $\dfrac{.01826}{10^4}$ **38.** $\dfrac{91.112}{10^0}$ **39.** $\dfrac{.6122}{10^3}$

40. $\dfrac{10.413}{10^5}$

41. Find the average of the numbers 86.7, 49.2, and 75.4 correct to the nearest tenth.

42. If a car averages 24.6 miles per gallon, how many miles will it go on 18 gallons of gas?

43. If a motorcycle averages 32.4 miles per gallon, how many miles will it go on 7 gallons of gas?

44. If a car travels 300 miles on 16 gallons of gas, how many miles will it travel per gallon?

45. If a bicyclist rode 250.6 miles in 13.2 hours, what was her average speed (to the nearest tenth)?

46. A quarter of beef can be bought cheaper than the same amount of meat purchased a few pounds at a time. What is the cost per pound if 150 pounds cost $187.50?

47. If you drive 9.5 hours at an average speed of 52.2 miles per hour, how far will you drive?

48. If new tires cost $56.50 per tire and tax is figured at 0.06 times the cost of each tire, what will you pay for 4 new tires?

49. If you bought 10 books for a total price of $225 plus tax at 0.06 times the price, what average amount did you pay per book including tax?

50. If the total price of a stereo was $312.70 including tax at 0.06 times the asking price, you can find the asking price by dividing the total price by 1.06. What was the asking price?

51. If the interest on a 30-year mortgage on a home loan of $60,000 is going to be $189,570, what will be the monthly payments (on the loan and interest)? (This does not include insurance or taxes.)

4.6   RATIO AND PROPORTION

We know that a rational number, such as $\frac{3}{4}$, can (1) indicate a certain number of equal parts of a whole, and (2) indicate division. A third purpose is simply to provide a comparison of two quantities. For example, if there are 30 students and 40 chairs in a room, we can compare the number of students to the number of chairs, as $\frac{30 \text{ students}}{40 \text{ chairs}}$. Since $\frac{30}{40} = \frac{3}{4}$, the comparison can also be written $\frac{3 \text{ students}}{4 \text{ chairs}}$. Such a comparison is called a ratio.

DEFINITION   A **ratio** is a comparison of two quantities.

During baseball season, major league players' batting averages are in the newspapers. These are ratios of hits to times at bat. For example, a player with an average of .300 averages 300 hits for 1000 times at bat, or 3 hits for 10 times at bat, since $\frac{300}{1000} = \frac{3}{10}$. A hitting percentage of .250 can be represented as $\frac{250}{1000}$ or $\frac{1}{4}$. This can also be written 1 to 4 or 1:4.

To avoid confusion, the units used in a ratio should be written down or otherwise explained in the problem. For example, 30 students to 40 chairs, or 3 hits to 10 times at bat, or 1 foot to 3 yards. In this last case, we can change the units so the numerator and denominator have the same units.

$$\frac{1 \text{ foot}}{3 \text{ yards}} = \frac{1 \text{ foot}}{9 \text{ feet}} = \frac{1}{9}$$

The last ratio is more meaningful than the first. **Common units should be used in the numerator and denominator of a ratio whenever possible.**

The statement $\frac{6}{8} = \frac{9}{12}$ is an equation that says the two ratios $\frac{6}{8}$ and $\frac{9}{12}$ are equal. Is this true? Is the equation $\frac{5}{8} = \frac{7}{10}$ true? These equations are called **proportions,** and we need some method to determine whether a proportion is true or false.

DEFINITION   A **proportion** is a statement that two ratios are equal. In symbols, $\frac{a}{b} = \frac{c}{d}$ is a proportion.

In a proportion, the numbers are called **terms** and are named as follows:

$$\text{first term} \longrightarrow \frac{a}{b} = \frac{c}{d} \longleftarrow \text{third term}$$
$$\text{second term} \longrightarrow b \qquad d \longleftarrow \text{fourth term}$$

Terms 1 and 4 (a and d) are called the **extremes.** Terms 2 and 3 (b and c) are called the **means.**

A proportion is true if the product of the extremes equals the product of the means.

$$\frac{a}{b} = \frac{c}{d} \quad \text{if and only if} \quad a \cdot d = b \cdot c \quad \text{where } b \neq 0 \text{ and } d \neq 0$$

Now we can say $\dfrac{6}{8} = \dfrac{9}{12}$ is true since $6 \cdot 12 = 8 \cdot 9$. Also, $\dfrac{5}{8} = \dfrac{7}{10}$ is false since $5 \cdot 10 \neq 8 \cdot 7$.

EXAMPLES

Determine whether the following proportions are true or false. [NOTE: The terms can be decimal numbers, fractions, or mixed numbers, as well as whole numbers. The rules are the same.]

1. $\dfrac{9}{13} = \dfrac{4.5}{6.5}$

   True because $9(6.5) = 58.5$ and $13(4.5) = 58.5$.

2. $\dfrac{3}{7} = \dfrac{6}{15}$

   False because $3 \cdot 15 = 45$ and $7 \cdot 6 = 42$ and $45 \neq 42$.

3. $\dfrac{\frac{1}{4}}{\frac{2}{3}} = \dfrac{9}{24}$

   True because $\dfrac{1}{4} \cdot 24 = 6$ and $\dfrac{2}{3} \cdot 9 = 6$. We could also simplify both sides:

   $$\frac{\frac{1}{4}}{\frac{2}{3}} = \frac{1}{4} \cdot \frac{3}{2} = \frac{3}{8} \quad \text{and} \quad \frac{9}{24} = \frac{\cancel{3} \cdot 3}{\cancel{3} \cdot 8} = \frac{3}{8}$$

   In either case, the proportion is true.

EXERCISES 4.6

Write the following comparisons as ratios reduced to lowest terms. Use common units in the numerator and denominator whenever possible.

1. 1 dime to 3 nickels

2. 30 chairs to 25 students

3. 2 yards to 5 feet

4. 18 inches to 2 feet

5. 3 bookshelves to 18 feet of lumber

6. 10 centimeters to 1 decimeter

7. 100 centimeters to 1 meter

8. 1000 millimeters to 1 meter

9. $525 to 100 stocks in the stock market

10. 38 miles to 2 gallons of gas

Determine whether the following proportions are true or false.

11. $\dfrac{5}{6} = \dfrac{10}{12}$　　12. $\dfrac{2}{7} = \dfrac{5}{17}$　　13. $\dfrac{7}{21} = \dfrac{4}{12}$　　14. $\dfrac{6}{15} = \dfrac{2}{5}$

15. $\dfrac{5}{8} = \dfrac{12}{17}$　　16. $\dfrac{12}{15} = \dfrac{20}{25}$　　17. $\dfrac{5}{3} = \dfrac{15}{9}$　　18. $\dfrac{6}{8} = \dfrac{15}{20}$

19. $\dfrac{2}{5} = \dfrac{4}{10}$　　20. $\dfrac{3}{5} = \dfrac{60}{100}$　　21. $\dfrac{125}{1000} = \dfrac{1}{8}$　　22. $\dfrac{3}{8} = \dfrac{375}{1000}$

23. $\dfrac{1}{4} = \dfrac{25}{100}$　　24. $\dfrac{7}{8} = \dfrac{875}{1000}$　　25. $\dfrac{3}{16} = \dfrac{9}{48}$　　26. $\dfrac{2}{3} = \dfrac{66}{100}$

27. $\dfrac{1}{3} = \dfrac{33}{100}$　　28. $\dfrac{14}{6} = \dfrac{21}{8}$　　29. $\dfrac{4}{9} = \dfrac{7}{12}$　　30. $\dfrac{19}{16} = \dfrac{20}{17}$

31. $\dfrac{3}{6} = \dfrac{4}{8}$　　32. $\dfrac{12}{18} = \dfrac{14}{21}$　　33. $\dfrac{5}{6} = \dfrac{7}{8}$　　34. $\dfrac{7.5}{10} = \dfrac{3}{4}$

35. $\dfrac{6.2}{3.1} = \dfrac{10.2}{5.1}$　　36. $\dfrac{8\frac{1}{2}}{2\frac{1}{3}} = \dfrac{4\frac{1}{4}}{1\frac{1}{6}}$　　37. $\dfrac{6\frac{1}{5}}{1\frac{1}{7}} = \dfrac{3\frac{1}{10}}{\frac{8}{14}}$

38. $\dfrac{6}{24} = \dfrac{10}{48}$　　39. $\dfrac{7}{16} = \dfrac{3\frac{1}{2}}{8}$　　40. $\dfrac{10}{17} = \dfrac{5}{8\frac{1}{2}}$

4.7 FINDING THE UNKNOWN TERM IN A PROPORTION

Before we try to solve word problems using proportions (Section 4.8), we must develop certain skills for solving proportions when one term is unknown. In such cases, three terms are known and only one term is unknown. This unknown term may be in any one of the four positions in a proportion.

The following examples are solved using certain basic steps. You should follow these steps in solving all proportions. Study these examples carefully.

EXAMPLES

1. Find x if $\dfrac{2}{4} = \dfrac{5}{x}$.

$$\frac{2}{4} = \frac{5}{x}$$

Basic Steps

A. Represent the unknown term with some letter, such as $x, y, w, A, B,$ etc.

$$2 \cdot x = 20$$

B. Write an equation stating that the product of the extremes equals the product of the means.

$$\frac{\cancel{2} \cdot x}{\cancel{2}} = \frac{20}{2}$$

C. The number next to the letter (unknown) is called its **coefficient.** Divide both sides by this coefficient and simplify.

$$x = 10$$

D. The resulting equation gives the missing value for the unknown. [IMPORTANT: Write one equation under the other, just as in the examples.]

You can check the results by substituting the solution in the proportion.

Check: $\dfrac{2}{4} = \dfrac{5}{10}$ is true since $2 \cdot 10 = 4 \cdot 5$.

2. Find y if $\dfrac{6}{16} = \dfrac{y}{24}$.

$$\frac{6}{16} = \frac{y}{24}$$

$$\frac{3}{8} = \frac{y}{24}$$

Reduce the ratio if possible. The numbers will be smaller.

$$3 \cdot 24 = 8 \cdot y$$

The unknown can be on the right as well as the left. The coefficient is 8.

$$\frac{72}{8} = \frac{\cancel{8} \cdot y}{\cancel{8}}$$ Divide both sides by the coefficient.

$$9 = y$$

3. Find w if $\dfrac{5}{\frac{3}{4}} = \dfrac{w}{21}$.

$$\frac{5}{\frac{3}{4}} = \frac{w}{21}$$ The terms can be fractions or decimals as well as whole numbers.

$$5 \cdot 21 = \frac{3}{4} \cdot w$$ The coefficient of w is $\dfrac{3}{4}$.

$$\frac{4}{3} \cdot 105 = \frac{4}{3} \cdot \frac{3}{4} \cdot w$$ Multiplying by $\dfrac{4}{3}$ is the same as dividing by $\dfrac{3}{4}$: $\dfrac{105}{\frac{3}{4}} = \dfrac{105}{1} \cdot \dfrac{4}{3}$

$$\frac{4}{\cancel{3}} \cdot \overset{35}{\cancel{105}} = \frac{\overset{1}{\cancel{4}}}{\cancel{3}} \cdot \frac{\overset{1}{\cancel{3}}}{\cancel{4}} \cdot w$$ Simplify.

$$140 = w$$

4. Find A if $\dfrac{A}{3} = \dfrac{7.5}{6}$.

$$\frac{A}{3} = \frac{7.5}{6}$$

$$6 \cdot A = 22.5$$ The coefficient of A is 6.

$$\frac{\cancel{6} \cdot A}{\cancel{6}} = \frac{22.5}{6}$$

$$A = 3.75$$ The unknown can be a decimal or a fraction as well as a whole number.

PRACTICE QUIZ

Solve the following proportions.

1. $\dfrac{3}{5} = \dfrac{R}{100}$

2. $\dfrac{2\frac{1}{2}}{6} = \dfrac{3}{y}$

ANSWERS

1. $R = 60$

2. $y = \dfrac{36}{5}$ or $7\dfrac{1}{5}$

EXERCISES 4.7

Be sure to study the examples carefully step by step before you do these exercises.

Solve the following proportions.

1. $\dfrac{3}{6} = \dfrac{6}{x}$ 2. $\dfrac{7}{21} = \dfrac{y}{6}$ 3. $\dfrac{5}{7} = \dfrac{z}{28}$ 4. $\dfrac{4}{10} = \dfrac{5}{x}$

5. $\dfrac{8}{B} = \dfrac{6}{30}$ 6. $\dfrac{7}{B} = \dfrac{5}{15}$ 7. $\dfrac{1}{2} = \dfrac{x}{100}$ 8. $\dfrac{3}{4} = \dfrac{x}{100}$

9. $\dfrac{A}{3} = \dfrac{7}{2}$ 10. $\dfrac{x}{100} = \dfrac{1}{20}$ 11. $\dfrac{3}{5} = \dfrac{60}{D}$ 12. $\dfrac{3}{16} = \dfrac{9}{x}$

13. $\dfrac{\frac{1}{2}}{x} = \dfrac{5}{10}$ 14. $\dfrac{\frac{2}{3}}{3} = \dfrac{y}{127}$ 15. $\dfrac{\frac{1}{3}}{x} = \dfrac{5}{9}$ 16. $\dfrac{\frac{3}{4}}{7} = \dfrac{3}{z}$

17. $\dfrac{\frac{1}{8}}{6} = \dfrac{\frac{1}{2}}{w}$ 18. $\dfrac{\frac{1}{6}}{5} = \dfrac{5}{w}$ 19. $\dfrac{1}{4} = \dfrac{1\frac{1}{2}}{y}$ 20. $\dfrac{1}{5} = \dfrac{x}{2\frac{1}{2}}$

21. $\dfrac{1}{5} = \dfrac{x}{7\frac{1}{2}}$ 22. $\dfrac{2}{5} = \dfrac{R}{100}$ 23. $\dfrac{3}{5} = \dfrac{R}{100}$ 24. $\dfrac{A}{4} = \dfrac{75}{100}$

25. $\dfrac{A}{4} = \dfrac{50}{100}$ 26. $\dfrac{20}{B} = \dfrac{1}{4}$ 27. $\dfrac{30}{B} = \dfrac{25}{100}$ 28. $\dfrac{A}{20} = \dfrac{15}{100}$

29. $\dfrac{1}{3} = \dfrac{R}{100}$ 30. $\dfrac{2}{3} = \dfrac{R}{100}$ 31. $\dfrac{9}{x} = \dfrac{4\frac{1}{2}}{11}$ 32. $\dfrac{y}{6} = \dfrac{2\frac{1}{2}}{12}$

33. $\dfrac{x}{4} = \dfrac{1\frac{1}{4}}{5}$ 34. $\dfrac{5}{x} = \dfrac{2\frac{1}{4}}{27}$ 35. $\dfrac{x}{3} = \dfrac{16}{3\frac{1}{5}}$ 36. $\dfrac{6.2}{5} = \dfrac{x}{15}$

37. $\dfrac{3.5}{2.6} = \dfrac{10.5}{B}$ 38. $\dfrac{4.1}{3.2} = \dfrac{x}{6.4}$ 39. $\dfrac{7.8}{1.3} = \dfrac{x}{.26}$ 40. $\dfrac{7.2}{y} = \dfrac{4.8}{14.4}$

4.8 SOLVING WORD PROBLEMS USING PROPORTIONS

A sale on lamps advertises three lamps for $127.50. But you want four of these lamps for your new home. What will be the price of four lamps? You can find the price of one lamp, then multiply this price by 4.

$$
\begin{array}{r}
\$42.50 \\
3{\overline{\smash{)}127.50}} \\
\underline{12} \\
07 \\
\underline{6} \\
1\,5 \\
\underline{1\,5} \\
00 \\
\underline{0}
\end{array}
$$

$$
\begin{array}{r}
\$\ 42.50 \\
\underline{4} \\
\$170.00 \quad \text{for four lamps}
\end{array}
$$

The price for four lamps would be $170.00.

You can also solve this problem using proportions. Be sure to set up the proportion using one of the following two patterns:

1. $\dfrac{3 \text{ lamps}}{4 \text{ lamps}} = \dfrac{\$127.50}{\$x}$

    Each ratio has the same units, the numerators correspond, and the denominators correspond.

2. $\dfrac{3 \text{ lamps}}{\$127.50} = \dfrac{4 \text{ lamps}}{\$x}$

    Each ratio has different units, but they are in the same order.

Solving either proportion gives the price of four lamps.

1. $\dfrac{3}{4} = \dfrac{127.50}{x}$

    $3 \cdot x = 510.00$

    $\dfrac{\cancel{3} \cdot x}{\cancel{3}} = \dfrac{510.00}{3}$

    $x = \$170.00$

2. $\dfrac{3}{127.50} = \dfrac{4}{x}$

    $3 \cdot x = 510.00$

    $\dfrac{\cancel{3} \cdot x}{\cancel{3}} = \dfrac{510.00}{3}$

    $x = \$170.00$

You **cannot** write either of the following proportions.

$\dfrac{3 \text{ lamps}}{4 \text{ lamps}} = \dfrac{\$x}{\$127.50}$ WRONG: Numerators and denominators do not correspond. (4 lamps do not cost $127.50.)

$\dfrac{3 \text{ lamps}}{\$127.50} = \dfrac{\$x}{4 \text{ lamps}}$ WRONG: The units are not in the same order in each ratio.

EXAMPLES   Solve the following problems using proportions.

1. An architect draws house plans using a scale of $\frac{1}{2}$ inch to represent 8 feet. How many inches would represent 20 feet?

Solution:

$$\frac{\frac{1}{2} \text{ inch}}{8 \text{ feet}} = \frac{x \text{ inches}}{20 \text{ feet}}$$

$$\frac{1}{2} \cdot 20 = 8 \cdot x$$

$$\frac{10}{8} = \frac{\cancel{8} \cdot x}{\cancel{8}}$$

$$\frac{5}{4} = x$$

So, $\frac{5}{4}$ inches (or $1\frac{1}{4}$ inches) represents 20 feet.

2. If 6 pencils cost $1.50, how many pencils could you buy with $2.75?

Solution:

$$\frac{6 \text{ pencils}}{\$1.50} = \frac{x \text{ pencils}}{\$2.75}$$

$$16.50 = 1.50 \cdot x$$

$$\frac{16.50}{1.50} = \frac{\cancel{1.50} \cdot x}{\cancel{1.50}}$$

$$11 = x$$

So, $2.75 will buy 11 pencils (not including tax).

Nurses and doctors sometimes work with proportions in prescribing medicine and in giving injections. Medical texts write proportions in the form $2:40::x:100$ instead of $\frac{2}{40} = \frac{x}{100}$. In either notation the solution is found by putting the product of the extremes equal to the product of the means.

EXAMPLE

Solve the proportion 2 ounces:40 grams::x ounces:100 grams

Solution:

2 ounces:40 grams::x ounces:100 grams

$$2 \cdot 100 = 40 \cdot x$$

$$\frac{200}{40} = \frac{\cancel{40} \cdot x}{\cancel{40}}$$

$$5 = x$$

or
$$\frac{2 \text{ ounces}}{40 \text{ grams}} = \frac{x \text{ ounces}}{100 \text{ grams}}$$

$$2 \cdot 100 = 40 \cdot x$$

$$\frac{200}{40} = \frac{\cancel{40} \cdot x}{\cancel{40}}$$

$$5 = x$$

The solution is $x = 5$ ounces.

EXERCISES 4.8

Study the text and examples carefully before working these exercises.

Solve the following word problems using proportions.

1. A map maker uses a scale of 2 inches to represent 30 miles. How many miles does 3 inches represent?

2. If gasoline sells for $1.35 per gallon, what will be the cost of 10 gallons?

3. If gasoline sells for $1.35 per gallon, how many gallons can be bought with $14.85?

4. If the odds on a horse are $5 to win on a $2 bet, what can be won with a $5 bet?

5. An investor thinks she should make $12 for every $100 she invests. How much would she expect to make on a $1500 investment?

6. If one dozen (12) eggs cost $1.09, what would three dozen eggs cost?

7. The price of a certain fabric is $1.75 per yard. How many yards can be bought with $35 (not including tax)?

8. An artist figures she can paint 3 portraits every two weeks. At this rate, how long will it take her to paint 18 portraits?

9. Two units of a certain gas weigh 175 grams. What is the weight of 5 units of this gas?

10. A baseball team bought 8 bats for $96. What would they pay for 10 bats?

11. A store owner expects to make a profit of $2 on an item that sells for

$10. How much profit will he expect to make on an item that sells for $60?

12. A saleswoman makes $8 out of every $100 worth of the product she sells. What will she make if she sells $5000 worth?

13. If property taxes are figured at $1.50 for every $100 in evaluation, what taxes will be paid on a home valued at $85,000?

14. A condominium owner pays property taxes of $2000 per year. If taxes are figured at a rate of $1.25 for every $100 in value, what is the value of his condominium?

15. Sales tax is figured at 6¢ for every $1.00 of merchandise purchased. What was the purchase price on an item that had a sales tax of $2.04?

16. An architect drew plans for a city park using a scale of $\frac{1}{4}$ inch to represent 25 feet. How many feet would 2 inches represent?

17. A building 14 stories high casts a shadow of 30 feet at a certain time of day. What is the length of the shadow of a 20-story building at the same time of day in the same city?

18. Two numbers are in the ratio of 4 to 3. The number 10 is in that same ratio to a fourth number. What is the fourth number?

19. A car is traveling at 45 miles per hour. Its speed is increased by 3 miles per hour every 2 seconds. By how much will its speed increase in 5 seconds? How fast will the car be traveling?

20. A truck is traveling at 55 miles per hour and is braked so that it slows down 2 miles per hour every 3 seconds. How long will it take for the truck to slow to a speed of 45 miles per hour?

21. A salesman figured he drove 560 miles every two weeks. How far would he drive in three months (12 weeks)?

22. Driving steadily, a woman made a trip of 200 miles in $4\frac{1}{2}$ hours. How long would she take to drive 500 miles at the same rate of speed?

23. If you can drive 286 miles in $5\frac{1}{2}$ hours, how long will it take you to drive 468 miles at the same rate of speed?

24. What will be the cost of 21 gallons of gasoline if gasoline costs $1.25 per gallon?

25. If diesel fuel costs $1.10 per gallon, how much diesel fuel will $24.53 buy?

26. An electric fan makes 180 revolutions per minute. How many revolutions will the fan make if it runs for 24 hours?

27. An investor made $144 in one year on a $1000 investment. What would she have earned if her investment had been $4500?

28. A typist can type 8 pages of manuscript in 56 minutes. How long will this typist take to type 300 pages?

29. On a map, $1\frac{1}{2}$ inches represent 40 miles. How many inches represent 50 miles?

30. In the metric system, there are 2.54 centimeters in one inch. How many centimeters are in one foot?

31. Find two numbers that are in the same ratio as 5:2.

32. An English teacher must read and grade 27 essays. If the teacher takes 20 minutes to read and grade 3 essays, how much time will the teacher need to grade all 27 essays?

33. If 2 cups of flour are needed to make 12 biscuits, how much flour will be needed to make 9 of the same kind of biscuits?

34. If 2 cups of flour are needed to make 12 biscuits, how many of the same kind of biscuits can be made with 3 cups of flour?

35. There are one thousand grams in one kilogram. How many grams are in four and seven tenths kilograms?

The following exercises are examples of proportions in medicine. Be sure to label you answers. (The abbreviations are from the metric system.)

36. $\dfrac{1 \text{ liter}}{1000 \text{ mL}} = \dfrac{x}{5000 \text{ mL}}$

37. $\dfrac{1 \text{ kg}}{1000 \text{ g}} = \dfrac{x}{2700 \text{ g}}$

38. $\dfrac{1 \text{ mg}}{1000 \text{ mcg}} = \dfrac{x}{4 \text{ mcg}}$

39. $\dfrac{1 \text{ g}}{1000 \text{ mg}} = \dfrac{0.5 \text{ g}}{x}$

40. $\dfrac{1 \text{ dram}}{60 \text{ grains}} = \dfrac{x}{90 \text{ grains}}$

41. $\dfrac{1 \text{ ounce}}{8 \text{ drams}} = \dfrac{0.5 \text{ ounce}}{x}$

42. $\dfrac{1 \text{ dram}}{60 \text{ minims}} = \dfrac{x}{180 \text{ minims}}$

43. $\dfrac{1 \text{ ounce}}{480 \text{ minims}} = \dfrac{4 \text{ ounces}}{x}$

44. $\dfrac{1 \text{ g}}{15 \text{ grains}} = \dfrac{x}{60 \text{ grains}}$

45. $\dfrac{1 \text{ grain}}{60 \text{ mg}} = \dfrac{x}{500 \text{ mg}}$

46. $\dfrac{1 \text{ ounce}}{30 \text{ g}} = \dfrac{2 \text{ ounces}}{x}$

47. $\dfrac{1 \text{ mL}}{15 \text{ minims}} = \dfrac{5 \text{ mL}}{x}$

48. $\dfrac{1 \text{ tsp}}{4 \text{ mL}} = \dfrac{x}{12 \text{ mL}}$

49. $\dfrac{1 \text{ ounce}}{30 \text{ mL}} = \dfrac{x}{15 \text{ mL}}$

50. $\dfrac{1 \text{ pint}}{500 \text{ mL}} = \dfrac{1.5 \text{ pints}}{x}$

SUMMARY: CHAPTER 4

DEFINITION   A **decimal number** is a rational number that has a power of ten as its denominator.

SPECIAL NOTES

A.   The **th** at the end of a word indicates a fraction part (a part to the right of the decimal point).

B.   The hyphen (-) indicates one word.

To **round off** a number means to find another number close to the original number.

RULE FOR ROUNDING OFF DECIMAL NUMBERS

1.   Look at the single digit just to the right of the digit that is in the place of desired accuracy.

2.   If this digit is 5 or greater, make the digit in the desired place of accuracy one larger and replace all digits to the right with zeros.

3.   If this digit is less than 5, leave the digit that is in the place of desired accuracy as it is and replace all digits to the right with zeros.

RULE FOR MULTIPLYING TWO DECIMAL NUMBERS

1.   Multiply the two numbers as if they were whole numbers.

2.   Count the total number of places to the right of the decimal points in both multipliers.

3.   This sum is the number of places to the right of the decimal point in the product.

Division by 10 moves the decimal point to the left one place.

Division by 100 moves the decimal point to the left two places.

Division by 1000 moves the decimal point to the left three places, and so on.

DEFINITION   A **ratio** is a comparison of two quantities.

DEFINITION   A **proportion** is a statement that two ratios are equal. In symbols, $\dfrac{a}{b} = \dfrac{c}{d}$ is a proportion.

A proportion is true if the product of the extremes equals the product of the means.

$$\frac{a}{b} = \frac{c}{d} \quad \text{if and only if} \quad a \cdot d = b \cdot c \quad \text{where } b \neq 0 \text{ and } d \neq 0$$

REVIEW QUESTIONS: CHAPTER 4

1. A decimal number is a rational number that has a power of _____ as its denominator.

Write the following decimal numbers in words.

2. 0.4   3. 7.08   4. 92.137   5. 18.5526

Write the following decimal numbers in mixed number form.

6. 81.47   7. 100.03   8. 9.592   9. 200.5

Write the following numbers in decimal notation.

10. two and seventeen hundredths

11. eighty-four and seventy-five thousandths

12. three thousand three and three thousandths

Round off as indicated.

13. 5863   (nearest hundred)

14. 7.649   (nearest tenth)

15. 0.0385   (nearest thousandth)

16. 2.069876   (nearest hundred-thousandth)

Add or subtract as indicated.

17. $5.4 + 7.34 + 14.08$

18. $3 + 7.86 + 52.891 + 0.4$

19. $34.967 + 40.8 + 9.451 + 8.2$

20. $32.5 - 14.71$

21. $16.92 - 7.9$

22. $5 - 1.0377$

23. Add   78.6
          9.683
        15.989

24. Subtract   42.008
             −19.3

Multiply.

25. $(.8)(.9)$

26. $(.2)(.1)$

27. $(.02)(.32)$

28. $100(2.35)$

29. $10(.17632)$

30. $10^3(5.9641)$

31. 2.4
    .05

32. 1.08
    .16

33. 36.5
    4.7

Divide. (Round off to the nearest hundredth.)

34. $4\overline{)2.83}$

35. $.06\overline{)52.832}$

36. $1.003\overline{)200.6}$

Divide by moving the decimal point the correct number of places.

37. $\dfrac{296.1}{100}$

38. $\dfrac{5.67}{10^3}$

39. $\dfrac{19.435}{10}$

40. Find the average (to the nearest tenth) of 16.5, 23.4, and 30.7.

41. In the proportion $\dfrac{2}{3} = \dfrac{x}{y}$, name the extremes and the means.

Determine whether the following proportions are true or false.

42. $\dfrac{3}{5} = \dfrac{9}{15}$

43. $\dfrac{15}{20} = \dfrac{18}{24}$

44. $\dfrac{6\frac{1}{2}}{14} = \dfrac{8}{16}$

Solve the following proportions.

45. $\dfrac{10}{12} = \dfrac{x}{6}$

46. $\dfrac{1.7}{5.1} = \dfrac{100}{y}$

47. $\dfrac{7\frac{1}{2}}{3\frac{1}{3}} = \dfrac{w}{2\frac{1}{4}}$

48. A motorcycle averages 42.8 miles per gallon of gas. How many miles can the motorcycle travel on 3.5 gallons of gas?

49. Find the difference between sixty-four and five hundred thirty-six ten-thousandths and fifty-nine and three thousand six hundred eighty-one ten-thousandths.

50. On a certain map, 1 inch represents 35.5 miles. What distance is represented by 4.7 inches?

51. A part-time clerk earned $420 the first month on a new job. This was 0.6 of what he had anticipated. How much had he anticipated making?

52. If a machine produces 5000 hairpins in 2 hours, how many will it produce in two 8-hour days?

53. An automobile was slowing down at the rate of 5 miles per hour (mph) for every 3 seconds. If the automobile was going 65 mph when it began to slow down, how fast was it going at the end of 12 seconds?

54. An architect draws house plans using a scale of $\frac{3}{4}$ inch to represent 10 feet. How many feet are represented by 2 inches?

55. If you can drive 200 miles in $4\frac{1}{2}$ hours, how far could you drive (at the same rate) in 6 hours?

CHAPTER TEST: CHAPTER 4

Write in words and in mixed number form.

1. 5692.4

2. 8.357

3. Write seventy-five and three hundred-thousandths in decimal notation.

Round off as indicated.

4. 9358   (nearest hundred)

5. 71.355   (nearest hundredths)

Add or subtract as indicated.

6. 52.536 + 46.849

7. 10 + 12.3 + 19.47

8. 19 − 3.08

9. 50.872 − 36.938

Multiply.

10. $10^3(13.85)$

11. (.25)(.56)

12. 41.62
.134

Divide. (Round off to the nearest thousandth.)

13. $17.23\overline{)5.6}$

14. $.052\overline{)364}$

15. $\dfrac{83.9}{10^3}$

16. Find the average (to the nearest hundredth) of 75.16, 84.22, and 93.35.

Determine whether the following proportions are true or false.

17. $\dfrac{1.5}{1.8} = \dfrac{3.6}{4.32}$

18. $\dfrac{1\frac{3}{4}}{2\frac{1}{2}} = \dfrac{\frac{6}{5}}{\frac{7}{3}}$

Solve the following proportions.

19. $\dfrac{x}{5} = \dfrac{3\frac{3}{5}}{7}$

20. $\dfrac{50}{z} = \dfrac{\frac{1}{2}}{\frac{3}{4}}$

21. $\dfrac{2.2}{3} = \dfrac{x}{1.53}$

22. The scale on a street map indicates that $1\frac{1}{4}$ inches represent $\frac{1}{2}$ mile. How many miles do 3 inches represent?

23. A new type of truck averaged 29.2 miles per gallon on a test track. How many miles did it travel on the track if it used 0.83 gallon of gas?

24. You borrowed $500 for 6 months from a loan company and paid $45 in interest. How much interest would you pay if you borrowed $600 for a year at the same interest rate?

25. A lighthouse light revolves once every 30 seconds. How many times will it revolve in 12 hours?

5
PERCENT

5.1   UNDERSTANDING PERCENT

The word **percent** comes from the Latin **per centum** meaning "per hundred." So, **percent means hundredths** or **the ratio of a number to 100.** The symbol for percent is %. For example,

$$\frac{50}{100} = 50\% \quad \text{and} \quad \frac{78}{100} = 78\%$$

We are affected daily by percents related to such ideas as profit and loss, discount, commission, taxes, interest, and batting or shooting percentages.

Consider the following situation. You have two ways to make some money:

A.   Make $150 profit by investing $300, or

B.   Make $200 profit by investing $500.

Which is the better investment for you? One way to answer this question is to use **percents.**

A. $\dfrac{\$150 \text{ profit}}{\$300 \text{ invested}} = \dfrac{\cancel{3} \cdot 50}{\cancel{3} \cdot 100} = \dfrac{50}{100} = 50\%$

B. $\dfrac{\$200 \text{ profit}}{\$500 \text{ invested}} = \dfrac{\cancel{5} \cdot 40}{\cancel{5} \cdot 100} = \dfrac{40}{100} = 40\%$

Since 50% is larger than 40%, A is the better investment. B makes more money ($200 is more than $150), but you must invest more. [Of course, the thing to do is invest $500 with A and make 50% of $500 or $250. We will study this procedure in Section 5.4.]

Since **percent means hundredths,** any fraction with denominator 100 can be written as a percent by writing the numerator and a percent sign (%).

EXAMPLES

Change the following fractions to percents.

1. $\dfrac{50}{100} = 50\%$ 2. $\dfrac{60}{100} = 60\%$ 3. $\dfrac{25}{100} = 25\%$

4. $\dfrac{17.3}{100} = 17.3\%$ 5. $\dfrac{100}{100} = 100\%$ 6. $\dfrac{164}{100} = 164\%$ (A number more than 1 is more than 100%.)

EXERCISES 5.1

Change the following fractions to percents.

1. $\dfrac{30}{100}$ 2. $\dfrac{20}{100}$ 3. $\dfrac{40}{100}$ 4. $\dfrac{50}{100}$ 5. $\dfrac{7}{100}$

6. $\dfrac{8}{100}$ 7. $\dfrac{90}{100}$ 8. $\dfrac{15}{100}$ 9. $\dfrac{25}{100}$ 10. $\dfrac{35}{100}$

11. $\dfrac{45}{100}$ 12. $\dfrac{65}{100}$ 13. $\dfrac{75}{100}$ 14. $\dfrac{42}{100}$ 15. $\dfrac{53}{100}$

16. $\dfrac{68}{100}$ 17. $\dfrac{77}{100}$ 18. $\dfrac{48}{100}$ 19. $\dfrac{125}{100}$ 20. $\dfrac{110}{100}$

21. $\dfrac{150}{100}$ 22. $\dfrac{175}{100}$ 23. $\dfrac{200}{100}$ 24. $\dfrac{250}{100}$ 25. $\dfrac{236}{100}$

26. $\dfrac{120}{100}$ 27. $\dfrac{16.3}{100}$ 28. $\dfrac{27.2}{100}$ 29. $\dfrac{13.4}{100}$ 30. $\dfrac{38.6}{100}$

31. $\dfrac{20.25}{100}$ 32. $\dfrac{93.5}{100}$ 33. $\dfrac{0.5}{100}$ 34. $\dfrac{1.5}{100}$ 35. $\dfrac{0.25}{100}$

36. $\dfrac{3\frac{1}{2}}{100}$ 37. $\dfrac{10\frac{1}{4}}{100}$ 38. $\dfrac{1\frac{1}{4}}{100}$ 39. $\dfrac{24\frac{1}{2}}{100}$ 40. $\dfrac{17\frac{3}{4}}{100}$

5.2 DECIMALS AND PERCENTS

To understand how decimals are related to percents, we can first relate decimals to fractions with denominator 100, then change to percents as in Section 5.1. For example,

$$0.42 = \frac{42}{100} = 42\%$$

$$0.76 = \frac{76}{100} = 76\%$$

$$0.253 = \frac{.253}{1} \cdot \frac{100}{100} = \frac{25.3}{100} = 25.3\%$$

$$0.905 = \frac{90.5}{100} = 90.5\%$$

decimal point moved two places to the right   % sign added

Moving the decimal point two places to the right corresponds to multiplying by 100. Writing the % sign corresponds to dividing by 100. Thus, we are changing the form of the decimal number but not its value.

> **To change a decimal to a percent,** move the decimal point two places to the right and write the % sign.

EXAMPLES   Change the following decimals to percents.

1. $0.2 = .20 = 20\%$ 2. $0.03 = 3\%$

3. $0.005 = 0.5\%$ 4. $1.5 = 150\%$

5. $0.07\frac{1}{2} = 7\frac{1}{2}\%$ or $0.07\frac{1}{2} = 0.075 = 7.5\%$

To change percents to decimals, we simply reverse the procedure for changing decimals to percents. Move the decimal point two places to the left and drop the % sign.

> **To change a percent to a decimal,** move the decimal point two places to the left and drop the % sign.

EXAMPLES   Change the following percents to decimals.

1. $56\% = 0.56$ 2. $18.5\% = 0.185$

3. $12\frac{1}{4}\% = 0.12\frac{1}{4}$ or 0.1225

4. $230\% = 2.30$ 5. $0.6\% = 0.006$

PRACTICE QUIZ	Change from decimals to percents.	ANSWERS
	1. 0.34	**1.** 34%
	2. 1.75	**2.** 175%
	Change from percents to decimals.	
	3. 100%	**3.** 1.00
	4. 1.5%	**4.** 0.015

EXERCISES 5.2

Change the following decimals to percents.

1. 0.02	**2.** 0.09	**3.** 0.1	**4.** 0.5	**5.** 0.7
6. 0.9	**7.** 0.36	**8.** 0.52	**9.** 0.83	**10.** 0.75
11. 0.25	**12.** 0.30	**13.** 0.40	**14.** 0.65	**15.** 0.025
16. 0.035	**17.** 0.046	**18.** 0.055	**19.** 0.003	**20.** 0.004
21. 1.10	**22.** 1.30	**23.** 1.25	**24.** 1.75	**25.** 2
26. 1.08	**27.** 1.05	**28.** 1.5	**29.** 2.3	**30.** 2.15

Change the following percents to decimals.

31. 2%	**32.** 7%	**33.** 10%	**34.** 18%
35. 15%	**36.** 20%	**37.** 25%	**38.** 30%
39. 35%	**40.** 80%	**41.** 10.1%	**42.** 11.5%
43. 13.2%	**44.** 17.3%	**45.** $5\frac{1}{4}\%$	**46.** $6\frac{1}{2}\%$
47. $13\frac{3}{4}\%$	**48.** $15\frac{1}{4}\%$	**49.** $20\frac{1}{4}\%$	**50.** $18\frac{1}{2}\%$
51. 0.25%	**52.** 1.25%	**53.** 0.17%	**54.** 0.50%
55. 125%	**56.** 150%	**57.** 130%	**58.** 120%
59. 222%	**60.** 215%		

5.3   FRACTIONS AND PERCENTS (CALCULATORS OPTIONAL)

If a fraction has a denominator that is a factor of 100 (2, 4, 5, 10, 20, 25, 50), then one can change that fraction to a percent by first writing it in its equivalent form with denominator 100. For example,

$$\frac{3}{4} = \frac{3}{4} \cdot \frac{25}{25} = \frac{75}{100} = 75\%$$

$$\frac{1}{2} = \frac{1}{2} \cdot \frac{50}{50} = \frac{50}{100} = 50\%$$

$$\frac{4}{5} = \frac{4}{5} \cdot \frac{20}{20} = \frac{80}{100} = 80\%$$

But this technique does not work well with fractions such as $\frac{5}{8}$ or $\frac{1}{3}$ in which the denominators are not factors of 100. A better technique in this case is to change the fraction to a decimal first, using long division; then change the decimal to a percent.

$$\frac{5}{8} \qquad 8\overline{)5.000} \;\; {\scriptstyle.625} \qquad \frac{5}{8} = .625 = 62.5\%$$

$$\begin{array}{r} .625 \\ 8\overline{)5.000} \\ \underline{4\,8} \\ 20 \\ \underline{16} \\ 40 \\ \underline{40} \\ 0 \end{array}$$

$$\frac{1}{3} \qquad \begin{array}{r} .33\frac{1}{3} \\ 3\overline{)1.00} \\ \underline{9} \\ 10 \\ \underline{9} \\ 1 \end{array} \qquad \frac{1}{3} = .33\frac{1}{3} = 33\frac{1}{3}\%$$

If the division will go beyond the third decimal place, write the remainder over the divisor in fraction form.

EXAMPLES

Change the following numbers to percents.

1. $\dfrac{13}{20}$

$$\frac{13}{20} = \frac{13}{20}\cdot\frac{5}{5} = \frac{65}{100} = 65\%$$

2. $\dfrac{1}{6}$

$$\frac{1}{6} \qquad \begin{array}{r} .16\frac{4}{6} = .16\frac{2}{3} \\ 6\overline{)1.00} \\ \underline{6} \\ 40 \\ \underline{36} \\ 4 \end{array} \qquad \frac{1}{6} = .16\frac{2}{3} = 16\frac{2}{3}\%$$

3. $2\dfrac{1}{3}$

$$2\frac{1}{3} = \frac{7}{3} \qquad \begin{array}{r} 2.33\frac{1}{3} \\ 3\overline{)7.00} \\ \underline{6} \\ 1\,0 \\ \underline{9} \\ 10 \\ \underline{9} \\ 1 \end{array} \qquad 2\frac{1}{3} = 2.33\frac{1}{3} = 233\frac{1}{3}\%$$

Calculator Section

If your instructor agrees that this is the appropriate time, you may choose to use a calculator to do the long division in changing a fraction to a decimal. You still should be able to change the decimal to a percent without the calculator.

Since most calculators give decimal answers accurate to 8 digits, we will make the following agreement concerning division with calculators:

> Decimal quotients that are exact with four decimal places or less will be written with four decimal places; otherwise, decimal quotients will be rounded off to the third decimal place (thousandths).

For example, $\dfrac{1}{3} = 0.3333333$ and $\dfrac{2}{3} = 0.6666667$ with a calculator. Our agreement gives

$$\frac{1}{3} = 0.333 = 33.3\% \quad \text{and} \quad \frac{2}{3} = 0.667 = 66.7\%$$

Thus, agreement to use a calculator automatically implies agreement to use some rounded off answers.

EXAMPLES

Change the following fractions to percents. Use a calculator to perform the long division.

1. $\dfrac{5}{8} = 0.625 = 62.5\%$

2. $\dfrac{3}{40} = 0.0075 = 0.75\%$

3. $\dfrac{1}{7} = 0.1428571 = 14.3\%$

To change a percent to a fraction, write the percent as a fraction with denominator 100, then reduce the fraction. The calculator will not be helpful here.

EXAMPLES

Change the following percents to fractions or mixed numbers.

1. 60%

$$60\% = \frac{60}{100} = \frac{3 \cdot \cancel{20}}{5 \cdot \cancel{20}} = \frac{3}{5}$$

2. 18%

$$18\% = \frac{18}{100} = \frac{9 \cdot \cancel{2}}{50 \cdot \cancel{2}} = \frac{9}{50}$$

3. $7\frac{1}{4}\%$

$$7\frac{1}{4}\% = \frac{7\frac{1}{4}}{100} = \frac{\frac{29}{4}}{100} = \frac{29}{4} \cdot \frac{1}{100} = \frac{29}{400}$$

4. 130%

$$130\% = \frac{130}{100} = \frac{13 \cdot \cancel{10}}{10 \cdot \cancel{10}} = \frac{13}{10} = 1\frac{3}{10}$$

PRACTICE QUIZ	Change to percents.	ANSWERS
	1. $\dfrac{3}{20}$	1. 15%
	2. $\dfrac{3}{8}$	2. 37.5%
	Change to fractions.	
	3. 35%	3. $\dfrac{7}{20}$
	4. 40%	4. $\dfrac{2}{5}$

EXERCISES 5.3

Change the following numbers to percents. A calculator may be used if your instructor thinks its use is appropriate at this time.

1. $\dfrac{3}{100}$ 2. $\dfrac{16}{100}$ 3. $\dfrac{7}{100}$ 4. $\dfrac{29}{100}$ 5. $\dfrac{1}{2}$ 6. $\dfrac{3}{4}$

7. $\dfrac{1}{4}$ 8. $\dfrac{1}{20}$ 9. $\dfrac{11}{20}$ 10. $\dfrac{7}{10}$ 11. $\dfrac{3}{10}$ 12. $\dfrac{3}{4}$

13. $\dfrac{1}{5}$ 14. $\dfrac{2}{5}$ 15. $\dfrac{4}{5}$ 16. $\dfrac{1}{50}$ 17. $\dfrac{13}{50}$ 18. $\dfrac{1}{25}$

19. $\dfrac{12}{25}$ 20. $\dfrac{24}{25}$ 21. $\dfrac{1}{8}$ 22. $\dfrac{5}{8}$ 23. $\dfrac{7}{8}$ 24. $\dfrac{1}{9}$

25. $\dfrac{5}{9}$ **26.** $\dfrac{2}{7}$ **27.** $\dfrac{3}{7}$ **28.** $\dfrac{5}{6}$ **29.** $\dfrac{7}{11}$ **30.** $\dfrac{5}{11}$

31. $1\dfrac{1}{14}$ **32.** $1\dfrac{1}{6}$ **33.** $1\dfrac{1}{20}$ **34.** $1\dfrac{1}{4}$ **35.** $1\dfrac{3}{4}$ **36.** $1\dfrac{1}{5}$

37. $1\dfrac{3}{8}$ **38.** $2\dfrac{1}{2}$ **39.** $2\dfrac{1}{10}$ **40.** $2\dfrac{1}{15}$

Change the following percents to fractions or mixed numbers.

41. 10%   **42.** 5%   **43.** 15%   **44.** 17%   **45.** 25%

46. 30%   **47.** 50%   **48.** $12\dfrac{1}{2}\%$ **49.** $37\dfrac{1}{2}\%$ **50.** $16\dfrac{2}{3}\%$

51. $33\dfrac{1}{3}\%$ **52.** $66\dfrac{2}{3}\%$ **53.** 33%   **54.** $\dfrac{1}{2}\%$ **55.** $\dfrac{1}{4}\%$

56. 1%   **57.** 100%   **58.** 125%   **59.** 120%   **60.** 150%

61. 0.3%   **62.** 2.5%   **63.** 62.5%   **64.** 0.2%   **65.** 0.75%

5.4   PROBLEMS INVOLVING PERCENT
(CALCULATORS OPTIONAL)

Many people have difficulty working with percent simply because they do
not know whether to add, subtract, multiply, or divide. There are **only
three** basic types of problems using percent. In this section, we will develop
a method for handling these problems that will be helpful in working with
word problems in Section 5.5.

Can you answer questions like the following?

What is 35% of 60?
25% of what number is 12?
19.2 is what percent of 96?

Each of the three questions is of a different type, but each can be answered
by using one basic proportion. This proportion is in the form of a formula,
or general rule.

Using A = amount, B = base, and R = percent, the following proportion can be used to solve any of the three basic types of percent problems.

$$\frac{A}{B} = \frac{R}{100}$$

B = **BASE** is the number we are finding the percent of.
A = **AMOUNT** is the result of multiplying the base by the percent.
R = **PERCENT** or **RATE** without the % sign.

For example, we know that

$$50\% \quad \text{of} \quad 30 \quad \text{is} \quad 15.$$
$$\uparrow \qquad\qquad \uparrow \qquad\qquad \uparrow$$
$$R \qquad\qquad\;\; B \qquad\qquad A$$

The proportion

$$\frac{A}{B} = \frac{R}{100} \quad \text{becomes} \quad \frac{15}{30} = \frac{50}{100}$$

$B = 30$ since we are finding 50% of 30.
$A = 15$ since 15 is the result of multiplying 50% times 30.
$R = 50$ since the percent or rate is 50%.

The three letters A, B, and R correspond to the three basic types of percent problems. In each problem, two of these quantities will be known and the third will be unknown. We solve the proportions using the techniques developed in Section 4.7.

PROBLEM TYPE 1

$$35\% \quad \text{of} \quad 60 \quad \text{is} \quad \underline{} \qquad R = 35, B = 60, A \textbf{ is unknown.}$$
$$\uparrow \qquad\; \uparrow \qquad\; \uparrow$$
$$R \qquad\; B \qquad A$$

$$\frac{A}{60} = \frac{35}{100}$$

$$100 \cdot A = 60 \cdot 35$$

$$\frac{\cancel{100} \cdot A}{\cancel{100}} = \frac{2100}{100}$$

$$A = 21$$

So, 35% of 60 is $\underline{21}$.

Check: Does $\dfrac{21}{60} = \dfrac{35}{100}$?

Yes, since $21 \cdot 100 = 2100$ and $60 \cdot 35 = 2100$.

PROBLEM
TYPE 2

25% of __ is 12 $R = 25$, $A = 12$, B **is unknown.**

$\underset{R}{\uparrow} \quad \underset{B}{\uparrow} \quad \underset{A}{\uparrow}$

$$\frac{12}{B} = \frac{25}{100} \qquad \text{First reduce } \frac{25}{100}.$$

$$\frac{12}{B} = \frac{1}{4}$$

$$12 \cdot 4 = 1 \cdot B$$

$$48 = B$$

So, 25% of $\underline{48}$ is 12.

Check: Does $\dfrac{12}{48} = \dfrac{25}{100}$?

Yes, since $\dfrac{12}{48} = \dfrac{1}{4}$ and $\dfrac{25}{100} = \dfrac{1}{4}$.

PROBLEM
TYPE 3

__% of 96 is 19.2 $B = 96$, $A = 19.2$, R **is unknown.**

$\underset{R}{\uparrow} \quad \underset{B}{\uparrow} \quad \underset{A}{\uparrow}$

$$\frac{19.2}{96} = \frac{R}{100}$$

$$19.2 \cdot 100 = 96 \cdot R$$

$$\frac{1920}{96} = \frac{\cancel{96} \cdot R}{\cancel{96}}$$

$$20 = R$$

So, $\underline{20}$% of 96 is 19.2.

Check: Does $\dfrac{19.2}{96} = \dfrac{20}{100}$?

Yes, since $19.2 \cdot 100 = 1920$ and $96 \cdot 20 = 1920$.

The key to the method using proportion is deciding what number is A and what number is B. **Remember that B is the number you are taking the percent of.** In many cases, B follows the word **of**.

Calculator Section

In Section 3.7, we discussed the fact that a fraction **of** a number means to multiply the fraction times the number. For example, $\frac{1}{4}$ of 80 means to multiply $\frac{1}{4} \cdot 80$. The same is true with percents. 25% of 80 means to multiply the decimal 0.25 times 80. This is a Type 1 percent problem (25% of 80 is _____.).

Type 1 percent problems (R and B are known) are easily solved by changing the percent to a decimal and multiplying this decimal times B. The multiplication can be done with a calculator.

EXAMPLES

1. Find 65% of 42.

$$42 = B$$
$$0.65 = R \text{ (as a decimal)}$$
$$A = 0.65(42) = 27.30$$

So, 65% of 42 is 27.3.

2. Find 18% of 244.

Using a calculator, $0.18(244) = 43.92$.

The proportion method has two advantages:

1. It works for all three types of problems, and
2. You don't have to decide whether to multiply or divide.

Also, once the proportion is properly set up, each multiplication or division can be done with a calculator. The proportion set-up assures you that you are multiplying or dividing with the correct numbers.

PRACTICE QUIZ	Solve each problem for the unknown quantity.	ANSWERS
	1. 5% of 70 is _____.	**1.** 3.5
	2. _____% of 80 is 9.6.	**2.** 12%
	3. 15% of _____ is 13.5.	**3.** 90

EXERCISES 5.4

Solve each problem for the unknown quantity. (A calculator may be used as an aid.)

1. 10% of 70 is _____.

2. 5% of 62 is _____.

3. 15% of 60 is _____.

4. 25% of 72 is _____.

5. 75% of 12 is _____.

6. 60% of 30 is _____.

7. 100% of 36 is _____.

8. 80% of 50 is _____.

9. 2% of _____ is 3.

10. 20% of _____ is 17.

11. 3% of _____ is 21.

12. 30% of _____ is 21.

13. 100% of _____ is 75.

14. 50% of _____ is 42.

15. 150% of _____ is 63.

16. 110% of _____ is 330.

17. _____% of 60 is 90.

18. _____% of 150 is 60.

19. _____% of 75 is 15.

20. _____% of 12 is 4.

21. _____% of 34 is 17.

22. _____% of 30 is 6.

23. _____% of 48 is 16.

24. _____% of 100 is 35.

Each of the following problems is one of the three types discussed, with slightly changed wording. Remember that B is the number you are finding the percent of.

25. _____ is 50% of 25.

26. _____ is 31% of 76.

27. 22 is 20% of _____.

28. 86 is 100% of _____.

29. 13 is _____% of 10.

30. 15 is _____% of 10.

31. 24 is $33\frac{1}{3}$% of _____.

32. 92.1 is 15% of _____.

33. 119.6 is 23% of _____.

34. 9.5 is 25% of _____.

35. 36 is _____% of 18.

36. 60 is _____% of 40.

37. _____ is 96% of 17.

38. _____ is 84% of 32.

39. _____ is 18% of 325.

40. _____ is 28% of 460.

5.5   APPLICATIONS: DISCOUNT, COMMISSION, SALES TAX, OTHERS (CALCULATORS RECOMMENDED)

You should be able to operate with decimals, round off decimals, change percents to decimals, and set up and solve the three basic types of percent problems. Since this section is concerned with the application of skills already learned, the author recommends the use of a calculator. A calculator is not meant to replace necessary skills and understanding but to enhance these abilities by providing answers rapidly and accurately. You must know from your own experience, knowledge, and understanding what numbers to work with and what operations to use.

GENERAL PLAN FOR SOLVING PERCENT PROBLEMS

1. Read the problem carefully.

2. Decide what is unknown (A, B, or R).

3. Write down the values you do know for A, B, or R. (You must know two of them.)

4. Set up and solve the proportion $\dfrac{A}{B} = \dfrac{R}{100}$.

5. Check to see that the answer is reasonable.

EXAMPLES

You may do any of the multiplying, dividing, adding, or subtracting with a calculator. In dealing with money, round off answers to the next highest cent.

1. A refrigerator that normally sells for $850 is on sale at a 30% discount. What is the amount of the discount? What is the sale price?

Solution:
There are two problems here: (a) find the discount, and (b) find the sale price.

a. Find the discount.

30% of $850 is _____ (Type 1 problem).

$$\begin{array}{r} \$850 \\ \times\,.30 \\ \hline \$255.00 \end{array}\ \text{discount}$$

The discount is $255.00.

b. Find the sale price by subtracting the amount of the discount from the original price.

$$\begin{array}{r} \$850.00 \\ -255.00 \\ \hline \$595.00 \end{array}\ \text{sale price}$$

The sale price is $595.00.

2. If the sales tax rate is 6%, what would be the final cost of the refrigerator in Example 1?

Solution:
First find the amount of the sales tax:

6% of $595 is _____ (Type 1 problem).

$$\begin{array}{r} \$595 \\ \times .06 \\ \hline \$35.70 \end{array} \text{ sales tax}$$

Second, add the sales tax to the price to get the final cost. (This involves your knowledge of how to compute sales tax.)

$$\begin{array}{r} \$595.00 \\ +35.70 \\ \hline \$630.70 \end{array} \begin{array}{l} \text{price} \\ \text{sales tax} \\ \text{final cost} \end{array}$$

The final cost is $630.70.

3. A salesman earns a salary of $700 a month plus a commission of 8% on whatever he sells after he has sold $5000 in merchandise. What did he earn the month that he sold $9600 in merchandise?

Solution:
A. First find the amount of his commission.

He earns 8% of what he sells **over** $5000.

$$\begin{array}{r} \$9600 \\ -5000 \\ \hline \$4600 \end{array} \text{ what his commission is based on}$$

8% of $4600 is _____ (Type 1 problem).

$$\begin{array}{r} \$4600 \\ \times .08 \\ \hline \$368.00 \end{array} \text{ commission}$$

B. His monthly pay is his salary plus his commission.

$$\begin{array}{r} \$700 \\ +368 \\ \hline \$1068 \end{array} \text{ Total earned for the month}$$

He earned $1068.

4. A woman paid $5684 for a used car. This was not the original price. She received a 2% discount off the original price because she paid cash. What was the original price?

Solution:

Do **not** take 2% of $5684. We do know that if 2% of something is gone, then 98% remains. That is, **by reasoning,** we know that $5684 represents 98% of the original price (100% − 2% = 98%).

98% of _____ is $5684   (Type 2 problem).

$$\frac{5684}{B} = \frac{98}{100}$$

$$\frac{568,400}{98} = \frac{98 \cdot B}{98}$$

$$\$5800 = B$$

The original price was $5800.

Percent of Profit (Profit = Selling Price − Cost)

The percent of profit can be found by writing a ratio, then changing the fraction to a percent. The ratio is either

$$\frac{\text{profit}}{\text{cost}} \qquad \text{profit based on cost}$$

or $$\frac{\text{profit}}{\text{selling price}} \qquad \text{profit based on selling price}$$

The amount of profit is the same in either case, but the percent will be different because the base (denominator) is different.

EXAMPLE

A company manufactures fittings that cost $6 each to produce. The fittings are sold for $8 each. What is the profit on each fitting? What is the percent of profit based on cost? What is the percent of profit based on the selling price?

Solution:

$$
\begin{array}{ll}
\$8.00 & \text{selling price} \\
-6.00 & \text{cost} \\
\hline
\$2.00 & \text{profit}
\end{array}
$$

Using ratios,

$$\frac{\$2}{\$6} \begin{array}{l} \text{profit} \\ \text{cost} \end{array} = \frac{1}{3} = .33\frac{1}{3} = 33\frac{1}{3}\% \quad \text{profit based on cost}$$

$$\frac{\$2}{\$8} \begin{array}{l} \text{profit} \\ \text{selling price} \end{array} = \frac{1}{4} = .25 = 25\% \quad \text{profit based on selling price}$$

Percent of profit **based on cost** is higher than percent of profit **based on selling price.** The business community reports whichever percent serves its purposes better. Your responsibility as an investor or consumer is to know which percent is reported and what it means to you.

EXERCISES 5.5

Some of the following problems involve several calculations. Write down the known information and the basic set-up to work each problem. Calculators are recommended for these exercises.

1. A realtor works on a 6% commission. What is his commission on a house he sold for $95,000?

2. A store owner received a 3% discount from the manufacturer because she bought $6500 worth of dresses. What was the amount of the discount? What did she pay for the dresses?

3. A sales clerk receives a monthly salary of $500 plus a commission of 6% on all sales over $2500. What did the clerk earn the month she sold $6000 in merchandise?

4. If a salesman works on a 10% commission only, how much will he have to sell to earn $1800 in one month?

5. A computer programmer was told she would be given a bonus of 5% of any money her programs could save the company. How much would she have to save the company to earn a bonus of $600?

6. The property taxes on a house were $750. What was the tax rate if the house was valued at $25,000?

7. Towels were on sale at a discount of 30%. If the sale price was $3.01, what was the original price?

8. Sheets were marked $12.50 and pillow cases were marked $4.50. What is the sale price of each item if each item is discounted 25% from the marked price?

9. One shoe salesman worked on a straight 9% commission. His friend worked on a salary of $300 per month plus a 5% commission. How much did each salesman make during the month in which each sold $4500 worth of shoes?

10. A student missed 3 problems on a math test and was given a grade of 85%. If all the problems were of equal value, how many problems were on the test?

11. In one season, a basketball player missed 15% of his free throws. How many free throws did he make if he attempted 180 free throws?

12. A basketball player made 120 of 300 shots she attempted. What percent of her shots did she make? What percent did she miss?

13. The discount on a fur coat was $150. This was a 20% discount. What was the original selling price of the coat? What was the sale price? What was paid for the coat if a 6% sales tax was added to the sale price?

14. If sales tax is figured at 6%, what was the tax on a purchase of $30.20? What was paid for the purchase?

15. Golf clubs were marked on sale for $320. This was a discount of 20% off the original selling price. What was the original selling price? The clubs cost the golf pro $240. What was his profit? What was his percent of profit based on cost? What was his percent of profit based on the sale price?

16. The discount on men's suits was $50, and they were on sale for $200. What was the original selling price? What was the rate of discount? If the suits cost the store owner $150 each, what was his percent of profit based on cost? What was his percent of profit based on the selling price?

17. In one year, Mr. James earned $15,000. He spent $4800 on rent, $5250 on food, and $1800 on taxes. What percent of his income did he spend on each of those items?

18. The author of a book was told she would have to cut the number of pages by 12% in order for the book to sell at a popular price and still show a profit. What percent of the pages were in the final form? If the book contained 220 pages in final form, how many pages were in it originally? How many pages were cut?

19. In order to get more subscribers, a book club offered three books whose total selling price was originally $17.55 for $7.02. What was the amount of the discount? Based on the original selling price, what was the rate of discount on these three books?

20. The cost of a television set to a store owner was $350, and he sold the set for $490. What was his profit? What was his percent of profit based on cost? What was his percent of profit based on selling price?

21. A car dealer bought a used car for $1500. He marked up the price so he would make a profit of 25% based on his cost. What was the selling

price? If the customer paid 8% of the selling price in taxes and fees, what did the customer pay for the car?

22. A man weighed 200 pounds. He lost 20 pounds in three months. What percent did he lose? Then he gained back 20 pounds two months later. What percent did he gain? The loss and the gain are the same, but the two percents are different. Why?

23. An auto supply store received a shipment of auto parts together with the bill for $845.30. Some of the parts were not as ordered, however, and were returned at once. The value of the parts returned was $175.50. The terms of the billing provided the store with a 2% discount if it paid cash within two weeks. What did the store finally pay for the parts it kept if it paid cash within two weeks?

24. Suppose you sell your home for $100,000 and you owe the Savings and Loan $60,000 on the first trust deed. You pay a real estate agent 6% of the selling price and other fees and taxes totaling $1200. How much cash do you have after the sale? (You may pay income taxes later unless you buy a new home.)

25. You purchase a new home for $98,000. The bank will loan you 80% of the purchase price. How much cash do you need if loan fees are 2% of the loan and other fees total $850?

26. Margorie enrolled in freshman calculus. She had the choice of buying the text in hardback form for $26 or in paperback for $19.50. Tax is figured at 6% of the selling price. If the bookstore buys back hardback books for 50% of the selling price and paperback books for 30% of the selling price, which book is the more economical buy for Margorie if she sells her book back to the bookstore at the end of the semester? How much will she save?

SUMMARY: CHAPTER 5

Percent means hundredths.

> **To change a decimal to a percent,** move the decimal point two places to the right and write the % sign.

> **To change a percent to a decimal,** move the decimal point two places to the left and drop the % sign.

Using A = amount, B = base, and R = percent, the following proportion can be used to solve any of the three basic types of percent problems.

$$\frac{A}{B} = \frac{R}{100}$$

B = **Base** is the number we are finding the percent of.
A = **Amount** is the result of multiplying the base by the percent.
R = **Percent** or **Rate** without the % sign.

GENERAL PLAN FOR SOLVING PERCENT PROBLEMS

1. Read the problem carefully.

2. Decide what is unknown (A, B, or R).

3. Write down the values you do know for A, B, or R. (You must know two of them.)

4. Set up and solve the proportion $\dfrac{A}{B} = \dfrac{R}{100}$.

5. Check to see that the answer is reasonable.

Percent of profit **based on cost** is higher than percent of profit **based on selling price.**

REVIEW QUESTIONS: CHAPTER 5

1. Percent means _____ .

Change the following fractions to percents.

2. $\dfrac{85}{100}$ 　　 3. $\dfrac{18}{100}$ 　　 4. $\dfrac{37}{100}$ 　　 5. $\dfrac{16\frac{1}{2}}{100}$

6. $\dfrac{15.2}{100}$ 　　 7. $\dfrac{115}{100}$

Change the following decimals to percents.

8. 0.06 　　 9. 0.3 　　 10. 0.67

11. 0.027 　　 12. 3 　　 13. 1.2

Change the following percents to decimals.

14. 35%

15. 4%

16. 0.25%

17. $\frac{1}{4}\%$

18. 7.1%

19. 132%

Change the following numbers to percents.

20. $\frac{6}{10}$

21. $\frac{3}{20}$

22. $\frac{4}{25}$

23. $\frac{3}{8}$

24. $\frac{5}{12}$

25. $1\frac{4}{15}$

Change the following percents to fractions or mixed numbers.

26. 14%

27. 40%

28. 66%

29. $12\frac{1}{2}\%$

30. 400%

31. $33\frac{1}{2}\%$

Solve each problem for the unknown quantity.

32. 30% of 52 is _____.

33. 15% of 17 is _____.

34. 3% of _____ is 7.

35. 42% of _____ is 18.

36. _____% of 36 is 7.2.

37. _____% of 48 is 16.

38. 75 is _____% of 300.

39. _____ is 6% of 18.25.

40. 5 is 10% of _____.

41. 14 is $5\frac{1}{2}\%$ of _____.

42. _____ is $6\frac{1}{2}\%$ of 15.

43. 62 is _____% of 31.

44. A shirt was marked 25% off. What would you pay for the shirt if the original price was $15 and you had to pay 6% sales tax?

45. Men's topcoats were on sale for $180. This was a discount of $30 from the original price. If the store owner paid $120 for the coats, what was his percent of profit based on his cost? Based on the selling price?

46. A student received a grade of 75% on a statistics test. If there were 32 problems on the test, all of equal value, how many problems did the student miss?

47. A salesman works on a 9% commission on his sales over $10,000 each month, plus a base salary of $600 per month. How much did he make the month he sold $25,000 in merchandise?

48. The property taxes on a house were $1800. What was the tax rate if the house was valued at $150,000?

49. Mary's allowance each week was $15. She saved for 6 weeks, then she spent $5 on a movie, $35 on clothes, and $20 on a gift for her parents' anniversary. What percent of her savings did she spend on each item?

50. The discount on a new car was $1500, including a rebate from the company. What was the original price of the car if the discount was 15% of the original price? What would be paid for the car if taxes and license fees totaled $650?

CHAPTER TEST: CHAPTER 5

1. Percent means _____ .

Change the following numbers to percents.

2. $\dfrac{63}{100}$ 3. $\dfrac{4.5}{100}$ 4. $\dfrac{9}{10}$ 5. $\dfrac{6}{25}$ 6. $1\dfrac{3}{8}$

Change the following decimals to percents.

7. 0.036   8. 0.54   9. 0.7   10. 2.6   11. 0.125

Change the following percents to decimals.

12. 16%   13. 8.1%   14. 95.2%   15. 183%   16. $11\dfrac{1}{4}\%$

Change the following percents to fractions or mixed numbers.

17. 9%   18. 55%   19. $37\dfrac{1}{2}\%$ 20. 350%   21. 6.2%

Solve each problem for the unknown quantity.

22. 22% of 60 is _____ .   23. 18% of 3000 is _____ .

24. _____% of 83 is 4.15.   25. _____% of 16 is 5.6.

26. 25% of _____ is 18.   27. 100% of _____ is 97.

28. A new refrigerator was marked at $750. For paying cash, the customer was allowed a 5% discount. What did the customer pay, including 6% sales tax, if she paid cash?

29. The bookstore sold used English books for $15. If the bookstore had paid $10 for the books, what was the percent of profit based on cost? Based on selling price?

30. A realtor works on a 6% commission. She must give 30% of her commission to the company she works for. What did she make on a house she sold for $85,000?

6

APPLICATIONS
WITH
CALCULATORS

A Note on Calculators

You are encouraged to use a calculator for all the problems in this chapter. A calculator is not necessary, but it will save you considerable time. Many of the problems have several steps. In these problems, be sure to write down, in an organized manner, the results of intermediate steps. Accuracy to three decimal places will be sufficient, but you should avoid rounding off if you can.

Be aware that some answers will differ slightly if you round off at different places in a calculation involving multiplication and/or division. For example, to calculate $\$500 \times 0.15 \times \frac{1}{3}$, you might write

$$\$500 \times 0.15 \times \frac{1}{3} = \frac{500 \times 0.15}{3} = \frac{75}{3} = \$25$$

or, using a decimal, $\frac{1}{3} = 0.333$ (to three decimal places),

$$\$500 \times 0.15 \times \frac{1}{3} = 500 \times 0.15 \times 0.333 = \$24.98$$

$25 is the correct answer, but if we are going to allow calculators we must allow for **round off errors,** and $24.98 will be accepted. Just be sure you understand that such slight errors can and do occur when decimals are rounded off.

6.1 SIMPLE INTEREST

[REMINDER: Even though you may use a calculator, be neat and organized and write down the results of intermediate steps.]

Interest is money paid for the use of money. The money that is invested or borrowed is called the **principal.** The **rate** is the **percent of interest** and is generally stated as an annual (yearly) rate. For example, if you borrow $80,000 at 15% interest from a bank to buy your home, the 15% interest is an annual interest rate. The $80,000 is the principal.

[NOTE: Interest rates have been changing at an unprecedented rate in the last few years. The author has chosen realistic interest rates at the time of writing the text. Regardless of the rates current as you study this material, the techniques of working with principal and interest are the same.]

Interest is either paid or earned, depending on whether you are the borrower or the lender. In either case, the concept of money paid for the use of money is the same. The purpose of this section is to explain how interest is calculated.

There are two kinds of interest, **simple interest** and **compound in-**

terest. If you put $500 in a savings account at 6% interest, you will earn $0.06 \times \$500 = \$30.00$ interest in one year. If you leave the $30 interest in your account, you will then earn **interest on your interest** or **compound interest.** During the second year, you would earn $0.06 \times \$530 = \$31.80$. We will discuss compound interest in detail in Section 6.2.

Many loans are based on simple interest. If you borrow $1000 for one year and pay 12% interest, and **you make no monthly payments,** then you are paying simple interest. At the end of the year, you would pay back the entire principal of $1000 plus the interest, $0.12 \times 1000 = \$120$.

Interest may be calculated for savings or loans that are made for only a few days or a few months with the following formula.

FORMULA FOR CALCULATING SIMPLE INTEREST

$I = P \times R \times T$ where I = interest
P = principal
R = rate
T = time (in years)

We will agree to use 360 days in one year (30 days in a month). This is common practice in business and banking. R **is a yearly rate** and T **is a fractional part of a year.**

EXAMPLES

1. A woman borrows $500 for 90 days at 14% interest. How much interest will she pay?

Solution:

$$I = P \times R \times T$$

$P = \$500$
$R = 14\% = 0.14$

$$T = 90 \text{ days} = \frac{90}{360} \text{ year} = \frac{1}{4} \text{ year}$$

$$I = 500 \times 0.14 \times \frac{1}{4} = \frac{90.00}{4} = \$22.50$$

or

$$I = 500 \times 0.14 \times 0.25 = 90 \times 0.25 = \$22.50$$

She will pay $22.50 in interest.

2. You loan $1000 to a friend for 6 months at an interest rate of 12%. How much will you be paid at the end of 6 months?

Solution:

You will be paid interest plus principal.

$$I = P \times R \times T$$

$$P = \$1000, \ R = 0.12, \ T = 6 \text{ months} = \frac{6}{12} = \frac{1}{2} \text{ year}$$

$$I = 1000 \times 0.12 \times \frac{1}{2} = \frac{120}{2} = \$60$$

You will be paid $1000 + $60 = $1060.

Suppose you would like to invest some money for 30 days, and you would like to make $100 interest. You know you can get 16% interest. How much do you need to invest?

You know the interest, rate, and time. You do not know the principal. Dividing both sides so that the unknown quantity is alone, we can rewrite the formula $I = P \times R \times T$ in three different forms.

$$P = \frac{I}{R \times T} \qquad\qquad R = \frac{I}{P \times T} \qquad\qquad T = \frac{I}{P \times R}$$

Now, let's solve the problem using $P = \dfrac{I}{R \times T}$.

$$P = \frac{100}{0.16 \times \dfrac{30}{360}} = \frac{100}{0.16 \times \dfrac{1}{12}} = \frac{100}{0.16} \times \frac{12}{1} = \frac{1200}{0.16} = \$7500$$

Or, changing to a decimal $\left(\dfrac{1}{12} = 0.083 \right)$,

$$P = \frac{100}{0.16 \times \dfrac{1}{12}} = \frac{100}{0.16 \times 0.083} = \frac{100}{0.01328} = \$7530.12$$

Or, if we round off 0.01328 to 0.013, we get

$$P = \frac{100}{0.013} = \$7692.31$$

The correct answer is $7500.

You must decide with your instructor just how much **round off error is acceptable. The more rounding off you do, the more error you are going to accumulate.** As a general rule, avoid rounding off by working with fractions as much as possible, then perform the final calculations with a calculator.

EXAMPLE

You want to borrow $1500 at 15%, and you are willing to pay $300 in interest. How long can you keep the money?

Solution:

$$T = \frac{I}{P \times R}$$

$I = \$300, P = \$1500, R = 0.15$

$$T = \frac{300}{1500 \times 0.15} = \frac{300}{225} = 1.333 \text{ or } 1\frac{1}{3} \text{ years}$$

(or 1 year 4 months)

EXERCISES 6.1

1. What will be the interest earned in one year on a savings account of $800 if the bank pays 6% interest?

2. If interest is paid at 10% for one year, what will a principal of $600 earn?

3. If a principal of $900 is invested at a rate of 14% for 90 days, what will be the interest earned?

4. A loan of $5000 is made at 11% for a period of 6 months. How much interest is paid?

5. If you borrow $750 for 30 days at 18%, how much interest will you pay?

6. How much interest is paid on a 60-day loan of $500 at 12%?

7. Find the simple interest paid on a savings account of $1800 for 120 days at 8%.

8. A savings account of $2300 is left for 90 days drawing interest at a rate of 7%. How much interest is earned? What is the amount in the account at the end of 90 days?

9. Every 6 months a stock pays 10% dividends (interest on investment). What will be the earnings of $14,600 invested for 6 months?

10. One thousand dollars worth of merchandise is charged at a local de-

partment store for 60 days at 18% interest. How much is owed at the end of 60 days?

11. You buy an oven on sale from $500 to $450, but you don't pay the bill for 60 days and are charged interest at a rate of 18%. How much do you pay for the oven by waiting 60 days to pay? How much did you save by buying the oven on sale?

12. A friend borrows $500 from you for a period of 8 months and pays you interest at 6%. How much interest are you paid? Suppose you ask 8% instead. Then how much interest are you paid?

13. How much would you have to invest at 8% for 60 days to earn interest of $500?

14. How many days must you leave $1000 in a savings account at $5\frac{1}{2}\%$ to earn $11.00?

15. What is the rate of interest charged if a loan of $2500 for 90 days is paid off with $2562.50?

16. Determine the missing item in each row.

PRINCIPAL	RATE	TIME	INTEREST
$ 400	16%	90 days	?
?	15%	120 days	$ 5.00
$ 560	12%	?	$ 5.60
$2700	?	40 days	$25.50

17. Determine the missing item in each row.

PRINCIPAL	RATE	TIME	INTEREST
$500	18%	30 days	?
$500	18%	?	$15.00
$500	?	90 days	$22.50
?	18%	30 days	$ 1.50

18. If you have a savings account of $25,000 drawing interest at 8%, how much interest will you earn in 6 months? How long must you leave the money in the account to earn $1500?

19. You have accumulated $50,000, and you want to live off the interest each year. If you need $800 a month to live on, what interest rate

must you be earning with your \$50,000? If you are earning $13\frac{1}{2}\%$, payable each month, what will your monthly interest be?

20. Your \$2500 savings account draws interest at $5\frac{1}{2}\%$. How many days will it take for you to earn \$68.75? If the interest rate is then raised to 6%, what will your money earn in the next 6 months?

6.2 COMPOUND INTEREST

Interest paid on interest is called **compound interest.** Suppose you deposit \$1000 in a savings account at 8% compounded quarterly. This means that interest is paid to your account every 3 months and that for the next 3 months, interest is paid on your new balance (principal plus interest). To calculate your yearly earnings, calculate simple interest four times. Each time use a new principal.

$$I = P \times R \times T$$

A. $I = \$1000 \times 0.08 \times \dfrac{1}{4}$ ($P = \$1000$)

$= 80 \times \dfrac{1}{4} = \$20$ interest

B. $I = \$1020 \times 0.08 \times \dfrac{1}{4}$ ($P = \$1000 + \$20 = \$1020$)

$= 81.60 \times \dfrac{1}{4} = \$20.40$ interest

C. $I = \$1040.40 \times 0.08 \times \dfrac{1}{4}$ ($P = \$1020 + \$20.40 = \$1040.40$)

$= 83.23 \times \dfrac{1}{4} = \$20.81$ interest

D. $I = \$1061.21 \times 0.08 \times \dfrac{1}{4}$ ($P = \$1040.40 + \$20.81 = \$1061.21$)

$= 84.90 \times \dfrac{1}{4} = \$21.23$ interest

$$
\begin{array}{l}
\$20.00 \\
20.40 \\
20.81 \\
\underline{21.23} \\
\$82.44 \quad \text{total interest}
\end{array}
$$

Simple interest on $1000 at 8% for one year is $I = 1000 \times 0.08 \times 1 = \$80$.

$$\begin{array}{ll} \$82.44 & \text{compound interest} \\ -80.00 & \text{simple interest} \\ \hline \$\ 2.44 & \text{more by compounding quarterly} \end{array}$$

Most savings and loan associations and banks pay interest compounded daily. That is, interest on your savings is figured daily and you are paid interest on your interest each day. To calculate interest compounded daily, we would need a computer or a large set of tables. Since the purpose here is to teach the concept of compound interest, we will compound annually (once a year simple interest), semiannually (twice a year), quarterly (four times a year), and monthly (twelve times a year).

EXAMPLES

1. $5000 is deposited in a savings account, and interest is compounded semiannually at 10%. How much interest will be earned in one year?

Solution:
There are two calculations because interest is accumulated every 6 months (or $\frac{1}{2}$ year).

A. $I = 5000 \times 0.10 \times \dfrac{1}{2} = \$250$

B. $I = 5250 \times 0.10 \times \dfrac{1}{2} = \$262.50$

$$\begin{array}{ll} \$250.00 & \\ +262.50 & \\ \hline \$512.50 & \text{interest in one year} \end{array}$$

2. Suppose your income is $1000 per month (or $12,000 per year) and you will receive a "cost of living raise" each year. If inflation is at 9% each year, in how many years will you be making $24,000?

Solution:
Compound annually at 9% until the principal (base salary) plus interest (raise) totals $24,000.

In table form (inflation at 9%),

YEAR	BASE	+	RAISE	=	TOTAL
1	$12,000.00		$ 0		$12,000.00
2	12,000.00		1080.00		13,080.00
3	13,080.00		1177.20		14,257.20
4	14,257.20		1283.15		15,540.35
5	15,540.35		1398.63		16,938.98
6	16,938.98		1524.51		18,463.49
7	18,463.49		1661.71		20,125.20
8	20,125.20		1811.27		21,936.47
9	21,936.47		1974.28		23,910.75

In 9 years, you would have almost doubled your salary, but your relative purchasing power would be the same as it was 9 years before. (Actually, you will be in a higher income tax bracket, and you will not be as well off as you were 9 years before.)

EXERCISES 6.2

1. If a bank compounds interest quarterly at 12% (on a certificate deposit), what will your $13,000 deposit be worth in 6 months? In one year?

2. An amount of $9000 is deposited in a savings and loan account, and interest is compounded monthly at 10%. What will be the balance of the account in 6 months?

3. If an account is compounded quarterly at a rate of 6%, what will be the interest earned on $5000 in one year? What will be the total amount in the account? How much more interest is earned in the first year because compounding is done quarterly rather than annually?

4. How much interest will be earned on a savings account of $4000 compounded semiannually at 6% in 4 years? What will be the balance of the account? If the saver took out the interest every 6 months for spending money, what would be the balance in 2 years?

5. How much interest will be earned on a savings account of $3000 in 2 years if interest is compounded annually at 6%? If interest is compounded semiannually? If interest is compounded quarterly?

6. If interest is calculated at 10% and compounded quarterly, what will be the value of $15,000 in $2\frac{1}{2}$ years?

7. You borrowed $4000 and agreed to make equal payments of $1000

each plus interest over the next 4 years. Interest is at a rate of 8% based only on what you owe. How much interest did you pay? How much interest would you have paid if you had not made the annual payments and paid only the $4000 plus interest compounded annually at the end of 4 years?

8. You borrow $600 from a friend and agree to pay 12% interest and pay in 60 days. But you can only pay $100 plus interest. Your friend agrees to let you pay $100 plus interest every 60 days until the debt is paid. How long will it take you to repay your friend? How much interest will you pay?

9. What will be the value of a $15,000 savings account at the end of 3 years if interest is calculated at 10% compounded annually? Suppose the interest is calculated at 5% compounded semiannually. Is the value the same? If not, what is the difference? Suppose the semiannual compounding is at 5% for 6 years. Now will the value be the same?

10. Calculate the interest earned in one year on $10,000 compounded monthly at 14%. What is the difference between this and simple interest at 14% for one year?

11. Calculate the interest you would pay on a loan of $6500 for one year if the interest were compounded every 3 months at 18% and you made no monthly payments. If you made payments of $1000 plus interest every 3 months and the balance plus interest at the end of the year, how much interest would you pay?

12. If a savings account of $10,000 draws interest at a rate of 10% compounded annually, how many years will it take for the $10,000 to double in value? (Make a table similar to the one in Example 2.)

13. Repeat Problem 12 with an interest rate of 6%.

14. Repeat Problem 12 with an interest rate of 12%.

6.3 INSTALLMENT PURCHASING

Have you ever bought a refrigerator, a car, furniture, or clothing and not paid the full purchase price in cash at the time of purchase? If you have, and most people have, you were buying on an **installment plan.** In an installment plan, the customer pays a certain percentage of the purchase price, then makes monthly payments until the remainder of the purchase price **plus interest** is paid.

The advantage to you, the customer, is that you have the item you want without having paid for it completely. The disadvantage is that you are essentially taking out a loan (the balance of the purchase price) at a very high interest rate.

You may have noticed that department store clerks, car dealers, and other merchants encourage you to charge items on a contract, revolving charge account, budget account, or some such installment plan. This is fine as long as you, the customer, understand exactly what you are doing and what you are paying. A few years ago, Congress passed the Truth-in-Lending Law so customers would understand just how much and at what annual interest rate they were paying when they bought "on time." Most people do not understand the total impact of a 1% per month (12% per year), a $1\frac{1}{2}$% per month (18% per year), or a 2% per month (24% per year) service charge or surcharge on their **unpaid balance.**

Consider the following example. You want to buy a used car for $5000. The dealer will allow you to pay $1000 down and monthly payments of $442.50 for one year (12 months). (The tax is $300 and the license fees are $200.) A good deal, right? You have the car and monthly payments you can afford. But just how much are you paying for the car?

In this case, the dealer calculated as follows:

$$
\begin{array}{rl}
\$5000 & \text{price} \\
300 & \text{tax} \\
+\ \ 200 & \text{fees} \\
\hline
\$5500 & \\
-1000 & \text{down payment} \\
\hline
\$4500 & \text{original balance}
\end{array}
$$

Now take 18% of the balance:

$$0.18 \times 4500 = \$810 \quad \text{interest}$$

$$
\begin{array}{rl}
\$5500 & \text{cost} \\
+\ \ 810 & \text{interest} \\
\hline
\$6310 & \text{total paid for the car}
\end{array}
$$

For the monthly payments:

$$
\begin{array}{rl}
\$4500 & \text{original balance} \\
+\ \ 810 & \text{interest} \\
\hline
\$5310 & \text{total in payments}
\end{array}
$$

Divide by 12:

$$\frac{5310}{12} = \$442.50 \quad \text{monthly payment}$$

But you are paying much more than 18% interest because you are making monthly payments and the balance of what you owe is decreasing (becoming less than $4500). For example, on your final payment of $442.50, the interest is $\frac{810}{12} = \$67.50$ and the principal (the balance you owe) is $442.50 - 67.50 = \$375.00$. The interest rate now is

$$R = \frac{I}{P \times T} = \frac{67.50}{375 \times \frac{1}{12}} = 2.16 = 216\%$$

A true 18% is figured at $1\frac{1}{2}\%$ of the **unpaid balance** each month, **not** the **original balance.**

Also, if the dealer paid, say, $4000 for the car, he probably borrowed the money from a bank at a much lower rate than 18%, say 12%, because he is a steady customer. So, he has your cash of $500 ($1000 − tax and fees) and you are paying him interest on money that the bank loaned him. Not a bad business.

Financially, depending on several factors such as income tax bracket and investments, the best procedure for most customers is to

1. Invest or save money (earn interest rather than pay it);

2. Pay cash for the item when it is on sale.

Unfortunately, this procedure involves a certain amount of patience and understanding of finances. So, if you are going to make installment purchases, at least understand just what you are paying.

EXAMPLES

1. Sam buys a coat for $100. He puts $10 down and agrees to pay a minimum payment (he may pay more) of $10 each month, which includes a service charge of $1\frac{1}{2}\%$ on the **unpaid balance.** What will he pay for the coat? How many payments will he make?

Solution:
Make a table.

$\left(1\frac{1}{2}\% = 0.015\right)$

INTEREST	BALANCE	+	INTEREST	−	PAYMENT	=	NEW BALANCE	PAYMENTS
0.015(90.00) = $1.35	$90.00	+	1.35	−	10.00	=	81.35	1st
0.015(81.35) = 1.22	81.35	+	1.22	−	10.00	=	72.57	2nd
0.015(72.57) = 1.09	72.57	+	1.09	−	10.00	=	63.66	3rd
0.015(63.66) = 0.95	63.66	+	0.95	−	10.00	=	54.61	4th
0.015(54.61) = 0.82	54.61	+	0.82	−	10.00	=	45.43	5th
0.015(45.43) = 0.68	45.43	+	0.68	−	10.00	=	36.11	6th
0.015(36.11) = 0.54	36.11	+	0.54	−	10.00	=	26.65	7th
0.015(26.65) = 0.40	26.65	+	0.40	−	10.00	=	17.05	8th
0.015(17.05) = 0.26	17.05	+	0.26	−	10.00	=	7.31	9th
0.015(7.31) = 0.11	7.31	+	0.11	−	7.42	=	Ø	10th
7.42					97.42			

+   10.00 deposit

$107.42

He made 10 payments and paid $107.42 for the coat.

2. June decided to buy a new car for $8500. She had to pay 6% sales tax and $200 in fees. Her down payment was $2000, and she had to pay off the remainder of the loan in 6 equal monthly payments with interest figured at 2% of the **original balance** each month. What were her monthly payments? What did she pay for the car?

Solution:

A. Sales tax on $8500:

$$\begin{array}{r} \$8500 \\ .06 \\ \hline \$510.00 \end{array}$$ sales tax

B. Fees: $200.00

C. Monthly payments:

$8500 - 2000 = \$6500$ balance

2% per month is a $12 \times 2\% = 24\%$ annual rate

$6500 \times 0.24 \times \dfrac{6}{12} = \$780.00$ simple interest paid in 6 months

$\$6500 + \$780 = \$7280$

$\dfrac{7280}{6} = \$1213.34$ per month

(Figure the total simple interest, add the cost, then divide by the number of payments.)

D. Total cost:

$$\begin{array}{ll} \$8500 & \text{price} \\ 510 & \text{sales tax} \\ 200 & \text{fees} \\ \underline{780} & \text{interest} \\ \$9990 & \text{total cost} \end{array}$$

In this case, the interest on June's last payment was $\dfrac{780}{6} = \$130.00$. Her principal (payment − interest) was $1213.34 - 130.00 = \$1083.34$. Her interest rate (on the last payment) was

$$R = \frac{I}{P \times T} = \frac{130.00}{1083.34 \times \dfrac{1}{12}} = 1.44 = 144\%$$

If June had chosen to have smaller monthly payments over a longer period of time, the amount of interest would have been greater and she would have paid more for the car.

When buying on any installment plan, be sure that you are paying interest only on the **unpaid balance** and **not** on the **original balance**.

EXERCISES 6.3

1. Tom bought a new refrigerator for $1200. He had to pay sales tax at 6% and a $300 down payment. He agreed to make payments of $100

per month, which included interest at $1\frac{1}{2}\%$ on the unpaid balance.
What did he pay for the refrigerator? How many monthly payments
did he make? How much interest did he pay? [Make a table similar to
the table in Example 1.]

2. A customer bought a new television set for $850 plus sales tax at 6%.
The down payment was $170. There were to be monthly payments of
$75 each month, which included interest at $1\frac{1}{2}\%$ on the unpaid bal-
ance. What was the total amount paid for the television set? How
many payments were made? [Make a table similar to the table in
Example 1.]

3. The Browns decided to buy a home computer. The computer was
priced at $3500 but was on sale at a 15% discount. They paid 6% sales
tax and $400 down. On the original balance (after the discount, sales
tax, and down payment), the retailer calculated interest at 1% per
month and allowed them to make 12 equal payments. What was the
amount of their monthly payments? What interest rate did they pay
on their last payment?

4. Mrs. Rose needed a new washer and dryer. The original selling price
for the pair was $850, but they were on sale for 25% off. She paid sales
tax of 6% and no down payment. Interest was figured at $1\frac{1}{2}\%$ of the
unpaid balance each month, and Mrs. Rose was to make monthly
payments of $150. How much did she actually pay for the washer and
dryer? How many payments did she make?

5. A woman decided to buy some new furniture that was priced at $5000.
Since she bought so much, the merchant gave her a 10% discount. She
had to pay 6% sales tax on the discount price. Then he offered her two
payment plans: (A) put down $1770, pay 2% per month on the original
balance, and make 12 equal monthly payments, or (B) make no down
payment and make payments of $600 per month including interest at
2% on the unpaid balance. What were her monthly payments under
plan A? Under which plan would she pay more for the furniture? How
much more?

6. Mike decided he needed a new stereo. The one he wanted was priced
at $1800 (plus 6% sales tax), and the dealer would give him $250 in
trade on his old stereo as a down payment. He could then pay the
balance in one of two ways: (A) no further down payment and 6 equal
monthly payments including $1\frac{1}{2}\%$ interest each month on the original balance, or (B) $500 more down and payments of $250 per month
including interest at 2% on the unpaid balance. How much would he

pay for the stereo under each plan? Which plan had the lower monthly payments?

7. Mr. and Mrs. Olson are going to have their home carpeted. They have two bids on the same carpet: (A) a cost of $2500 plus 6% sales tax, no down payment, and 12 equal monthly payments including $1\frac{1}{2}\%$ per month interest on the original balance, or (B) a cost of $2800 plus 6% sales tax, $500 down payment, and 8 equal monthly payments including 1% per month interest on the original balance. What is the monthly payment under each plan? Which plan is cheaper? How much cheaper? What interest rate will they be paying on the last payment under each plan?

8. You have decided to buy a microwave oven. The price is $375. You have agreed to make the minimum payment of $40 per month, with interest charged at $1\frac{1}{2}\%$ on the unpaid balance. You also paid 6% sales tax and $50 down. How many payments will you make? How much interest will you pay? What will the total cost of the oven be? [HINT: Make a table similar to that in Example 1.]

9. You want to buy a new piano. The price is $1500 plus tax at 6%. The store manager has the following plan: You pay $150 down and $150 per month, which includes interest on the unpaid balance at 1% per month. How many payments will you make? How much will you pay for the piano, including tax? How much interest will you pay? [HINT: Make a table similar to that in Example 1.]

10. Mr. and Mrs. Willis decided to buy a new car. The purchase price was $12,000 plus 6% sales tax plus $500 in license fees. The dealer gave them $2500 in trade for their used car. They agreed to pay the balance in equal monthly payments including $1\frac{1}{2}\%$ interest on the unpaid balance until the balance is paid. The monthly payments will be $600. How many payments will they make? How much interest will they pay? What will be the total cost of the car?

6.4   REAL ESTATE

The term **real estate** refers to land or buildings or both. A person who deals in real estate, trying to bring buyers and sellers together and get them to agree on terms of a sale, is called a **realtor.** The purpose of this section is to give a brief introduction to some of the kinds of expenses people can expect to have when they buy or sell a home or other real estate.

The following presentation is not meant to be complete, as the laws and systems for handling real estate vary from state to state. This discussion simply highlights some of the basic ideas, which apply regardless of where you live.

In the sale of a home, the expenses for the buyer and seller are not usually the same. That is, buyers and sellers do not simply share the expenses equally. The seller must pay such items as

1. a realtors commission (a flat fee or a percentage of the selling price)

2. title insurance (insurance that guarantees the title to the property)

3. property taxes to date (taxes are prorated to the date of sale, and the seller will get a refund)

4. mortgage pay-off (amount owed to bank or loan company)

5. penalty fees, if any (for paying off mortgage ahead of schedule)

6. termite inspection and any necessary repairs

7. legal fees for necessary paperwork

8. recording fees to state, county, or city (records of ownership must be kept up to date).

The seller will get some refunds on

1. property taxes

2. insurance premiums (paid for the year).

The buyer must pay such items as

1. the down payment

2. loan fees, if any (many lenders charge a percentage of the loan as a fee for giving the loan; the rate of interest and terms of payment are separate)

3. fire insurance

4. property taxes (taxes are prorated to date of sale)

5. legal fees for necessary paperwork

6. recording fees to state, county, or city (records of ownership must be kept up to date).

The amounts of these fees and payments can vary a great deal and depend largely on the selling price of the property and the laws that govern the sale. We will try to make the figures reasonable, but there are too many variables in a sale to be very specific. Each transaction will involve different amounts.

REAL ESTATE SALE FORM

| SELLER | | Statement of Sale of Property at: | BUYER | |
Charges	Credits	(Address or Legal Description of Property in This Sale)	Charges	Credits
_____	_____	A. Sales Price	_____	_____
_____	_____	B. Realtor's Fee (if any)	_____	_____
_____	_____	C. Cash Down Payment	_____	_____
_____	_____	D. First Trust Deed (mortgage owned by lender)	_____	_____
_____	_____	E. Taxes (pro-rated)	_____	_____
_____	_____	F. Fire Insurance Premium	_____	_____
_____	_____	G. Termite Report (if required)	_____	_____
_____	_____	H. Termite Repairs (if any)	_____	_____
_____	_____	I. Title Insurance	_____	_____
_____	_____	J. Legal Fees	_____	_____
_____	_____	K. Recording Fees (city, county, or state, as required)	_____	_____
_____	_____	L. Prepayment Fee (if any)	_____	_____
_____	_____	M. New Loan Fee (if any)	_____	_____
_____	_____	TOTALS	_____	_____
_____		BALANCE Due to SELLER		
		BALANCE Due from BUYER	_____	

Figure 6.1

Forms such as that shown in Figure 6.1 are usually completed by a lawyer or designated company. (In California, such companies are called escrow companies.) A copy is given to both the seller and buyer so that each can see just how much money is needed and for what.

EXAMPLE

A house is sold for $150,000. The seller must pay the following:

6% realtors fee
$80,000 mortgage
$530 for title insurance
$1500 prepayment penalty to the lender

REAL ESTATE SALE FORM

| SELLER | | Statement of Sale of Property at: | BUYER | |
Charges	Credits	(Address or Legal Description of Property in This Sale)	Charges	Credits
	$150,000	A. Sales Price		
$9,000		B. Realtor's Fee (if any)		
		C. Cash Down Payment	$30,000	
$80,000		D. First Trust Deed (mortgage owned by lender)		[$120,000]
	$350.	E. Taxes (pro-rated)	$350	
	$200	F. Fire Insurance Premium	$500	
$45		G. Termite Report (if required)		
$200		H. Termite Repairs (if any)		
$530		I. Title Insurance		
$310		J. Legal Fees	$310	
$180		K. Recording Fees (city, county, or state, as required)	$50	
$1,500		L. Prepayment Fee (if any)		
		M. New Loan Fee (if any)	$2400	
$91,765	$150,550	TOTALS	$33,610	[$120,000]
$58,785		BALANCE Due to SELLER		
		BALANCE Due from BUYER	$33,610	

Figure 6.2

$310 for legal fees
$180 for recording fees
$45 for a termite report
$200 for termite repairs

The seller gets a refund of $200 on fire insurance (the premium was paid for a whole year) and a refund of $350 on property taxes (taxes were paid for a 6-month period).

The buyer must pay the following:

$30,000 down payment (20% of the selling price)
$2400 loan fee to the lender (2% of the new mortgage of $120,000)

$500 premium for fire insurance
$350 for taxes
$310 for legal fees
$50 for recording fees

How much cash will the seller get? How much cash must the buyer provide?

Solution:

Complete the Sale Form, entering all the charges and credits for both the buyer and the seller (see Figure 6.2). On this particular form, a very simplified version of sale forms, the selling price is a credit for the seller. The seller will receive the difference: total credits minus total charges. The new trust deed (loan from the lender) is considered a credit for the buyer. The buyer will pay the sum of the charges.

EXERCISES 6.4

Photocopy the Real Estate Sale Form in Figure 6.1 and complete the form with the information given in each of the following problems. (Use Figure 6.2 as a guide.)

1. The sale price is $85,000. The seller pays a 6% realtor's commission, $40 for termite report (no repairs), $250 for title insurance, $110 for recording fees, $195 for legal fees, a mortgage of $52,000, and a prepayment penalty of $500. The seller gets refunds of $75 on fire insurance and $170 on taxes. The buyer pays a $17,000 down payment, a loan fee of $1360 (the loan or first trust deed is for $68,000), $250 premium for fire insurance, $35 for recording fees, $170 for taxes, and $195 for legal fees. What will the seller receive? How much does the buyer owe?

2. The sale price is $62,500. The seller pays a 6% realtor's commission, $35 for termite report plus $100 in repairs, $175 for title insurance, $95 for recording fees, $170 for legal fees, a mortgage of $45,000, and no prepayment penalty. The seller gets refunds of $50 on fire insurance and $80 on taxes. The buyer pays $19,750 down payment (25% of selling price), a loan fee of $427.50, a $200 premium for fire insurance, $50 for recording fees, $80 for taxes, and $170 for legal fees. What will the seller receive? How much does the buyer owe?

3. The selling price on a house is $98,000, and no realtor is involved. (The seller and buyer made their own negotiations.) There is no termite inspection. The seller pays $420 for title insurance, $320 for recording fees, $250 for legal fees, and a mortgage of $62,500 plus a $625 pay-off penalty. The seller gets refunds of $75 on fire insurance and $320 on taxes. The buyer pays only $9800 down (10% of the selling price) but pays a loan fee of $2\frac{1}{2}\%$ of the new mortgage of $88,200, no recording fees (the seller agreed to pay these fees), $250 for legal fees, $320 for

taxes, and $425 for fire insurance. What is the balance due to the seller? What does the buyer owe?

4. The selling price is $80,000. The seller pays a 5% realtor's commission, $45 for termite report and $375 for repairs, $85 for recording fees, $235 for title insurance, $125 for legal fees, a mortgage of $43,000, and a prepayment penalty of $800. The seller gets refunds of $35 on fire insurance and $200 on taxes. The buyer pays 30% of the selling price as a down payment, $50 in loan fees, $480 premium for fire insurance, $200 for taxes, and $125 for legal fees. How much is due the seller? How much does the buyer owe?

5. A house is sold for $125,000. The seller is to pay a 6% realtor's commission, a $65 termite report and $325 in repairs, $516 for title insurance, $450 for recording fees, $345 for legal fees, and a mortgage of $68,500, with a pay-off penalty of $685. The seller gets refunds of $180 on fire insurance and $450 on taxes. The buyer is to pay $25,000 as a down payment, a loan fee of $2000, $220 for recording fees, $345 for legal fees, $450 for taxes, and $520 for fire insurance. What is the balance due the seller? What does the buyer owe?

6. A 4-unit apartment building is sold for $250,000. The seller pays a 4% realtor's commission, $100 for a termite report and $520 for repairs, $810 for title insurance, $615 for recording fees, $550 for legal fees, and a mortgage of $165,000, with a pay-off penalty of $1500. The seller gets refunds of $450 on fire insurance and $875 on taxes. The buyer pays a down payment of $40,000, a $1\frac{1}{2}\%$ fee on the $210,000 mortgage, $875 for taxes, $550 for legal fees, and $1530 for fire insurance. What amount is due the seller? How much cash does the buyer owe?

7. A home is sold for $45,500. The seller must pay a flat fee of $1000 to a realtor, $35 for a termite report (no repairs), $95 for title insurance, $80 for recording fees, $125 for legal fees, a mortgage of $28,000 with no prepayment penalties. The seller will get refunds of $40 on fire insurance and $180 on taxes. The buyer will pay a down payment of $4550, a loan fee of $819 (2% of the new loan), $220 premium for fire insurance, $35 for recording fees, $125 for legal fees, and $180 for taxes. What will be the balance due to the seller? How much cash does the buyer have to provide?

8. The sale price is $103,000. The seller pays a 6% realtor's commission, $60 for a termite report and $150 for repairs, $255 for title insurance, $120 for recording fees, $250 for legal fees, and a mortgage of $65,750 with a prepayment penalty of $500. The seller is to get refunds of $55 on fire insurance and $310 on taxes. The buyer will pay a down payment of $25,750, a loan fee of $750, $580 for fire insurance, $45 for recording fees, $250 for legal fees, and $310 on taxes. How much will be due the seller? What does the buyer owe in cash?

6.5 READING GRAPHS AND CHARTS

Graphs and charts are "pictures" of numerical information. Familiar types of graphs seen in newspapers, magazines, textbooks, and corporate reports are bar graphs, circle graphs (or pie graphs), line graphs (or jagged line charts), and pictographs. Figure 6.3 shows an example of each.

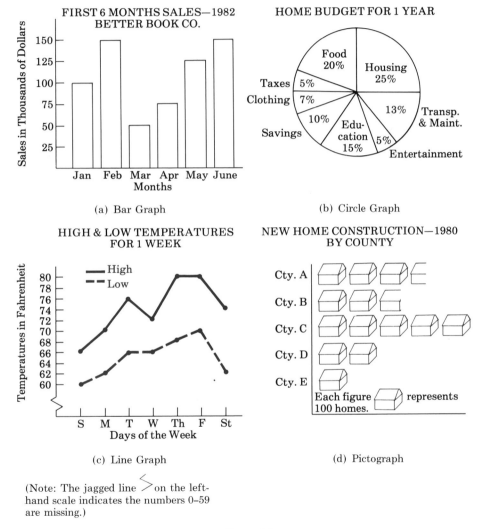

(a) Bar Graph

(b) Circle Graph

(c) Line Graph

(d) Pictograph

(Note: The jagged line ⟩ on the left-hand scale indicates the numbers 0–59 are missing.)

Figure 6.3

By scanning a graph quickly, you can get a general understanding of how certain quantities are related. For example, you can tell at a glance that, in Figure 6.3 (a), the most sales were in February and June; in Figure 6.3 (b), housing is the largest item in the budget; in Figure 6.3 (c), the lowest temperature was on Sunday; and in Figure 6.3 (d), County E had the least amount of home construction.

If you want more specific information, you must study the graph more carefully. Let's investigate Figure 6.3 (a) in detail. First, the scale on the left indicates that sales for January were not $100 but $100,000 because the label reads "Sales in Thousands." Also, only rounded off figures are given. The sales for February might have been $152,500, but this much accuracy cannot be indicated on the graph. For more exact numbers, you would study a table of sales figures prepared by an accountant.

Still in Figure 6.3 (a), what was the amount of decrease between February and March? What was the percent of decrease? What was the amount and percent of increase between March and April? Reading the figures from the graph and doing some simple calculations, we get:

$$\begin{array}{ll} \$150,000 & \text{February} \\ -\ \ 50,000 & \text{March} \\ \hline \$100,000 & \text{decrease} \end{array} \qquad \frac{\$100,000}{\$150,000} = \frac{2}{3} = .67 = 67\% \quad \text{decrease}$$

$$\begin{array}{ll} \$75,000 & \text{April} \\ -50,000 & \text{March} \\ \hline \$25,000 & \text{increase} \end{array} \qquad \frac{\$25,000}{\$50,000} = \frac{1}{2} = .50 = 50\% \quad \text{increase}$$

In this section you will not be asked to draw or construct your own graphs from information. The graphs will be given. You will read these given graphs and answer questions related to how you read each graph, some of which will involve calculations.

EXAMPLES

These examples refer to the graphs in Figure 6.3.

1. Using Figure 6.3 (b), suppose a family has an annual income of $15,000. How much will be allocated to each of the items indicated?
    Solution:

    Multiply each indicated percent times the annual income, $15,000.

Housing:	$0.25 \times 15,000 = \$3750.00$
Food:	$0.20 \times 15,000 = \$3000.00$
Taxes:	$0.05 \times 15,000 = \$\ 750.00$
Clothing:	$0.07 \times 15,000 = \$1050.00$
Savings:	$0.10 \times 15,000 = \$1500.00$
Education:	$0.15 \times 15,000 = \$2250.00$
Entertainment:	$0.05 \times 15,000 = \$\ 750.00$
Transportation & Maintenance:	$0.13 \times 15,000 = \$1950.00$

2. Using Figure 6.3 (c), find the average difference between the high and low temperatures for the week shown.

Solution:

First find the differences, then average these differences.

Sunday: $66 - 60 = 6°$
Monday: $70 - 62 = 8°$
Tuesday: $76 - 66 = 10°$
Wednesday: $72 - 66 = 6°$
Thursday: $80 - 66 = 14°$
Friday: $80 - 70 = 10°$
Saturday: $74 - 62 = \underline{12°}$
                        $66°$ total

$$
\begin{array}{r}
9.4° \quad \text{average difference}\\
7\overline{)66.0}\\
63\\
\overline{30}\\
28\\
\overline{2}
\end{array}
$$

3. Using Figure 6.3 (d), answer the following questions.
   (a) Which county had the most number of new homes built? How many?
   (b) Which county had the least number of new homes built? How many?
   (c) What was the difference between home construction in County A and in County D?

Solution:

(a) County C, 500 homes
(b) County E, 100 homes
(c) $350 - 200 = 150$ homes difference

EXERCISES 6.5

1.

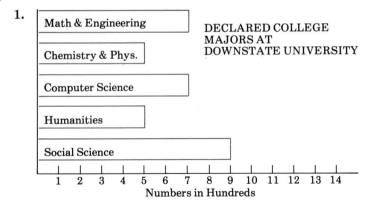

Which field of study has the largest number of declared majors? Which field of study has the smallest number of declared majors? How many declared majors are indicated in the entire graph? What percent are computer science majors?

2. The school budget shown is based on a total budget of $12,500,000. What amount will be spent on each category? How much more will be spent on teachers' salaries than on administration salaries? What percent will be spent on items other than salaries? How much?

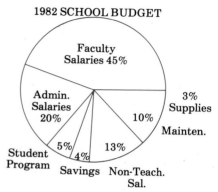

1982 SCHOOL BUDGET

3.

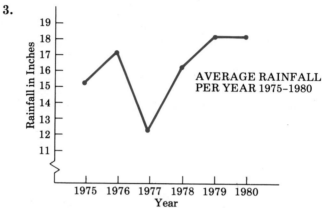

AVERAGE RAINFALL PER YEAR 1975–1980

In what year was the rainfall least? What was the most rainfall in a year? What year? What was the average rainfall over the 6-year period?

4.

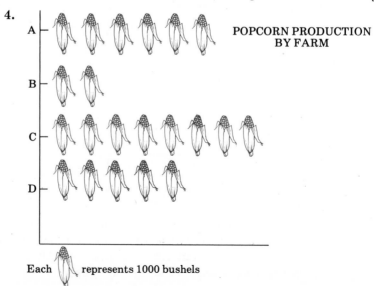

POPCORN PRODUCTION BY FARM

Each 🌽 represents 1000 bushels

Which farm showed the most corn production? How much was this? What percent of the total of the production of the four farms was this?

5. Daily programming is only for 20 hours. (Station XYZ is off the air from 2 A.M. to 6 A.M.) Sports are not shown here because they are considered special events. In the 20-hour period shown, how much time (in minutes) is devoted to each category?

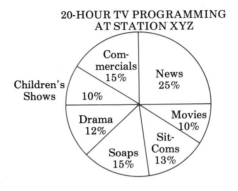

20-HOUR TV PROGRAMMING
AT STATION XYZ

6.

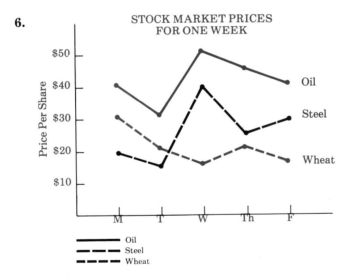

STOCK MARKET PRICES
FOR ONE WEEK

If on Monday morning you had 100 shares in each of the three stocks shown (oil, steel, wheat), and you held the stock all week, in which stock would you have lost money? How much? In which stock would you have gained money? How much? In which stock could you have made the most money if you had sold at the best time? How much? In which stock could you have lost the most money if you had sold at the worst time? How much?

7.

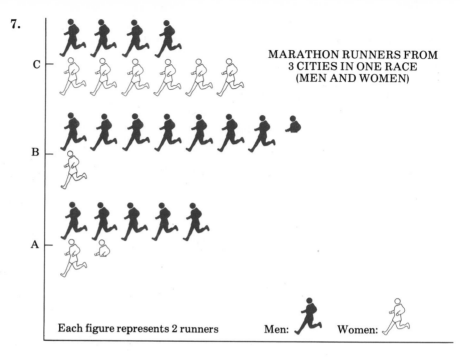

MARATHON RUNNERS FROM
3 CITIES IN ONE RACE
(MEN AND WOMEN)

Each figure represents 2 runners   Men: Women:

Which city had the greatest number of marathon runners? How
many? What percent was this of the total number in the race? Which
city had the greatest number of women runners? How many? What
percent of the total number of women runners was this?

8.

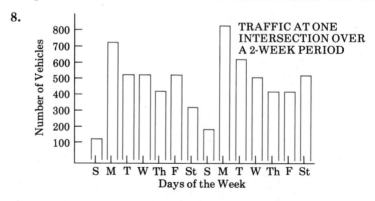

TRAFFIC AT ONE
INTERSECTION OVER
A 2-WEEK PERIOD

What was the total number of vehicles that crossed this intersection
in the 2 weeks? Which day averaged the highest number? How many?
What percent of the traffic was counted on Sundays? On Mondays?

9.

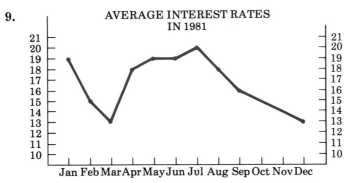

AVERAGE INTEREST RATES
IN 1981

During what month or months in 1981 were interest rates highest? Lowest? What was the average of the interest rates over the entire year?

10.

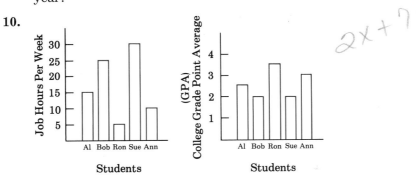

Assume that all five students (Al, Bob, Ron, Sue, Ann) had the same GPA from the same high school. What would be a major difficulty in putting the two graphs shown here together on one graph? Who worked the most hours per week? Who had the lowest GPA? If Bob spent 30 hours per week studying for his classes, what percent of his total work week (part-time work plus study time) did he spend studying? Ann also spent 30 hours per week studying. What percent of her work week did she spend studying? Does there seem to be any relationship between GPA and job hours?

11.

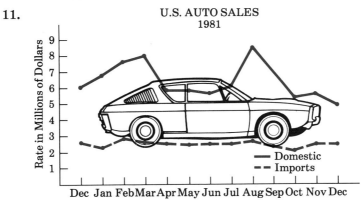

U.S. AUTO SALES
1981

During which month were domestic sales highest? How much higher were they than for the lowest month? What was the difference in import sales for the same two months? What was the difference between domestic and import sales in March? What percent of sales were imports in December 1980? In December 1981? (Answers will be approximate.)

12.

GROWTH OF
THE INFORMATION ECONOMY

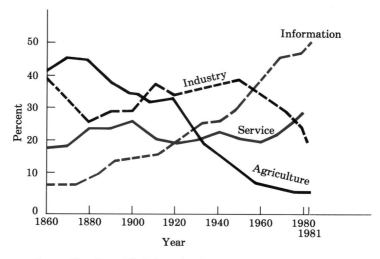

Source: Marc Porat, *The Information Economy: Definition and Measurement,*
Office of Telecommunications Special Publication 77-12 (1), May 1977.

What percent of workers were in each of the four areas in 1860? In 1980? Which area of work seems to have had the most stable percent of workers between 1860 and 1980? What is the difference between the highest and lowest percents for this area? Which area has had the most growth? What was its lowest percent and when? What was its highest percent and when? Which area has had the most decline?

SUMMARY: CHAPTER 6

Interest is money paid for the use of money. The **principal** is the money that is invested or borrowed. The **rate** is the **percent of interest** and is generally an annual rate.

Simple interest is interest paid on principal only. **Compound interest** is interest paid on interest.

FORMULA FOR CALCULATING SIMPLE INTEREST

$I = P \times R \times T$ where I = interest

$\qquad\qquad\qquad P$ = principal

$\qquad\qquad\qquad R$ = rate

$\qquad\qquad\qquad T$ = time (in years)

Other useful versions of the same formula:

$$P = \frac{I}{R \times T} \qquad\qquad R = \frac{I}{P \times T} \qquad\qquad T = \frac{I}{P \times R}$$

Be careful of round off error when using a calculator.

An **installment plan** of purchasing involves making payments at regular periods after an initial down payment, if required, is made. In many installment plans, interest is paid on the unpaid balance each period.

The term **real estate** refers to land or buildings or both.

Four kinds of graphs are presented in this chapter: bar graphs, circle graphs, line graphs, and pictographs. **Any graph should be well labeled so the reader will have a complete and accurate impression of the information indicated.**

REVIEW QUESTIONS: CHAPTER 6

1. What will be the interest earned in one year on a savings account of $1500 if the bank pays $6\frac{1}{2}\%$ interest?

2. If a stock pays 12% dividends every 6 months, what will be the dividend paid on an investment of $13,450 in 6 months?

3. If a principal of $1000 is invested at a rate of 9% for 30 days, what will be the interest earned?

4. You made $50 on an investment at 10% for 3 months. What principal did you invest?

5.   Determine the missing items in each row.

PRINCIPAL	RATE	TIME	INTEREST
$ 200	12%	180 days	?
$ 300	18%	?	$ 81
$1000	?	1 year	$ 85
?	9%	18 months	$270

6.   How much interest will be earned on a savings account of $6500 in 2 years if interest is compounded annually at 6%? If interest is compounded semiannually? If interest is compounded quarterly?

7.   Make a table based on an initial principal of $10,000 and an inflation rate of 8% that indicates the principal, interest, and total for 9 years. (See Example 2 on pages 178 and 179.)

8.   Mr. and Mrs. Atkinson decided to buy a new car. The purchase price was $8900 plus 6% sales tax plus $450 in license fees. They received a $1500 credit for trading in their used car. They agreed to pay another $2000 in cash and then the balance in payments of $450 per month including interest based on 1% of the unpaid balance. How many payments will they make? What is the total amount paid for the car?

9.   Complete a copy of the real estate sale form in Figure 6.1, using the following information about the sale of a house. The sale price is $75,000. The seller pays a 5% realtor's commission, $35 for a termite report and $250 for repairs, $85 for recording fees, $175 for legal fees, $250 for title insurance, a mortgage of $30,000, and a prepayment penalty of $300. The seller gets refunds of $50 on fire insurance and $275 on taxes. The buyer pays a $15,000 down payment, a loan fee of $600 (the first trust deed is for $60,000), a $350 premium for fire insurance, $60 for recording fees, $175 for legal fees, and $425 for taxes. What is the balance due to the seller? What does the buyer owe?

10.   A home budget is shown in the circle graph. What amount will be spent on each category if the family income is $35,000? How much more will be spent on food than on clothing?

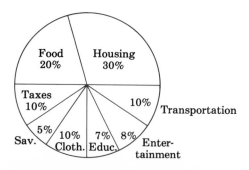

CHAPTER TEST: CHAPTER 6

1. You loan $2500 to a friend for one year at 13% interest. What will he pay you at the end of the year?

2. Calculate the interest you would receive in 6 months on a savings account of $1200 at 7% if interest were compounded monthly. What is the difference between this amount and the amount you would earn if simple interest were calculated for 6 months?

3. John bought a new boat based on the following terms. The price was $8500 plus 6% sales tax. He paid $3000 as a down payment and monthly payments of $800 for 5 months including $1\frac{1}{2}$% of the unpaid balance.

   At the end of the sixth month, he paid the entire balance (including interest). What total amount did he pay for the boat?

4. Complete a copy of the form in Figure 6.1, using the following information. A home is sold for $160,000. The seller pays a 5% realtor's commission, no termite report or repair fees, $170 for recording fees, $500 for title insurance, $290 for legal fees, a mortgage of $93,000, and no prepayment penalty. Seller gets refunds of $280 on fire insurance and $500 on taxes. Buyer pays 25% of the selling price as a down payment, $1500 in loan fees, $800 premium for fire insurance, $1200 for taxes, and $290 for legal fees. How much is due the seller? How much does the buyer owe?

5.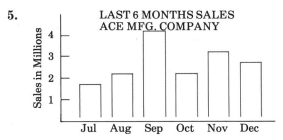

   During what month were sales highest? What were the sales during July? What was the difference in sales in October and November? What percent of the total sales for the six months were the December sales? What was the growth percent between August and September? What was the drop percent between September and October?

7

MEASUREMENT: THE METRIC SYSTEM

About 90% of the people in the world use the metric system. The United States is the only major industrialized country still committed to the U.S. customary system (formerly called the English system). Even in the United States, the metric system has been used for years in such fields as medicine, science, and military activities. Full adoption of the metric system by the United States now has strong backing from many organizations and is generally considered to be a distinct probability.

The main purpose of this chapter is to introduce the metric units of measurement and to help the student gain skill in manipulating within the metric system. Geometrical concepts are also considered in Sections 7.1, 7.3 and 7.4, and conversions between the metric system and the U.S. customary system are discussed in Section 7.5.

7.1 LENGTH

In the metric system, the basic unit of length is the **meter,** the basic unit of mass (or weight) is the **gram** (or sometimes the **kilogram**), and the basic unit of volume is the **liter.*** Smaller and larger units are named by putting a prefix in front of the basic unit. The prefixes† we will use are, from smallest to largest unit size,

milli, centi, deci, deka, hecto, kilo

These prefixes must be memorized, and memorized in order. (See Table 7.1.)

If the measurement is length, then the basic unit is the meter. (Your instructor may have a meter stick for demonstration purposes.) The units

TABLE 7.1 METRIC PREFIXES AND THEIR VALUES

PREFIX	VALUE	
milli	0.001	—thousandths
centi	0.01	—hundredths
deci	0.1	—tenths
basic unit	1	—ones
deka	10	—tens
hecto	100	—hundreds
kilo	1000	—thousands

* Optional spellings for meter and liter are metre and litre, respectively.
† Other prefixes that indicate extremely small units are micro, nano, pico, femto, and atto. Prefixes that indicate extremely large units are mega, giga, and tera. These prefixes will not be used in this text.

of measure for length and their abbreviations are listed in Table 7.2. [NOTE: There are **no periods** in the abbreviations in the metric system.]

TABLE 7.2　MEASURES OF LENGTH

1 **milli**meter	(mm)	= 0.001 meter
1 **centi**meter	(cm)	= 0.01 meter
1 **deci**meter	(dm)	= 0.1 meter
1 meter	(m)	= 1.0 meter
1 **deka**meter	(dam)	= 10 meters
1 **hecto**meter	(hm)	= 100 meters
1 **kilo**meter	(km)	= 1000 meters

The rule in Figure 7.1 is marked along the top in centimeters and along the bottom in millimeters. The ruler shows that 10 mm = 1 cm. That is, it takes 10 mm to equal 1 cm, 20 mm to equal 2 cm, 30 mm to equal 3 cm, and so on.

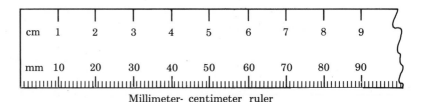

Millimeter- centimeter ruler

Figure 7.1

In the metric system, it takes 10 of any unit to equal 1 of the next higher unit. That is why, to be able to change from one unit to another, you must memorize the order of the prefixes. Tables 7.3 and 7.4 show some of the relationships you should know. Try to reproduce these tables with the book closed.

TABLE 7.3　RELATIONSHIPS
BETWEEN SMALLER AND LARGER
MEASURES OF LENGTH

10 mm = 1 cm	or	0.1 cm = 1 mm
10 cm = 1 dm		0.1 dm = 1 cm
10 dm = 1 m		0.1 m = 1 dm
10 m = 1 dam		0.1 dam = 1 m
10 dam = 1 hm		0.1 hm = 1 dam
10 hm = 1 km		0.1 km = 1 hm

TABLE 7.4 EQUIVALENT MEASURES OF LENGTH

				1 mm
			1 cm =	10 mm
		1 dm =	10 cm =	100 mm
	1 m =	10 dm =	100 cm =	1000 mm
				1 m
			1 dam =	10 m
		1 hm =	10 dam =	100 m
	1 km =	10 hm =	100 dam =	1000 m

To change units of measure, either multiply or divide by a power of 10. That is, multiply or divide by 10, 100, 1000, and so on.

To change to a measure

one unit smaller, multiply by 10	3 cm = 30 mm
two units smaller, multiply by 100	5 m = 500 cm
three units smaller, multiply by 1000	14 m = 14 000 mm
and so on	

To change to a measure

one unit larger, divide by 10	50 cm = 5 dm
two units larger, divide by 100	50 cm = 0.5 m
three units larger, divide by 1000	13 mm = 0.013 m
and so on	

A chart like the one in Figure 7.2 is very helpful for changing units in the metric system. In this chart, (a) list each unit across the top, (b) write only one digit in a column, and (c) mark decimal points on the vertical lines.

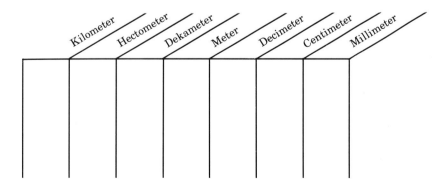

Figure 7.2

For example, 15 m = 15 000 mm since there is a three-unit change from meters to millimeters. This corresponds to moving the decimal point three places to the right, as shown in Figure 7.3.

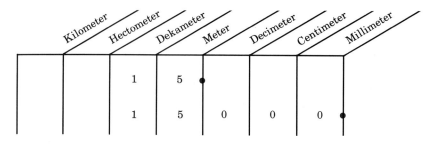

Figure 7.3

We need to use the following procedure when using the chart method for changing units.

TO CHANGE FROM ONE UNIT TO ANOTHER USING THE CHART METHOD

1.   Enter the number so that each digit is in one column and the decimal point is on the given unit line.

2.   Move the decimal point to the desired unit line.

3.   Fill in all spaces with 0's.

EXAMPLES

Use the chart method to change units as indicated.

1.   5 m = _____ km

2.   3 m = _____ cm

3.   16 cm = _____ m

4.   7.6 cm = _____ mm

5.   8 km = _____ dam

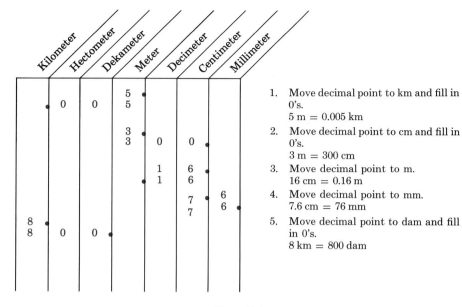

1. Move decimal point to km and fill in 0's.
    5 m = 0.005 km
2. Move decimal point to cm and fill in 0's.
    3 m = 300 cm
3. Move decimal point to m.
    16 cm = 0.16 m
4. Move decimal point to mm.
    7.6 cm = 76 mm
5. Move decimal point to dam and fill in 0's.
    8 km = 800 dam

Figure 7.4

SPECIAL NOTE: In the metric system,

1. A 0 is written to the left of the decimal point if there is no whole number part. (0.25 m)

2. No commas are used in writing numbers. If a number has more than four digits (left or right of the decimal point), the digits are grouped in threes from the decimal point with a space between the groups. (14 000 m)

PRACTICE QUIZ	Draw a metric chart and change the following units as indicated.	ANSWERS
	1. 4 m = ____ mm	1. 4000 mm
	2. 3.1 cm = ____ mm	2. 31 mm
	3. 50 cm = ____ m	3. 0.5 m
	4. 18 km = ____ m	4. 18 000 m

The **perimeter** of a geometric figure is the total distance around the figure. We can use the measures of length to find perimeters by applying the correct formula as shown under the corresponding figure. For circles, the perimeter is called the **circumference.** The distance from the center to a point on a circle is called its **radius.** The distance from one point on a circle to another point on the circle measured through the center is called its **diameter.** The diameter is twice the radius.

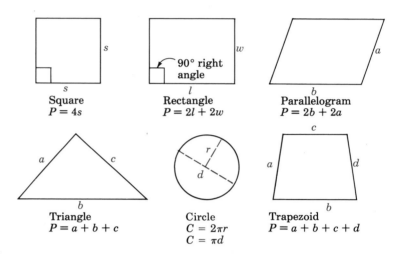

Square
$P = 4s$

Rectangle
$P = 2l + 2w$

Parallelogram
$P = 2b + 2a$

Triangle
$P = a + b + c$

Circle
$C = 2\pi r$
$C = \pi d$

Trapezoid
$P = a + b + c + d$

[NOTE: π is the symbol used for the constant number 3.1415926535. . . . This constant is an infinite decimal with no pattern to its digits. For our purposes, we will use $\pi \approx 3.14$, but you should be aware that 3.14 is only an approximation to π.]

EXAMPLES

1. Find the perimeter of a rectangle with length 20 cm and width 15 cm.

    Solution:
    Sketch the figure first.

    $$P = 2l + 2w$$
    $$P = 2 \cdot 20 + 2 \cdot 15$$
    $$= 40 + 30 = 70 \text{ cm}$$

    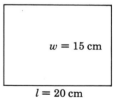

    $w = 15 \text{ cm}$

    $l = 20 \text{ cm}$

    The perimeter is 70 cm.

2. Find the circumference of a circle with diameter 3 m.

    Solution:
    Sketch the figure first.

    $$C = \pi d$$
    $$C = 3.14(3) = 9.42 \text{ m}$$

    3 m

    The circumference is 9.42 m.

EXERCISES 7.1

1. Write the six prefixes discussed in this section in order from smallest to largest unit size.

Change the following units as indicated. Use the chart method until you are accustomed to the metric system.

2. $1 \, m = \underline{\hspace{1cm}} cm$ 3. $5 \, m = \underline{\hspace{1cm}} cm$

4. $12 \, m = \underline{\hspace{1cm}} cm$ 5. $6 \, m = \underline{\hspace{1cm}} cm$

6. $2 \, m = \underline{\hspace{1cm}} mm$ 7. $0.3 \, m = \underline{\hspace{1cm}} mm$

8. $0.7 \, m = \underline{\hspace{1cm}} mm$ 9. $1.4 \, m = \underline{\hspace{1cm}} mm$

10. $1.6 \, cm = \underline{\hspace{1cm}} mm$ 11. $1.8 \, cm = \underline{\hspace{1cm}} mm$

12. $25 \, cm = \underline{\hspace{1cm}} mm$ 13. $35 \, cm = \underline{\hspace{1cm}} mm$

14. $4 \, m = \underline{\hspace{1cm}} dm$ 15. $16 \, m = \underline{\hspace{1cm}} dm$

16. $7 \, dm = \underline{\hspace{1cm}} cm$ 17. $21 \, dm = \underline{\hspace{1cm}} cm$

18. $3 \, km = \underline{\hspace{1cm}} m$ 19. $5 \, km = \underline{\hspace{1cm}} m$

20. $5.28 \, km = \underline{\hspace{1cm}} m$ 21. $6.4 \, km = \underline{\hspace{1cm}} m$

22. $11 \, mm = \underline{\hspace{1cm}} cm$ 23. $26 \, mm = \underline{\hspace{1cm}} cm$

24. $72 \, mm = \underline{\hspace{1cm}} cm$ 25. $48 \, mm = \underline{\hspace{1cm}} cm$

26. $6 \, mm = \underline{\hspace{1cm}} dm$ 27. $12 \, mm = \underline{\hspace{1cm}} dm$

28. $20 \, mm = \underline{\hspace{1cm}} m$ 29. $30 \, mm = \underline{\hspace{1cm}} m$

30. $145 \, mm = \underline{\hspace{1cm}} m$ 31. $256 \, mm = \underline{\hspace{1cm}} m$

32. $25 \, cm = \underline{\hspace{1cm}} m$ 33. $32 \, cm = \underline{\hspace{1cm}} m$

34. $150 \, cm = \underline{\hspace{1cm}} m$ 35. $170 \, cm = \underline{\hspace{1cm}} m$

36. $3000 \, m = \underline{\hspace{1cm}} km$ 37. $2400 \, m = \underline{\hspace{1cm}} km$

38. $500 \, m = \underline{\hspace{1cm}} km$ 39. $400 \, m = \underline{\hspace{1cm}} km$

40. $3.45 \, m = \underline{\hspace{1cm}} cm$ 41. $4.62 \, m = \underline{\hspace{1cm}} cm$

42. $6.3 \, cm = \underline{\hspace{1cm}} m$ 43. $5.2 \, cm = \underline{\hspace{1cm}} m$

44. $3.25 \, m = \underline{\hspace{1cm}} mm$ 45. $6.41 \, m = \underline{\hspace{1cm}} mm$

46. $3 \, mm = \underline{\hspace{1cm}} cm$ 47. $5 \, mm = \underline{\hspace{1cm}} cm$

48. $32 \, mm = \underline{\hspace{1cm}} m$ 49. $57 \, mm = \underline{\hspace{1cm}} m$

50. $20\,000 \, m = \underline{\hspace{1cm}} km$ 51. $35\,000 \, m = \underline{\hspace{1cm}} km$

52. $1.5 \, km = \underline{\hspace{1cm}} m$ 53. $2.3 \, km = \underline{\hspace{1cm}} m$

54. $0.5 \, m = \underline{\hspace{1cm}} km$ 55. $0.3 \, m = \underline{\hspace{1cm}} km$

56. Find the perimeter of a triangle with sides of 4 cm, 8.3 cm, and 6.1 cm.

57. Find the perimeter of a rectangle with length 35 mm and width 17 mm.

58. Find the perimeter of a square with sides of 13.3 m.

59. Find the circumference of a circle with radius 5 cm (use $\pi \approx 3.14$).

60. Find the perimeter of a parallelogram with one side 43 cm and another side 20 mm. Write your answer in millimeters.

61. Find the circumference of a circle with diameter 6.2 cm (use $\pi \approx 3.14$).

62. Find the perimeter of a rectangle with length 50 m and width 50 dm. Write your answer in meters.

63. Find the perimeter of a triangle with sides of 5 cm, 55 mm, and 0.3 dm. Write your answer in centimeters.

64. Find the circumference of a circle in meters if its radius is 70 cm (use $\pi \approx 3.14$).

65. Find the perimeter of a square in meters if one side is 4 km long.

7.2 MASS (WEIGHT)

Mass is the amount of material in an object. Regardless of where the object is in space, its mass remains the same. (See Figure 7.5.) **Weight** is the force of the Earth's gravitational pull on an object. The further an object is from Earth, the less the gravitational pull of the Earth. Thus, astronauts experience weightlessness in space, but their mass is unchanged.

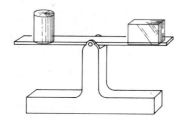

The two objects have the same *mass* and balance on an equal arm balance, regardless of their location in space.

Figure 7.5

Because most of us do not stray far from the Earth's surface, in this text weight and mass will be used interchangeably. Thus, a **mass** of 20 kilograms will be said to **weigh** 20 kilograms.

The basic unit of mass in the metric system is the **kilogram,**[*] about 2.2 pounds. In some fields, such as medicine, the **gram** (about the mass of a paper clip) is more convenient as a basic unit than the kilogram.

Large masses, such as loaded trucks and railroad cars, are measured by the **metric ton** (1000 kilograms or about 2200 pounds).

TABLE 7.5   MEASURES OF MASS

---

1 **milli**gram   (mg) = 0.001 gram
1 gram   (g) = 1.0 gram
1 **kilo**gram   (kg) = 1000 grams
1 metric ton   (t) = 1000 kilograms

The centigram, decigram, dekagram, and hectogram have little practical use and so have been omitted from Table 7.5. As with length and the meter, it takes 10 of any unit to equal 1 of the next higher unit. (See Table 7.6.)

TABLE 7.6   RELATIONSHIPS BETWEEN SMALLER AND LARGER MEASURES OF MASS

1000 mg = 1 g	0.001 g = 1 mg
1000 g = 1 kg	0.001 kg = 1 g
1000 kg = 1 t	0.001 t = 1 kg
1 t = 1000 kg = 1 000 000 g = 1 000 000 000 mg	

Since we are going to deal only with milligrams, grams, kilograms, and tons, we only need to multiply or divide by powers of 1000 to change units.

To change from

grams to milligrams, multiply by 1000   5 g = 5000 mg
kilograms to grams, multiply by 1000   14 kg = 14 000 g
tons to kilograms, multiply by 1000   6 t = 6000 kg

---

[*]Technically, a kilogram is the mass of a certain cylinder of platinum-iridium alloy kept by the International Bureau of Weights and Measures in Paris.

Originally, the basic unit was a gram, defined to be the mass of 1 cm^3 of distilled water at 4°Celsius. This mass is still considered accurate for many purposes, so that

1 cm^3 of water has a mass of 1 g.
1 dm^3 of water has a mass of 1 kg.
1 m^3 of water has a mass of 1000 kg, or 1 metric ton.

Cubic units indicated by cm^3, dm^3, and m^3 will be discussed in Section 7.4.

To change from

milligrams to grams, divide by 1000   500 mg = 0.5 g
grams to kilograms, divide by 1000   32 g   = 0.032 kg
kilograms to tons, divide by 1000   176 kg = 0.176 t

EXAMPLES

A chart may also be used to change units of mass. The headings are parts of a gram. Enter one digit per column and move the decimal point to the new unit line. Fill in any spaces with 0's.

1.   23 mg = _____ g

2.   6 g = _____ mg

3.   49 kg = _____ g

4.   5 t = _____ kg

5.   70 kg = _____ t

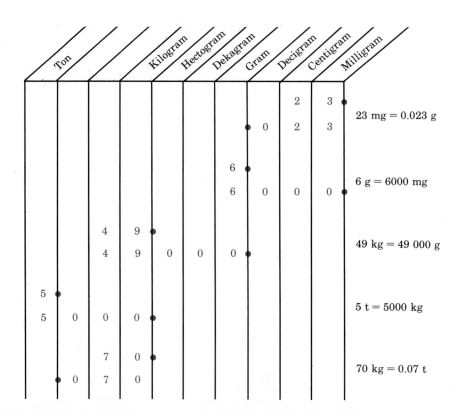

Figure 7.6

Change the following units as indicated without a chart.

6. 60 mg = _____ g
    60 mg = 0.06 g

7. 135 mg = _____ g
    135 mg = 0.135 g

8. 5700 kg = _____ t
    5700 kg = 5.7 t

9. 100 g = _____ kg
    100 g = 0.1 kg

10. 78 g = _____ mg
    78 g = 78 000 mg

PRACTICE QUIZ	Change the following units as indicated.	ANSWERS
	1. 500 mg = ___ g	1. 0.5 g
	2. 500 kg = ___ t	2. 0.5 t
	3. 43 g = ___ mg	3. 43 000 mg
	4. 62 g = ___ kg	4. 0.062 kg

EXERCISES 7.2

Change the following units as indicated.

1. 7 g = ___ mg   2. 2 kg = ___ g

3. 34.5 mg = ___ g   4. 3700 kg = ___ t

5. 4000 kg = ___ t   6. 5600 g = ___ kg

7. 73 kg = ___ mg   8. 91 kg = ___ t

9. 0.54 g = ___ mg   10. 0.7 g = ___ mg

11. 5 t = ___ kg   12. 17 t = ___ kg

13. 2 t = ___ kg   14. 896 mg = ___ g

15. 896 g = ___ mg   16. 342 kg = ___ g

17. 75 000 g = ___ kg   18. 3000 mg = ___ g

19. 7 t = ___ g   20. 0.4 t = ___ g

21. 0.34 g = ___ kg   22. 0.78 g = ___ mg

23. 16 mg = ___ g   24. 2.5 g = ___ mg

25. $3.94\text{ g} = \underline{\hspace{1cm}}\text{ mg}$ **26.** $92.3\text{ g} = \underline{\hspace{1cm}}\text{ kg}$

27. $5.6\text{ t} = \underline{\hspace{1cm}}\text{ kg}$ **28.** $7.58\text{ t} = \underline{\hspace{1cm}}\text{ kg}$

29. $3547\text{ kg} = \underline{\hspace{1cm}}\text{ t}$ **30.** $2963\text{ kg} = \underline{\hspace{1cm}}\text{ t}$

7.3 AREA

Area is a measure of the interior, or enclosure, of a surface. For example, the two rectangles in Figure 7.7 have different areas because they have different amounts of interior space, or different amounts of space are enclosed by the sides of the figures.

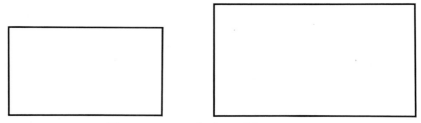

These two rectangles have different **areas**

Figure 7.7

Area is measured in square units. A square that is 1 centimeter long on each side is said to have an area of 1 square centimeter, or the area is 1 cm^2. A rectangle that is 7 cm on one side and 4 cm on the other side encloses 28 squares that have area 1 cm^2. So the rectangle is said to have an area of 28 square centimeters or 28 cm^2. (See Figure 7.8.)

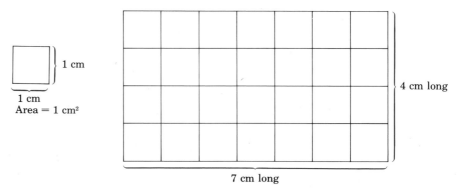

Area $= 7\text{ cm} \times 4\text{ cm} = 28\text{ cm}^2$

There are 28 squares that are each 1 cm^2 in the large rectangle

Figure 7.8

In the metric system, a square that is 1 centimeter on a side encloses 100 square millimeters, as shown in Figure 7.9.

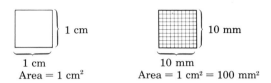

1 cm
Area = 1 cm²

10 mm
Area = 1 cm² = 100 mm²

Figure 7.9

If a square is 10 cm on a side, then it contains 10 cm × 10 cm = 100 cm². But this same square is 1 dm on a side. So, it contains 1 dm². This means that

$$1 \text{ dm}^2 = 100 \text{ cm}^2$$

We already know that each square centimeter contains 100 square millimeters. So,

$$1 \text{ dm}^2 = 100 \text{ cm}^2 = 10\ 000 \text{ mm}^2$$

This relationship is illustrated in Figure 7.10 on page 216.

Square millimeters, square centimeters, square decimeters, and square meters are useful measures for relatively small areas. For example, the area of the floor of your classroom could be measured in square meters, and the area of this page of paper could be measured in square centimeters or square millimeters.

TABLE 7.7 MEASURES OF AREA

$$1 \text{ cm}^2 = 100 \text{ mm}^2$$
$$1 \text{ dm}^2 = \quad 100 \text{ cm}^2 = 10\ 000 \text{ mm}^2$$
$$1 \text{ m}^2 = 100 \text{ dm}^2 = 10\ 000 \text{ cm}^2 = 1\ 000\ 000 \text{ mm}^2$$

[NOTE: Each smaller unit of area is 100 times the previous unit of area—**not** just 10 times.]

In the U.S. customary system, large areas of land are measured in acres. In the metric system, large areas are measured in ares and hectares. (**Are** is pronounced "air" and abbreviated a.) A square with each side 10 m long has an area of 1 are, which is equal to 10 m × 10 m = 100 m². (See Figure 7.11.) A hectare (ha) is 100 ares. That is, 1 ha = 100 a.

1 dm = 10 cm = 100 mm

1 dm = 10 cm = 100 mm

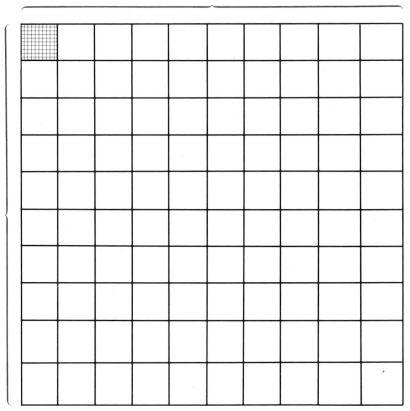

Figure 7.10

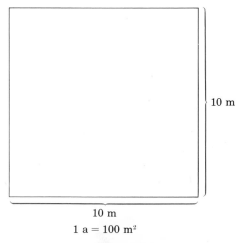

10 m

10 m

1 a = 100 m²

Figure 7.11

TABLE 7.8 MEASURES OF LAND AREA

$$1 \text{ a} = 100 \text{ m}^2$$
$$1 \text{ ha} = 100 \text{ a} = 10\,000 \text{ m}^2$$

EXAMPLES

Change the following units as indicated. Use Tables 7.7 and 7.8.

1. $5 \text{ cm}^2 = $ _____ mm^2
    $5 \text{ cm}^2 = 500 \text{ mm}^2$

2. $3 \text{ dm}^2 = $ _____ $\text{cm}^2 = $ _____ mm^2
    $3 \text{ dm}^2 = 300 \text{ cm}^2 = 30\,000 \text{ mm}^2$

3. $1.4 \text{ m}^2 = $ _____ $\text{dm}^2 = $ _____ $\text{cm}^2 = $ _____ mm^2
    $1.4 \text{ m}^2 = 140 \text{ dm}^2 = 14\,000 \text{ cm}^2 = 1\,400\,000 \text{ mm}^2$

4. $3.2 \text{ a} = $ _____ m^2
    $3.2 \text{ a} = 320 \text{ m}^2$

5. $7.63 \text{ ha} = $ _____ $\text{a} = $ _____ m^2
    $7.63 \text{ ha} = 763 \text{ a} = 76\,300 \text{ m}^2$

6. How many ares are in 1 km^2?

    [NOTE: One km is about 0.6 mile, so 1 km^2 is about $0.6 \times 0.6 = 0.36$ square mile.]

    Remember that $1 \text{ km} = 1000 \text{ m}$, so

    $$1 \text{ km}^2 = (1000 \text{ m}) \times (1000 \text{ m})$$
    $$= 1\,000\,000 \text{ m}^2$$
    $$= 10\,000 \text{ a} \qquad \text{(Divide m}^2 \text{ by 100 to get}$$

    (Divide m² by 100 to get ares because every 100 m² is equal to 1 are.)

7. A farmer plants corn and beans as shown in the figure. How many ares and how many hectares are planted in corn? In beans? (From Example 6 we know $1 \text{ km}^2 = 10\,000 \text{ a}$.)

    Corn: $(2 \text{ km}) \cdot (1 \text{ km}) = 2 \text{ km}^2 = 20\,000 \text{ a}$
    $\phantom{Corn: (2 \text{ km}) \cdot (1 \text{ km}) = 2 \text{ km}^2} = 200 \text{ ha}$

    Beans: $(0.5 \text{ km}) \cdot (1 \text{ km}) = 0.5 \text{ km}^2 = 5000 \text{ a}$
    $\phantom{Beans: (0.5 \text{ km}) \cdot (1 \text{ km}) = 0.5 \text{ km}^2} = 50 \text{ ha}$

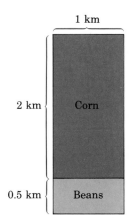

PRACTICE QUIZ	Change the following units as indicated.	ANSWERS
	1. $22 \text{ cm}^2 = \underline{\quad} \text{ mm}^2$	1. 2200 mm^2
	2. $500 \text{ mm}^2 = \underline{\quad} \text{ cm}^2$	2. 5 cm^2
	3. $3.7 \text{ dm}^2 = \underline{\quad} \text{ cm}^2 = \underline{\quad} \text{ mm}^2$	3. $370 \text{ cm}^2 = 37\,000 \text{ mm}^2$
	4. $3.6 \text{ a} = \underline{\quad} \text{ m}^2$	4. 360 m^2
	5. $0.73 \text{ ha} = \underline{\quad} \text{ a} = \underline{\quad} \text{ m}^2$	5. $73 \text{ a} = 7300 \text{ m}^2$

The formulas for the areas of common geometric figures are given here for reference and for use in the exercises.

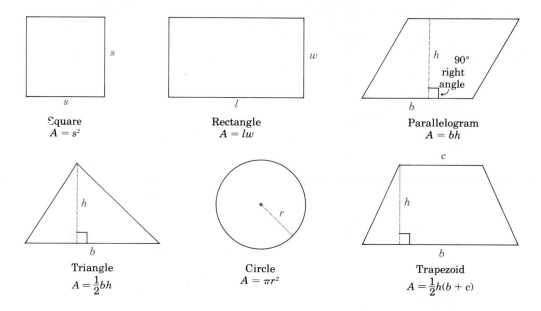

Square
$A = s^2$

Rectangle
$A = lw$

Parallelogram
$A = bh$

Triangle
$A = \frac{1}{2}bh$

Circle
$A = \pi r^2$

Trapezoid
$A = \frac{1}{2}h(b + c)$

EXAMPLES   1. Find the area of the figure shown here with the indicated dimensions.

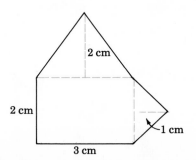

Solution:
There are two triangles and one rectangle.

Rectangle	Larger Triangle	Smaller Triangle
$A = lw$	$A = \dfrac{1}{2}bh$	$A = \dfrac{1}{2}bh$
$A = 2 \cdot 3 = 6 \text{ cm}^2$	$A = \dfrac{1}{2} \cdot 3 \cdot 2 = 3 \text{ cm}^2$	$A = \dfrac{1}{2} \cdot 2 \cdot 1 = 1 \text{ cm}^2$

$$\text{Total area} = 6 \text{ cm}^2 + 3 \text{ cm}^2 + 1 \text{ cm}^2$$
$$= 10 \text{ cm}^2$$

2. Find the area of the washer (shaded portion) with dimensions as shown (use $\pi \approx 3.14$).

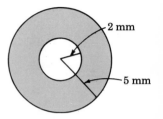

Solution:
Subtract the area of the inside (smaller) circle from the area of the outside (larger) circle.

Larger Circle	Smaller Circle
$A = \pi r^2$	$A = \pi r^2$
$A = 3.14(5^2)$	$A = 3.14(2^2)$
$= 3.14(25)$	$= 3.14(4)$
$= 78.50 \text{ mm}^2$	$= 12.56 \text{ mm}^2$

Washer

$$\begin{array}{r} 78.50 \text{ mm}^2 \\ -12.56 \text{ mm}^2 \\ \hline 65.94 \text{ mm}^2 \end{array} \quad \text{area of washer}$$

EXERCISES 7.3

Change the following units as indicated.

1. $3 \text{ cm}^2 = \underline{\hspace{1cm}} \text{ mm}^2$

2. $5.6 \text{ cm}^2 = \underline{\hspace{1cm}} \text{ mm}^2$

3. $8.7 \text{ cm}^2 = \underline{\hspace{1cm}} \text{ mm}^2$

4. $3.61 \text{ cm}^2 = \underline{\hspace{1cm}} \text{ mm}^2$

5. $600 \text{ mm}^2 = \underline{\hspace{1cm}} \text{ cm}^2$

6. $28 \text{ mm}^2 = \underline{\hspace{1cm}} \text{ cm}^2$

7. $1400 \text{ mm}^2 = \underline{\hspace{1cm}} \text{ cm}^2$

8. $20\,000 \text{ mm}^2 = \underline{\hspace{1cm}} \text{ cm}^2$

9. $4 \text{ dm}^2 = \underline{\hspace{1cm}} \text{ cm}^2 = \underline{\hspace{1cm}} \text{ mm}^2$

10. $7.3 \text{ dm}^2 = \underline{\hspace{1cm}} \text{ cm}^2 = \underline{\hspace{1cm}} \text{ mm}^2$

11. $57 \text{ dm}^2 = \underline{\hspace{1cm}} \text{ cm}^2 = \underline{\hspace{1cm}} \text{ mm}^2$

12. $0.6 \, \text{dm}^2 = \underline{\quad} \text{cm}^2 = \underline{\quad} \text{mm}^2$

13. $17 \, \text{m}^2 = \underline{\quad} \text{dm}^2 = \underline{\quad} \text{cm}^2 = \underline{\quad} \text{mm}^2$

14. $2.9 \, \text{m}^2 = \underline{\quad} \text{dm}^2 = \underline{\quad} \text{cm}^2 = \underline{\quad} \text{mm}^2$

15. $0.03 \, \text{m}^2 = \underline{\quad} \text{dm}^2 = \underline{\quad} \text{cm}^2 = \underline{\quad} \text{mm}^2$

16. $0.5 \, \text{m}^2 = \underline{\quad} \text{dm}^2 = \underline{\quad} \text{cm}^2 = \underline{\quad} \text{mm}^2$

17. $142 \, \text{mm}^2 = \underline{\quad} \text{cm}^2$ **18.** $5800 \, \text{mm}^2 = \underline{\quad} \text{cm}^2$

19. $200 \, \text{dm}^2 = \underline{\quad} \text{m}^2$ **20.** $35 \, \text{dm}^2 = \underline{\quad} \text{m}^2$

21. $7.8 \, \text{a} = \underline{\quad} \text{m}^2$ **22.** $300 \, \text{a} = \underline{\quad} \text{m}^2$

23. $0.04 \, \text{a} = \underline{\quad} \text{m}^2$ **24.** $0.53 \, \text{a} = \underline{\quad} \text{m}^2$

25. $8.69 \, \text{ha} = \underline{\quad} \text{a} = \underline{\quad} \text{m}^2$ **26.** $7.81 \, \text{ha} = \underline{\quad} \text{a} = \underline{\quad} \text{m}^2$

27. $0.16 \, \text{ha} = \underline{\quad} \text{a} = \underline{\quad} \text{m}^2$ **28.** $0.02 \, \text{ha} = \underline{\quad} \text{a} = \underline{\quad} \text{m}^2$

29. $1 \, \text{a} = \underline{\quad} \text{ha}$ **30.** $15 \, \text{a} = \underline{\quad} \text{ha}$

31. $5 \, \text{km}^2 = \underline{\quad} \text{a} = \underline{\quad} \text{ha}$ **32.** $4.76 \, \text{km}^2 = \underline{\quad} \text{a} = \underline{\quad} \text{ha}$

33. $0.3 \, \text{km}^2 = \underline{\quad} \text{a} = \underline{\quad} \text{ha}$ **34.** $0.532 \, \text{km}^2 = \underline{\quad} \text{a} = \underline{\quad} \text{ha}$

Find the area of each of the following figures.

35.   a rectangle 35 cm long and 25 cm wide

36.   a triangle with base 2 cm and altitude 6 cm

37.   a triangle with base 5 mm and altitude 8 mm

38.   a circle of radius 5 m (use $\pi \approx 3.14$)

39.   a circle of radius 1.5 cm (use $\pi \approx 3.14$)

40.   a trapezoid with parallel sides of 8 cm and 10 cm and altitude of 35 cm

41.   a trapezoid with parallel sides of 3.5 mm and 4.2 mm and altitude of 1 cm

42.   a parallelogram of altitude 10 cm to a base of 5 mm

Find the area of each of the following figures with the indicated dimensions (use $\pi \approx 3.14$).

43.

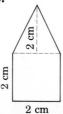

2 cm

2 cm

2 cm

44.

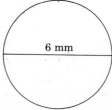

6 mm

45.

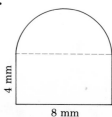

4 mm

8 mm

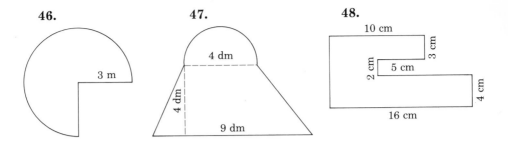

46. **47.** **48.**

Find the areas of the shaded portions in ares (use $\pi \approx 3.14$).

49. **50.**

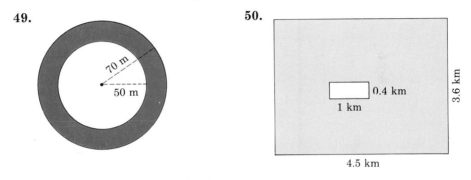

51. Find the area of a circle of radius 1 m.

52. Find the area of a circle with diameter 1 m.

53. Find the area of a rectangle 2 m long and 60 cm wide. Write your answer in both square meters and square centimeters.

54. Find the area of a square with sides of 50 cm. Write your answer in both cm² and m².

55. Find the area of a rectangle 0.5 m long and 35 cm wide. Write your answer in both cm² and m².

7.4 VOLUME

Volume is a measure of the space enclosed by a three-dimensional figure. The volume or space contained within a cube that is 1 cm on each edge is **one cubic centimeter,** or 1 cm³, as shown in Figure 7.12. A cubic centimeter is about the size of a sugar cube.

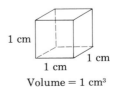

1 cm

1 cm

1 cm

Volume = 1 cm³

Figure 7.12

A rectangular solid that has edges of 3 cm and 2 cm and 5 cm has a volume of 3 cm $\times$ 2 cm $\times$ 5 cm = 30 cm^3. We can think of the rectangular solid as being three layers of ten cubic centimeters, as shown in Figure 7.13.

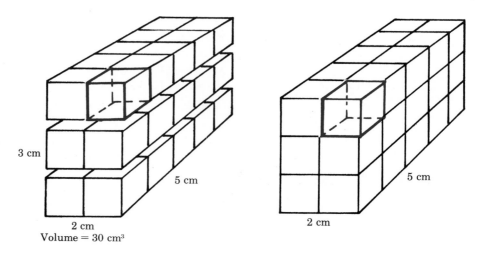

3 cm

5 cm

2 cm

Volume = 30 cm^3

5 cm

2 cm

Figure 7.13

A **liter** (abbreviated L) is the volume enclosed in a cube that is 10 cm on each edge. So, 1 liter is equal to

$$10 \text{ cm} \times 10 \text{ cm} \times 10 \text{ cm} = 1000 \text{ cm}^3$$

That is, 1 liter is equal to the volume of a box that holds about 1000 sugar cubes. (See Figure 7.14.) Also, since 10 cm = 1 dm,

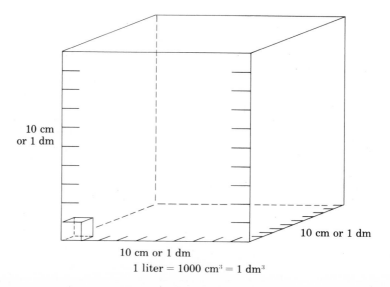

10 cm
or 1 dm

10 cm or 1 dm

10 cm or 1 dm

1 liter = 1000 cm^3 = 1 dm^3

Figure 7.14

$$10 \text{ cm} \times 10 \text{ cm} \times 10 \text{ cm} = 1 \text{ dm} \times 1 \text{ dm} \times 1 \text{ dm} = 1 \text{ dm}^3$$

Thus, $1 \text{ L} = 1 \text{ dm}^3 = 1000 \text{ cm}^3$

The prefixes kilo, hecto, deka, deci, centi, and milli all indicate the same parts of a liter as they do of the meter and the gram. The same type of chart used before will be helpful for changing units. However, the centiliter, deciliter, and dekaliter are not commonly used and so are not included in any tables or exercises. [NOTE: Liter will be abbreviated L.]

TABLE 7.9 MEASURES OF VOLUME	
1 **milli**liter	(mL) = 0.001 liter
1 liter	(L) = 1.0 liter
1 **hecto**liter	(hL) = 100 liters
1 **kilo**liter	(kL) = 1000 liters

TABLE 7.10 EQUIVALENT MEASURES OF VOLUME			
1000 mL = 1 L		1 mL = 1 cm^3	
1000 L = 1 kL		1 L = 1 dm^3	
10 hL = 1 kL		1 kL = 1 m^3	

The chart in Figure 7.15 shows several unit changes.

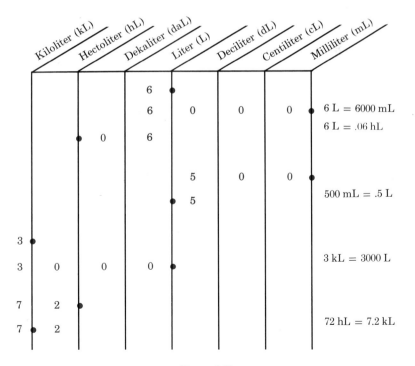

Figure 7.15

EXAMPLES

Change the following units as indicated. Use Tables 7.9 and 7.10.

1. 5000 mL = _____ L
    5000 mL = 5 L

2. 3.2 L = _____ mL
    3.2 L = 3200 mL

3. 60 hL = _____ kL
    60 hL = 6 kL

4. 637 mL = _____ L
    637 mL = 0.637 L

NOTE: Since 1 L = 1000 mL and 1 L = 1000 cm³,

$$1 \text{ cm}^3 = 1 \text{ mL}$$

Also, 1 kL = 1000 L = 1 000 000 cm³ and 1 000 000 cm³ = 1 m³. So,

$$1 \text{ kL} = 1 \text{ m}^3$$

5. 70 mL = _____ cm³
    70 mL = 70 cm³

6. 3.8 kL = _____ m³
    3.8 kL = 3.8 m³

PRACTICE QUIZ	Change the following units as indicated.	ANSWERS
	1. 2 mL = ___ L	**1.** 0.002 L
	2. 3.6 kL = ___ L	**2.** 3600 L
	3. 500 mL = ___ L	**3.** 0.5 L
	4. 500 mL = ___ cm³	**4.** 500 cm³
	5. 42 hL = ___ kL	**5.** 4.2 kL

The formulas for the volumes of various geometric solids are given here for reference and for use in the exercises.

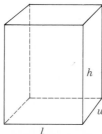

Rectangular solid
$V = lwh$

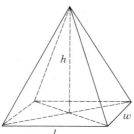

Rectangular pyramid
$V = \frac{1}{3}lwh$

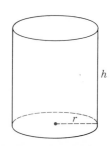

Right circular cylinder
$V = \pi r^2 h$

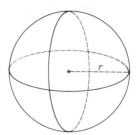

Right circular cone
$V = \frac{1}{3}\pi r^2 h$

Sphere
$V = \frac{4}{3}\pi r^3$

EXAMPLE

Find the volume of the solid with the dimensions indicated (use $\pi \approx 3.14$).

Solution:
On top of the cylinder is a hemisphere (half a sphere). Find the volume of the cylinder and the hemisphere and add the results.

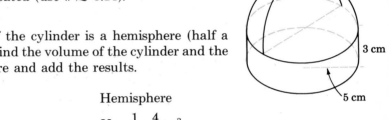

3 cm

5 cm

Cylinder

$V = \pi r^2 h$

$V = 3.14(5^2)(3)$

$= 235.5 \text{ cm}^3$

Hemisphere

$V = \frac{1}{2} \cdot \frac{4}{3}\pi r^3$

$V = \frac{2}{3}(3.14)(5^3)$

$= 261.67 \text{ cm}^3$

Total Volume
235.50 cm³
261.67 cm³
497.17 cm³ (or 497.17 mL)   total volume

EXERCISES 7.4

Copy and complete the following tables.

1.
1 cm³ = ____ mm³	
1 dm³ = ____ cm³	
1 m³ = ____ dm³	
1 km³ = ____ m³	

2.
1 dm = ____ cm	
1 dm = ____ mm	
1 dm² = ____ cm²	
1 dm² = ____ mm²	
1 dm³ = ____ cm³	
1 dm³ = ____ mm³	

3.

$$1 \text{ m} = \underline{\hspace{1cm}} \text{ dm}$$
$$1 \text{ m} = \underline{\hspace{1cm}} \text{ cm}$$
$$1 \text{ m}^2 = \underline{\hspace{1cm}} \text{ dm}^2$$
$$1 \text{ m}^2 = \underline{\hspace{1cm}} \text{ cm}^2$$
$$1 \text{ m}^3 = \underline{\hspace{1cm}} \text{ dm}^3$$
$$1 \text{ m}^3 = \underline{\hspace{1cm}} \text{ cm}^3$$

4.

$$1 \text{ km} = \underline{\hspace{1cm}} \text{ m}$$
$$1 \text{ km}^2 = \underline{\hspace{1cm}} \text{ m}^2$$
$$1 \text{ km}^3 = \underline{\hspace{1cm}} \text{ m}^3$$
$$1 \text{ km} = \underline{\hspace{1cm}} \text{ dm}$$
$$1 \text{ km}^2 = \underline{\hspace{1cm}} \text{ ha}$$
$$1 \text{ km}^3 = \underline{\hspace{1cm}} \text{ kL}$$

Change the following units as indicated.

5. $73 \text{ kL} = \underline{\hspace{1cm}} \text{ L}$

6. $0.9 \text{ kL} = \underline{\hspace{1cm}} \text{ L}$

7. $400 \text{ mL} = \underline{\hspace{1cm}} \text{ L}$

8. $525 \text{ mL} = \underline{\hspace{1cm}} \text{ L}$

9. $63 \text{ L} = \underline{\hspace{1cm}} \text{ mL}$

10. $8.7 \text{ L} = \underline{\hspace{1cm}} \text{ mL}$

11. $5 \text{ hL} = \underline{\hspace{1cm}} \text{ kL}$

12. $69 \text{ hL} = \underline{\hspace{1cm}} \text{ kL}$

13. $19 \text{ mL} = \underline{\hspace{1cm}} \text{ cm}^3$

Change the following units as indicated.

14. $5 \text{ cm}^3 = \underline{\hspace{1cm}} \text{ mm}^3$

15. $2 \text{ dm}^3 = \underline{\hspace{1cm}} \text{ cm}^3$

16. $76.4 \text{ mL} = \underline{\hspace{1cm}} \text{ L}$

17. $5.3 \text{ L} = \underline{\hspace{1cm}} \text{ mL}$

18. $30 \text{ cm}^3 = \underline{\hspace{1cm}} \text{ mL}$

19. $30 \text{ cm}^3 = \underline{\hspace{1cm}} \text{ L}$

20. $5.3 \text{ mL} = \underline{\hspace{1cm}} \text{ L}$

21. $48 \text{ kL} = \underline{\hspace{1cm}} \text{ L}$

22. $72\,000 \text{ L} = \underline{\hspace{1cm}} \text{ kL}$

23. $32 \text{ L} = \underline{\hspace{1cm}} \text{ hL}$

24. $80 \text{ L} = \underline{\hspace{1cm}} \text{ mL}$

25. $290 \text{ L} = \underline{\hspace{1cm}} \text{ kL}$

26. $569 \text{ mL} = \underline{\hspace{1cm}} \text{ L}$

27. $72 \text{ hL} = \underline{\hspace{1cm}} \text{ mL}$

28. $7 \text{ L} = \underline{\hspace{1cm}} \text{ mL}$

29. $95 \text{ hL} = \underline{\hspace{1cm}} \text{ L}$

30. $72 \text{ L} = \underline{\hspace{1cm}} \text{ hL}$

Find the volume of each of the following solids in a convenient unit from Table 7.9 (use $\pi \approx 3.14$).

31. a rectangular solid with a length of 5 dm, a width of 2 dm, and a height of 7 dm

32. a right circular cylinder 15 cm high and 1 dm in diameter (A diameter of a circle is a segment through the center with end points on the circle.)

33. a sphere with radius 4.5 cm

34. a sphere with diameter 12 dm

35. a right circular cone 3 dm high with a 2-dm radius

36. a rectangular pyramid with a length of 8 cm, a width of 10 mm, and a height of 3 dm

Find the volume of each of the solids with the dimensions indicated.

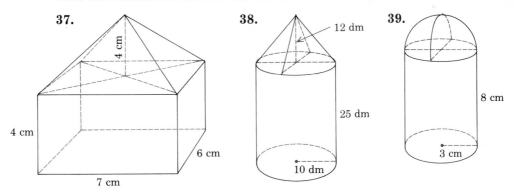

7.5 U.S. CUSTOMARY AND METRIC EQUIVALENTS

The U.S. measure of temperature is in **degrees Fahrenheit (°F)** and the metric measure of temperature is in **degrees Celsius (°C).** The two scales are shown on thermometers in Figure 7.16. By holding the edge of a piece of paper or a ruler horizontally across the page, you can get approximate conversions from one scale to the other.

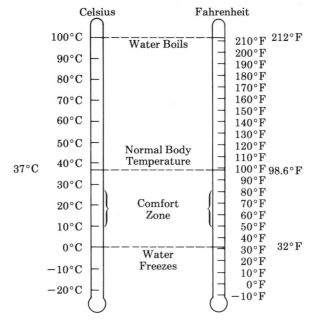

Figure 7.16

Hold a straightedge horizontally across the two thermometers in Figure 7.16, and you will read the following approximate equivalencies.

$$100°C = 212°F \qquad \text{(water boils at sea level)}$$
$$40°C = 105°F \qquad \text{(hot day in the desert)}$$
$$20°C = 68°F \qquad \text{(comfortable room temperature)}$$

The following two formulas give exact conversions from one scale to the other [F = Fahrenheit temperature and C = Celsius temperature].

$$C = \frac{5(F - 32)}{9} \qquad\qquad F = \frac{9 \cdot C}{5} + 32$$

Let $F = 86°$. $\qquad\qquad$ Let $C = 40°$.

$$C = \frac{5(86 - 32)}{9} \qquad\qquad F = \frac{9 \cdot 40}{5} + 32$$

$$= \frac{5(54)}{9} \qquad\qquad = 72 + 32$$

$$= 30 \qquad\qquad = 104$$

Thus, $86°F = 30°C$. $\qquad$ Thus, $40°C = 104°F$.

A calculator will give answers accurate to 8 digits. Answers that are not exact may be rounded off to whatever place of accuracy you choose.

In Tables 7.11 to 7.14 the equivalent measures are rounded off. Any calculations with these measures (with or without a calculator) cannot be any more accurate than the measure in the table.

TABLE 7.11   LENGTH EQUIVALENTS

U.S. TO METRIC	METRIC TO U.S.
1 in. = 2.54 cm (exact)	1 cm = 0.394 in.
1 ft = 0.305 m	1 m = 3.28 ft
1 yd = 0.915 m	1 m = 1.09 yd
1 mi = 1.61 km	1 km = 0.62 mi

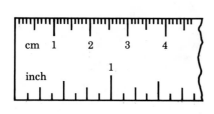

1 inch = 2.54 cm

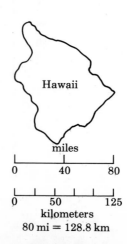

Hawaii

miles

0   40   80

0   50   125
kilometers
80 mi = 128.8 km

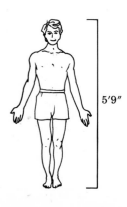

5′9″

5 ft 9 in. = 175 cm

EXAMPLES

Use Table 7.11 to convert the measures as indicated.

1. 6 ft = _____ cm
   6 ft = 72 in. = 72(2.54 cm) = 183 cm   (rounded off)

   or

   6 ft = 6(0.305 m) = 1.83 m = 183 cm

2. 25 mi = _____ km
   25 mi = 25(1.61 km) = 40.25 km

3. 30 m = _____ ft
   30 m = 30(3.28 ft) = 98.4 ft

4. 10 km = _____ mi
   10 km = 10(0.62 mi) = 6.2 mi

TABLE 7.12   AREA EQUIVALENTS

U.S. TO METRIC	METRIC TO U.S.
$1 \text{ in.}^2 = 6.45 \text{ cm}^2$	$1 \text{ cm}^2 = 0.155 \text{ in.}^2$
$1 \text{ ft}^2 = 0.093 \text{ m}^2$	$1 \text{ m}^2 = 10.764 \text{ ft}^2$
$1 \text{ yd}^2 = 0.836 \text{ m}^2$	$1 \text{ m}^2 = 1.196 \text{ yd}^2$
$1 \text{ acre} = 0.405 \text{ ha}$	$1 \text{ ha} = 2.47 \text{ acres}$

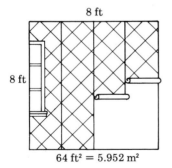

8 ft

8 ft

$64 \text{ ft}^2 = 5.952 \text{ m}^2$

1 in.

0.875 in.

$0.875 \text{ in.}^2 = 5.64 \text{ cm}^2$

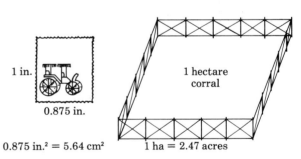

1 hectare corral

$1 \text{ ha} = 2.47 \text{ acres}$

EXAMPLES

Use Table 7.12 to convert the measures as indicated.

1. $40 \text{ yd}^2 = $ _____ m^2
   $40 \text{ yd}^2 = 40(0.836 \text{ m}^2) = 33.44 \text{ m}^2$

2. 5 acres = _____ ha
   5 acres = 5(0.405 ha) = 2.2 ha   (rounded off)

3. 5 ha = _____ acres
   5 ha = 5(2.47 acres) = 12.35 acres

4. $100 \text{ cm}^2 = $ _____ in.^2
   $100 \text{ cm}^2 = 100(0.155 \text{ in.}^2) = 15.5 \text{ in.}^2$

TABLE 7.13   VOLUME EQUIVALENTS

U.S. TO METRIC	METRIC TO U.S.
$1 \text{ in.}^3 = 16.387 \text{ cm}^3$	$1 \text{ cm}^3 = 0.06 \text{ in.}^3$
$1 \text{ ft}^3 = 0.028 \text{ m}^3$	$1 \text{ m}^3 = 35.315 \text{ ft}^3$
$1 \text{ qt} = 0.946 \text{ L}$	$1 \text{ L} = 1.06 \text{ qt}$
$1 \text{ gal} = 3.785 \text{ L}$	$1 \text{ L} = 0.264 \text{ gal}$

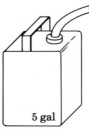

5 gal   1 liter   1 in.³ ice

5 gal = 18.925 L   1 L = 1.06 qt   3 in.³ = 49.161 cm³

EXAMPLES

Use Table 7.13 to convert the measures as indicated.

1. 20 gal = _____ L
    20 gal = 20(3.785 L) = 75.7 L

2. 42 L = _____ gal
    42 L = 42(0.264 gal) = 11.088 gal

    or

    42 L = 11.1 gal   (rounded off)

3. 6 qt = _____ L
    6 qt = 6(0.946 L) = 5.676

    or

    6 qt = 5.7 L   (rounded off)

4. $10 \text{ cm}^3 = $ _____ in.^3
    $10 \text{ cm}^3 = 10(0.06 \text{ in.}^3) = 0.6 \text{ in.}^3$

TABLE 7.14   MASS EQUIVALENTS

U.S. TO METRIC	METRIC TO U.S.
1 oz = 28.35 g	1 g = 0.035 oz
1 lb = 0.454 kg	1 kg = 2.205 lb

25 lb = 11.35 kg   9 kg = 19.85 lb

EXAMPLES   Use Table 7.14 to convert the measures as indicated.

1. 5 lb = _____ kg
    5 lb = 5(0.454 kg) = 2.27 kg

2. 15 kg = _____ lb
    15 kg = 15(2.205 lb) = 33.075 lb

    or

    15 kg = 33.1 lb   (rounded off)

EXERCISES 7.5

Use the appropriate formula to convert the degrees as indicated.

1. 25°C = ____°F   2. 80°C = ____°F   3. 10°C = ____°F

4. 35°C = ____°F   5. 50°F = ____°C   6. 100°F = ____°C

7. 32°F = ____°C   8. 41°F = ____°C

Use the appropriate table to convert the following measures as indicated.

9. 5 ft 2 in. = ____ cm   10. 6 ft 3 in. = ____ cm

11. 3 yd = ____ m   12. 5 yd = ____ m

13. 60 mi = ____ km   14. 100 mi = ____ km

15. 400 mi = ____ km   16. 350 mi = ____ km

17. 200 km = ____ mi   18. 65 km = ____ mi

19. 35 km = ____ mi   20. 450 km = ____ mi

21. 50 cm = ____ in.   22. 100 cm = ____ in.

23. 3 in.^{2} = ____ cm^2 24. 16 in.^{2} = ____ cm^2

25. 600 ft^2 = ____ m^2 26. 300 ft^2 = ____ m^2

27. 100 yd^2 = ____ m^2 28. 250 yd^2 = ____ m^2

29. 1000 acres = ____ ha   30. 250 acres = ____ ha

31. 300 ha = ____ acres

32. 400 ha = ____ acres

33. 5 m² = ____ ft²

34. 10 m² = ____ yd²

35. 30 cm² = ____ in.²

36. 50 cm² = ____ in.²

37. 10 qt = ____ L

38. 20 qt = ____ L

39. 25 gal = ____ L

40. 18 gal = ____ L

41. 10 L = ____ qt

42. 25 L = ____ qt

43. 42 L = ____ gal

44. 50 L = ____ gal

45. 200 in.³ = ____ cm³

46. 10 m³ = ____ ft³

47. 10 lb = ____ kg

48. 16 oz = ____ g

49. 500 kg = ____ lb

50. 100 g = ____ oz

SUMMARY: CHAPTER 7

The metric equivalents are given in tables throughout Chapter 7 and are not reproduced here. You should be familiar with all of them.

The prefixes used, from smallest to largest unit size, are milli, centi, deci, deka, hecto, kilo.

Abbreviations such as mm, cm, and dm do not have periods. Zero is written to the left of the decimal point if there is no whole number part. No commas are used in writing numbers.

The basic units are meter (length), gram (mass), are (area), liter (volume). Temperature is read from a Celsius scale.

Some useful formulas related to geometric figures:

FIGURE	PERIMETER	AREA
square	$P = 4s$	$A = s^2$
rectangle	$P = 2l + 2w$	$A = lw$
parallelogram	$P = 2b + 2a$	$A = bh$
triangle	$P = a + b + c$	$A = \frac{1}{2}bh$
circle	$C = 2\pi r$ $C = \pi d$	$A = \pi r^2$
trapezoid	$P = a + b + c + d$	$A = \frac{1}{2}h(b + c)$

FIGURE	VOLUME
rectangular solid	$V = lwh$
rectangular pyramid	$V = \dfrac{1}{3} lwh$
right circular cylinder	$V = \pi r^2 h$
right circular cone	$V = \dfrac{1}{3} \pi r^2 h$
sphere	$V = \dfrac{4}{3} \pi r^2$

For exact conversions between Fahrenheit and Celsius,

$$C = \frac{5(F - 32)}{9} \qquad\qquad F = \frac{9 \cdot C}{5} + 32$$

REVIEW QUESTIONS: CHAPTER 7

Change the following units as indicated.

1. $15\,\text{m} = $ _____ cm

2. $35\,\text{mm} = $ _____ dm

3. $37\,\text{cm}^2 = $ _____ mm^2

4. $17\,\text{mm}^2 = $ _____ cm^2

5. $3\,\text{ha} = $ _____ a

6. $3\,\text{ha} = $ _____ m^2

7. $5\,\text{L} = $ _____ cm^3

8. $36\,\text{L} = $ _____ mL

9. $13\,\text{dm}^3 = $ _____ cm^3

10. $68\,\text{cm}^3 = $ _____ mm^3

11. $5\,\text{kg} = $ _____ g

12. $3.4\,\text{g} = $ _____ mg

13. $6.71\,\text{t} = $ _____ kg

14. $19\,\text{mg} = $ _____ g

15. $8\,\text{kg} = $ _____ g

16. $4290\,\text{g} = $ _____ kg

Find the perimeter and area of each of the following figures with the dimensions indicated.

17.

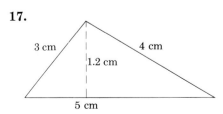

3 cm   4 cm   1.2 cm   5 cm

18.

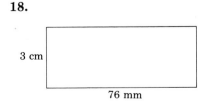

3 cm   76 mm

19.

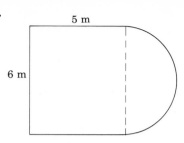

20.

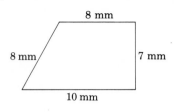

Find the volume in liters of each of the following solids with dimensions indicated (use $\pi \approx 3.14$).

21.

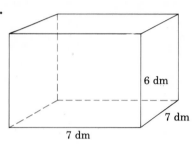

22.

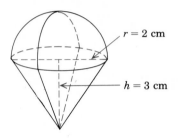

23.   Use the formula $C = \dfrac{5(F - 32)}{9}$ to convert to Celsius:

   a. $50°F = $ ____ $°C$ **b.** $122°F = $ ____ $°C$

24.   Convert the units as indicated.

   a.   3 in. = ____ cm   **b.**   2 mi = ____ km

   c.   3 acres = ____ ha   **d.**   4 qt = ____ L

   e.   5 lb = ____ kg

CHAPTER TEST: CHAPTER 7

1. Which is longer, 20 mm or 20 cm? How much longer?

2. Which is heavier, 10 g or 10 kg? How much heavier?

3. Which has the greater volume, 15 mL or 15 cm³? How much greater?

Change the following units as indicated.

4. 37 cm = _____ m

5. 23 m = _____ cm

6. 2 L = _____ cm³

7. 1200 g = _____ kg

8. 5.6 t = _____ kg

9. 75 a = _____ m²

10. 11 000 mm = _____ m

11. 4 cm³ = _____ mm³

12. 960 mm² = _____ cm²

13. 83.5 mg = _____ g

Find the perimeter and area of each figure.

14.

15.

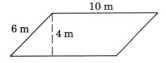

16. Find the volume in liters of a right circular cylinder with radius 3 cm and height 12 cm.

17. Use the formula $F = \dfrac{9 \cdot C}{5} + 32$ to convert 80°C to Fahrenheit.

18. Convert the units as indicated.

a. 5.08 cm = _____ in.

b. 2.12 L = _____ qt

8
NEGATIVE NUMBERS

8.1 NUMBER LINES AND ABSOLUTE VALUE

Draw a horizontal line, choose any point on the line, and label it with the number 0. (See Figure 8.1.)

Figure 8.1

Now choose another point on the same line to the right of 0 and label it with the number 1 (Figure 8.2).

Figure 8.2

The line in Figure 8.2 is called a **number line.** The points corresponding to the whole numbers are determined by the distance between 0 and 1. The point corresponding to 2 is the same distance to the right of 1 as 1 is from 0, 3 is the same distance to the right of 2, and so on (Figure 8.3).

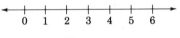

Figure 8.3

The **graph** of a number is indicated with a shaded dot at the corresponding point on the number line. The point or dot is the graph of the number, and the number is the **coordinate** of the point. We will use the terms **number** and **point** interchangeably. Thus, "number 3" and "point 3" are considered to have the same meaning.

The numbers 0, 1, and 5 form a set. Sets are indicated with braces and are named with capital letters. Thus, we can say $A = \{0, 1, 5\}$. The graph of A is shown in Figure 8.4.

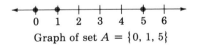

Graph of set $A = \{0, 1, 5\}$

Figure 8.4

The symbol -1 is read **"the opposite of 1"** or **"negative 1"** and represents the point 1 unit to the left of 0. The symbol -2 is read **"the opposite of 2"** or **"negative 2"** and represents the point 2 units to the left of 0. The opposite of 3 is -3, the opposite of 4 is -4, and so on. (See Figure 8.5.)

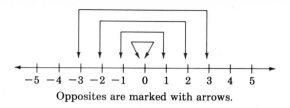

Opposites are marked with arrows.

Figure 8.5

The whole numbers and their opposites form a set called the set of **integers.** The counting numbers are called **positive integers,** and their opposite are called **negative integers. The number 0 is neither positive nor negative, and 0 is its own opposite.** Note also that the opposite of a negative integer is a positive integer.

Integers	$\ldots, -3, -2, -1, 0, 1, 2, 3, \ldots$
Positive integers	$1, 2, 3, 4, 5, \ldots$
Negative integers	$\ldots, -3, -2, -1$

(The three dots, . . . , indicate a continuing pattern without end.)

EXAMPLES

1. Find the opposite of 8.

   Answer: -8

2. Find the opposite of -2.

   Answer: $-(-2)$ read "the opposite of negative 2"

   or $+2$ read "positive 2"

   or 2 A number is understood to be positive if there is no sign in front of it.

3. Graph the set $B = \{-2, -1, 3\}$.

   Answer: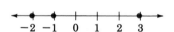

4. Graph the set $\{-4, -2, 0, 2, 4, \ldots\}$.

   Answer: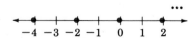

The **absolute value** of a number is its distance from 0 on a number line, regardless of the direction. For example, -5 and $+5$ are both 5 units from 0, and both have the same absolute value, namely 5. (See Figure 8.6.)

The symbol for absolute value is two vertical bars, $|\ |$. So we can write $|-5| = 5$ and $|+5| = 5$.

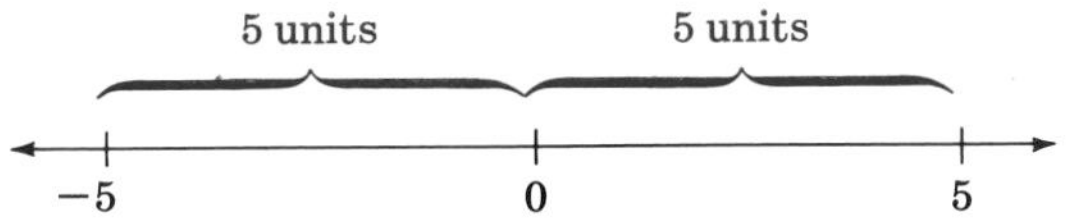

Figure 8.6

EXAMPLES

1. $|+3| = 3$ and $|-3| = 3$

2. $|-7| = |+7| = 7$

3. $|0| = 0$ The absolute value of 0 is 0.

4. Which number has the larger absolute value, -8 or -6?

Answer: -8 has the larger absolute value since $|-8| = 8$ and $|-6| = 6$, and 8 is larger than 6.

EXERCISES 8.1

In each of the following exercises, draw a number line and graph the given set of integers.

1. $\{0, 1, 2\}$ 2. $\{0, 2, 4\}$

3. $\{-3, -1, 1\}$ 4. $\{-3, -2, 0\}$

5. $\{-10, -9, -8, -7\}$ 6. $\{-5, -4, -2, -1\}$

7. $\{-5, 0, 5\}$ 8. $\{-3, -1, 0, 1, 3\}$

9. $\{1, 2, 3, \ldots, 10\}$ 10. $\{-2, -1, 0, \ldots, 5\}$

11. $\{1, 3, 5, 7, \ldots\}$ 12. $\{0, 2, 4, 6, \ldots\}$

13. $\{\ldots, -5, -4, -3\}$ 14. $\{\ldots, -10, -9, -8\}$

15. $\{-3, 0, 3, 6, 9, \ldots\}$ 16. $\{\ldots, -8, -4, 0, 4\}$

17. $\{|-7|, |-2|, 0\}$ 18. $\{|-3|, 0, 1\}$

19. $\{-3, 0, |-3|, |-5|\}$ 20. $\{-2, -1, |-1|, |-2|\}$

Find the opposite of each number.

21. -10 22. -9 23. 14 24. 12 25. -6

26. -3 27. 30 28. 40 29. 0 30. -7

Find each absolute value as indicated.

31. $|-6|$ 32. $|-10|$ 33. $|+24|$ 34. $|+16|$

35. $|-20|$ 36. $|-50|$ 37. $|0|$ 38. $|27|$

State which number in each pair has the larger absolute value.

39. $10, 13$ 40. $16, 18$ 41. $-13, -10$ 42. $-18, -16$

43. 20, 30   **44.** 10, 15   **45.** −9, 5   **46.** −6, 3

47. −8, 9   **48.** −3, 7   **49.** −11, 7   **50.** −12, 10

8.2 ADDING INTEGERS

The sum of two integers can be indicated with a plus (+) sign between the two numbers. For example,

$$(+2) \quad + \quad (+7)$$

$$\uparrow \qquad \uparrow \qquad \uparrow$$

positive 2   plus   positive 7

The sum in this example is obviously +9 since we already know how to add 2 and 7:

$$(+2) + (+7) = +9$$

But the sums

$$(+2) + (-7) = ?$$

$$(-2) + (+7) = ?$$

and $\qquad (-2) + (-7) = ?$

are not so obvious.

By using a number line, we can develop an intuitive idea of how to add integers. Start at the first number to be added, then

1. Move right if the second number is positive, or

2. Move left if the second number is negative.

The distance moved is the absolute value of the second number.

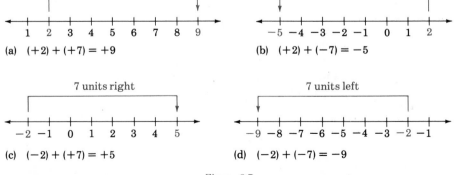

(a) $(+2) + (+7) = +9$

(b) $(+2) + (-7) = -5$

(c) $(-2) + (+7) = +5$

(d) $(-2) + (-7) = -9$

Figure 8.7

EXAMPLES   Find the following sums, using a number line as illustrated in Figure 8.7.

1. $(-5) + (+4) = ?$

Answer:

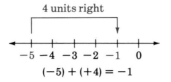

$(-5) + (+4) = -1$

2. $(-3) + (-8) = ?$

Answer:

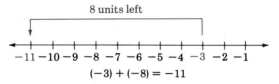

$(-3) + (-8) = -11$

3. $(+6) + (-10) = ?$

Answer:

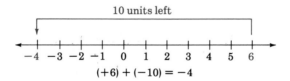

$(+6) + (-10) = -4$

4. $(-7) + (+7) = ?$

Answer:

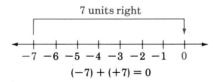

$(-7) + (+7) = 0$

[NOTE: The sum of two opposites will always be 0.]

Integers can be added mentally, without use of number lines, once you know the following rules.

1. **To add two integers with like signs,** add their absolute values and use the common sign.

$$(+6) + (+5) = +[|+6| + |+5|] = +[6 + 5] = +11$$
$$(-2) + (-8) = -[|-2| + |-8|] = -[2 + 8] = -10$$

2. **To add two integers with unlike signs,** subtract their absolute values (smaller from larger) and use the sign of the number with the larger absolute value.

$$(+10) + (-15) = -[|-15| - |+10|] = -[15 - 10] = -5$$
$$(+12) + (-9) = +[|+12| - |-9|] = +[12 - 9] = +3$$

EXAMPLES

1. $(+4) + (-1) = +(4 - 1) = +3$ (unlike signs)

2. $(-5) + (-3) = -(5 + 3) = -8$ (like signs)

3. $(+7) + (+10) = +(7 + 10) = +17$ (like signs)

4. $(+6) + (-19) = -(19 - 6) = -13$ (unlike signs)

5. $(+30) + (-30) = (30 - 30) = 0$ (opposites)

If more than two numbers are to be added, add any two, then add their sum to one of the other numbers until all numbers have been added.

In algebra equations are written horizontally, so the ability to work with sums written horizontally is very important. However, there are some situations (as in long division) where sums are written vertically. The rules for adding are the same whether the numbers are written horizontally or vertically.

EXAMPLES

1. $(+9) + (+2) + (-6) = +[9 + 2] + (-6)$
$$= +[11] + (-6)$$
$$= +5$$

2. $\begin{array}{r} -12 \\ +8 \\ \hline -4 \end{array}$

3. $\begin{array}{r} -5 \\ 6 \\ -14 \\ \hline -13 \end{array}$

PRACTICE QUIZ	Find each sum.		ANSWERS	
	1. $(-10) + (+3) =$		**1.** -7	
	2. $(-5) + (+9) =$		**2.** $+4$	
	3. $(-2) + (-3) =$		**3.** -5	
	4. $(+4) + (-6) =$		**4.** -2	

EXERCISES 8.2

Find each sum.

1. $(+6) + (-4)$ **2.** $(+8) + (-7)$ **3.** $(4) + (+6)$

4. $(5) + (-8)$ **5.** $(16) + (+3)$ **6.** $(-8) + (-2)$

7. $(-3) + (-6)$ **8.** $(-2) + (+2)$ **9.** $(+4) + (-4)$

10. $(13) + (12)$ **11.** $(6) + (-10)$ **12.** $(14) + (-17)$

13. $(+5) + (-3)$ **14.** $(+15) + (-18)$ **15.** $(-4) + (-12)$

16. $(-8) + (+8)$ **17.** $(+2) + (-6)$ **18.** $(-9) + (+5)$

19. $(-16) + (+3) + (+13)$ **20.** $(-5) + (+5) + (14)$

21. $(-1) + (-2) + (+7)$ **22.** $(+3) + (-4) + (-5)$

23. $(+6) + (+3) + (+5)$ **24.** $(-18) + (-5) + (-7)$

25. $(-1) + (+2) + (-4) + (+2)$

26. $\begin{array}{r} -4 \\ +8 \\ \hline \end{array}$	**27.** $\begin{array}{r} -5 \\ -10 \\ \hline \end{array}$	**28.** $\begin{array}{r} -13 \\ -6 \\ \hline \end{array}$	**29.** $\begin{array}{r} +16 \\ +25 \\ \hline \end{array}$	**30.** $\begin{array}{r} +14 \\ -8 \\ \hline \end{array}$
31. $\begin{array}{r} +20 \\ -7 \\ \hline \end{array}$	**32.** $\begin{array}{r} +2 \\ -5 \\ -3 \\ \hline \end{array}$	**33.** $\begin{array}{r} +8 \\ +3 \\ -1 \\ \hline \end{array}$	**34.** $\begin{array}{r} +10 \\ -4 \\ +2 \\ \hline \end{array}$	**35.** $\begin{array}{r} -16 \\ -8 \\ +12 \\ \hline \end{array}$
36. $\begin{array}{r} -15 \\ -20 \\ -6 \\ \hline \end{array}$	**37.** $\begin{array}{r} -4 \\ -17 \\ +11 \\ \hline \end{array}$	**38.** $\begin{array}{r} +13 \\ -5 \\ +17 \\ -25 \\ \hline \end{array}$	**39.** $\begin{array}{r} +14 \\ -14 \\ +37 \\ -37 \\ \hline \end{array}$	**40.** $\begin{array}{r} -8 \\ -5 \\ -13 \\ -22 \\ \hline \end{array}$
41. $\begin{array}{r} -100 \\ -50 \\ -85 \\ \hline \end{array}$	**42.** $\begin{array}{r} -96 \\ +14 \\ -83 \\ \hline \end{array}$	**43.** $\begin{array}{r} +750 \\ -632 \\ -198 \\ +200 \\ \hline \end{array}$	**44.** $\begin{array}{r} -300 \\ +450 \\ +325 \\ +500 \\ \hline \end{array}$	**45.** $\begin{array}{r} -25 \\ -95 \\ -48 \\ -20 \\ +67 \\ \hline \end{array}$

8.3   SUBTRACTING INTEGERS

In the intuitive discussion of addition using number lines, we

1.   Moved right if the second number was positive, or

2.   Moved left if the second number was negative.

With **subtraction,** we simply reverse the moves. That is, we

1.   Move left if the second number is positive, or

2.   Move right if the second number is negative.

In other words, in subtraction, move in the opposite direction from the direction indicated by the number being subtracted. (See Figure 8.8.)

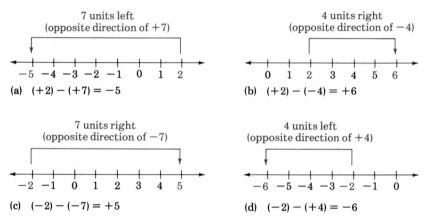

Figure 8.8

We can restate the previous discussion and the moves outlined in Figure 8.8 in terms of addition. **In subtraction, add the opposite of the number being subtracted.** For example,

$$(+4) \quad - \quad (+6) \; = \; (+4) \quad + \quad (-6) \; = \; -2$$

positive 4   minus   positive 6   positive 4   plus   opposite of $+6$

$$(+5) \quad - \quad (-8) \; = \; (+5) \quad + \quad (+8) \; = \; +13$$

positive 5   minus   negative 8   positive 5   plus   opposite of -8

DEFINITION   The **difference** of two integers a and b is the sum of a and the opposite of b. Symbolically, $(a) - (b) = (a) + (-b)$.

EXAMPLES

1. $(+2) - (-6) = (+2) + (+6) = +8$

     ↑   ↑ ↑

    minus   plus opposite of -6

2. $(-3) - (-7) = (-3) + (+7) = +4$

                   ↑

         opposite of -7

3. $(-5) - (+2) = (-5) + (-2) = -7$

                   ↑

         opposite of $+2$

4. $(-9) - (-9) = (-9) + (+9) = 0$

                   ↑

         opposite of -9

The numbers may also be written vertically, one underneath the other. In this case, change the sign of the number being subtracted (the bottom number), then add.

EXAMPLES

	To subtract:	Add:
1.	-10	-10
	$\underline{-3}$	$\underline{+3}$
		-7
2.	$+14$	$+14$
	$\underline{+9}$	$\underline{-9}$
		5

PRACTICE QUIZ

Find each difference.

1. $(+8) - (-3) =$

2. $(-4) - (-5) =$

3. $(-10) - (+3) =$

ANSWERS

1. 11

2. $+1$

3. -13

EXERCISES 8.3

Find each difference.

1. $(+5) - (+2)$ 2. $(+16) - (+3)$ 3. $(+8) - (-3)$

4. $(+12) - (-4)$ 5. $(-5) - (+2)$ 6. $(-10) - (+3)$

7. $(-10) - (-1)$ 8. $(-15) - (-1)$ 9. $(-3) - (-7)$

10. $(-2) - (-12)$ **11.** $(-4) - (+6)$ **12.** $(-9) - (+13)$

13. $(-13) - (-14)$ **14.** $(-12) - (-15)$ **15.** $(+9) - (-9)$

16. $(+11) - (-11)$ **17.** $(+15) - (-2)$ **18.** $(+20) - (-3)$

19. $(-17) - (+14)$ **20.** $(-16) - (+10)$ **21.** $(+3) - (+8)$

22. $(+1) - (+5)$ **23.** $(-5) - (-5)$ **24.** $(-7) - (-7)$

25. $(+7) - (+12)$

Subtract the second number from the first.

26.	**27.**	**28.**	**29.**	**30.**
18	24	-8	-13	-4
-12	16	-12	-18	$+5$

31.	**32.**	**33.**	**34.**	**35.**
32	-6	-25	-45	28
-48	-30	-13	-16	-15

8.4 ANOTHER NOTATION

We have been using a plus (+) sign for addition, a minus (−) sign for subtraction, and parentheses around each number being added or subtracted. In another notation more commonly used in algebra, the parentheses and the plus (+) and minus (−) signs are dropped. **The numbers are written horizontally, and the problem is considered as adding positive and negative numbers.** All positive and negative signs between numbers must be included.

If the first number to the left is positive, the + sign may be omitted with the understanding that the number is positive. For example,

$$9 - 12 \text{ is the same as } (+9) + (-12)$$

So,

$$9 - 12 = (+9) + (-12) = -3$$

After some practice, you will be able to do the second step mentally and will simply write

$$9 - 12 = -3$$

EXAMPLES   1. $7 - 3 = (+7) + (-3) = 4$
or $7 - 3 = 4$

2. $-13 + 6 = (-13) + (+6) = -7$
    or $-13 + 6 = -7$

3. $-4 - 8 = (-4) + (-8) = -12$
    or $-4 - 8 = -12$

4. $6 - 9 = -3$

5. $-20 + 12 = -8$

6. $-3 - 14 = -17$

7. $17 - 4 - 8 = 13 - 8 = 5$

8. $-14 + 30 - 5 + 1 = 16 - 5 + 1 = 11 + 1 = 12$

PRACTICE QUIZ	Evaluate each of the following expressions.	ANSWERS
	1. $-10 + 4$	**1.** -6
	2. $-12 - 8$	**2.** -20
	3. $13 - 5 + 6$	**3.** 14

EXERCISES 8.4

Evaluate each of the following expressions.

1. $6 + 2$ **2.** $4 + 8$ **3.** $7 - 1$ **4.** $9 - 4$

5. $4 + 6$ **6.** $8 + 9$ **7.** $-3 - 1$ **8.** $-2 - 6$

9. $12 - 6$ **10.** $9 - 3$ **11.** $-13 + 4$ **12.** $-20 + 14$

13. $-10 + 9$ **14.** $-18 + 3$ **15.** $24 - 32$ **16.** $14 - 17$

17. $-12 - 6$ **18.** $-2 - 8$ **19.** $-15 + 18$ **20.** $-25 + 30$

21. $-20 + 21$ **22.** $-30 + 32$ **23.** $-7 + 7$ **24.** $-6 + 6$

25. $18 - 3$ **26.** $-4 + 16 - 8$ **27.** $-5 + 12 - 3$

28. $-20 - 2 + 6$ **29.** $14 - 5 - 12$ **30.** $13 + 15 - 6$

31. $-6 - 8 - 13$ **32.** $-4 - 10 - 7$ **33.** $30 + 12 - 18$

34. $16 + 4 - 20$ **35.** $15 + 6 - 21$ **36.** $13 - 4 + 6 - 5$

37. $16 - 3 - 7 - 1$ **38.** $19 - 5 - 8 - 6$ **39.** $-4 + 10 - 12 + 1$

40. $-8 + 14 - 10 + 3$

8.5   MULTIPLYING INTEGERS

Multiplication with whole numbers is shorthand for repeated addition. For example,

$$7 + 7 + 7 + 7 = 4 \cdot 7 = 28$$
$$3 + 3 + 3 + 3 + 3 + 3 = 6 \cdot 3 = 18$$

Multiplication with integers can also be considered shorthand for repeated addition with integers.

$$(-7) + (-7) + (-7) + (-7) = 4(-7) = -28$$
$$(-3) + (-3) + (-3) + (-3) + (-3) + (-3) = 6(-3)$$
$$= -18$$

Repeated addition of a negative integer results in the product of a positive integer and a negative integer. Since the sum of negative integers is negative, **the product of a positive integer and a negative integer will be negative.**

EXAMPLES

1. $4(-3) = -12$ 2. $5(-2) = -10$

3. $7(-8) = -56$ 4. $-9(5) = -45$

The product of two negative integers is not related to repeated addition. Instead, we will develop a rule based on intuition related to patterns in the answers. Consider the following patterns of products and try to decide what the missing products should be.

-5	-5	-5	-5	-5	-5	-5
$+3$	$+2$	$+1$	0	-1	-2	-3
-15	-10	-5	0	$?$	$?$	$?$

Did you decide

-5	-5	-5
-1	-2	-3
$+5$	$+10$	$+15$

Try again with the numbers written horizontally.

$$(+3)(-2) = -6$$
$$(+2)(-2) = -4$$
$$(+1)(-2) = -2$$
$$(0)(-2) = 0$$
$$(-1)(-2) = ?$$
$$(-2)(-2) = ?$$
$$(-3)(-2) = ?$$

You should have

$$(-1)(-2) = +2$$
$$(-2)(-2) = +4$$
$$(-3)(-2) = +6$$

The following statement is true. **The product of two negative integers is positive.**

EXAMPLES

1. $(-6)(-4) = +24$

2. $(-7)(-9) = 63$

3. $-2(-3) = 6$

4. $-8(-20) = 160$

SUMMARY OF RULES FOR MULTIPLYING TWO INTEGERS

1. $(+ \text{ integer}) \cdot (+ \text{ integer}) = + \text{ integer}$

2. $(+ \text{ integer}) \cdot (- \text{ integer}) = - \text{ integer}$

3. $(- \text{ integer}) \cdot (- \text{ integer}) = + \text{ integer}$

Or, using symbols,

If a and b are positive integers,

1. $a \cdot b = ab$

2. $a(-b) = -ab$

3. $(-a)(-b) = ab$

Or, in words,

---

1. The product of two positive integers is positive.

2. The product of a positive integer and a negative integer is negative.

3. The product of two negative integers is positive.

If a product involves more than two integers, multiply any two, then continue to multiply that product by the next integer until all the integers have been multiplied.

EXAMPLES

1. $(-3)(4)(-2) = [(-3)(4)](-2)$
$$= [-12](-2)$$
$$= 24$$

2. $(-5)(-3)(-10) = [(-5)(-3)](-10)$
$$= [+15](-10)$$
$$= -150$$

3. $(-2)(+3)(-2)(-6) = [(-2)(+3)](-2)(-6)$
$$= [-6](-2)(-6)$$
$$= [(-6)(-2)](-6)$$
$$= [+12](-6)$$
$$= -72$$

After some practice, you may note that a product with an odd number (1, 3, 5, 7, and so on) of negative factors will be negative; and a product with an even number (0, 2, 4, 6, and so on) of negative factors will be positive. **If 0 is a factor, of course the product will be 0.**

PRACTICE QUIZ

Find the following products.

ANSWERS

1. $(-5)(5)$ 1. -25

2. $(-8)(-2)$ 2. $+16$

3. $(-3)(-3)(0)$ 3. 0

4. $(-2)(-2)(-2)$ 4. -8

EXERCISES 8.5

Find the following products.

1. $5(-3)$	2. $4(-6)$	3. $-6(-4)$
4. $-2(-7)$	5. $-5(4)$	6. $-8(3)$
7. $14(2)$	8. $13(3)$	9. $-10(5)$
10. $-11(3)$	11. $(-7)3$	12. $(-2)9$
13. $6(-8)$	14. $9(-4)$	15. $-7(-9)$
16. $-8(-9)$	17. $0(-6)$	18. $0(-4)$

19. $(-6)(-5)(3)$	**20.** $(-2)(-1)(7)$	**21.** $4(-2)(-3)$
22. $5(-6)(-1)$	**23.** $(-5)(3)(-4)$	**24.** $(-3)(7)(-5)$
25. $(-7)(-2)(-3)$	**26.** $(-4)(-4)(-4)$	**27.** $(-3)(-3)(-5)$
28. $(-2)(-2)(-8)$	**29.** $(-3)(-4)(-5)$	**30.** $(-2)(-5)(-7)$
31. $(-5)(0)(-6)$	**32.** $(-6)(0)(-2)$	**33.** $(-1)^3$
34. $(-3)^3$	**35.** $(-2)^4$	**36.** $(-4)^3$

37. $(-1)(-4)(-7)(+3)$ **38.** $(-5)(-2)(-1)(+5)$

39. $(-2)(-3)(-10)(-5)$ **40.** $(-11)(-2)(-4)(-1)$

41. $(-2)^2(-3)^3$ **42.** $(-5)(-4)^3$

43. $(|-7|)(|-5|)(|-8|)$ **44.** $(|-10|)(|-11|)(|-3|)$

45. $(|-2|)(|6|)(-3)^2$ **46.** $(|-5|)^2(-5)$

47. $(|4|)^2(|-9|)(|-8|)$ **48.** $(|-1|)^3(|-2|)(|5|)^2$

49. $(|-3|)^2(|-4|)(|-9|)^2$ **50.** $(|-5|)^2(|-7|)(|-2|)^3$

8.6 DIVIDING INTEGERS

Division with whole numbers is closely related to multiplication. For example,

$$\frac{42}{6} = 7 \quad \text{because} \quad 42 = 6 \cdot 7$$

This same relationship is true for integers and is stated as a definition for division.

DEFINITION   If a and b are integers and $b \neq 0$,

$$\frac{a}{b} = x \quad \text{means} \quad a = b \cdot x$$

If a is an integer, then $\dfrac{a}{0}$ is undefined.

We can develop the rules for division with integers simply by applying the definition to each of the possible cases.

EXAMPLES   1. $\dfrac{+28}{+4} = +7$ because $+28 = (+4)(+7)$.

2. $\dfrac{-28}{+4} = -7$ because $-28 = (+4)(-7)$.

3. $\dfrac{+28}{-4} = -7$ because $+28 = (-4)(-7)$.

4. $\dfrac{-28}{-4} = +7$ because $-28 = (-4)(+7)$.

These four examples illustrate all the possibilities that occur when we divide with positive and negative integers.

SUMMARY OF RULES FOR DIVISION WITH INTEGERS

1. $\dfrac{+ \text{ integer}}{+ \text{ integer}} = +$ number

2. $\dfrac{- \text{ integer}}{+ \text{ integer}} = -$ number

3. $\dfrac{+ \text{ integer}}{- \text{ integer}} = -$ number

4. $\dfrac{- \text{ integer}}{- \text{ integer}} = +$ number

Or, using symbols,

If a and b are positive integers,

1. $\dfrac{+a}{+b} = +\dfrac{a}{b}$

2. $\dfrac{-a}{+b} = -\dfrac{a}{b}$

3. $\dfrac{+a}{-b} = -\dfrac{a}{b}$

4. $\dfrac{-a}{-b} = +\dfrac{a}{b}$

Or, in words,

---

1. The quotient of two positive integers is positive.

2, 3. The quotient of a positive integer and a negative integer is negative.

4. The quotient of two negative integers is positive.

The quotient of two integers may not always be an integer. The quotient may be a fraction. However, in this section the problems will be set up so that any quotients will be integers.

EXAMPLES

1. $\dfrac{+16}{-2} = -8$

2. $\dfrac{-10}{+2} = -5$

3. $\dfrac{-40}{-4} = 10$

4. $\dfrac{23}{0} = $ undefined

PRACTICE QUIZ

Find the following quotients.

1. $\dfrac{-20}{10}$

2. $\dfrac{-40}{-2}$

3. $\dfrac{0}{-8}$

4. $\dfrac{36}{-2}$

ANSWERS

1. -2

2. 20

3. 0

4. -18

EXERCISES 8.6

Find the following quotients.

1. $\dfrac{-12}{4}$ 2. $\dfrac{-18}{2}$ 3. $\dfrac{-14}{7}$ 4. $\dfrac{-28}{7}$ 5. $\dfrac{-20}{-5}$

6. $\dfrac{-30}{-3}$ 7. $\dfrac{-50}{-10}$ 8. $\dfrac{-30}{-5}$ 9. $\dfrac{30}{-6}$ 10. $\dfrac{40}{-8}$

11. $\dfrac{75}{-25}$ 12. $\dfrac{80}{-4}$ 13. $\dfrac{12}{6}$ 14. $\dfrac{24}{8}$ 15. $\dfrac{36}{9}$

16. $\dfrac{22}{11}$ 17. $\dfrac{-39}{13}$ 18. $\dfrac{27}{-9}$ 19. $\dfrac{32}{-4}$ 20. $\dfrac{23}{-23}$

21. $\dfrac{-34}{-17}$ 22. $\dfrac{-60}{-15}$ 23. $\dfrac{-8}{-8}$ 24. $\dfrac{26}{-13}$ 25. $\dfrac{-31}{0}$

26. $\dfrac{17}{0}$ 27. $\dfrac{0}{-20}$ 28. $\dfrac{0}{-16}$ 29. $\dfrac{35}{0}$ 30. $\dfrac{0}{25}$

Fill in the blank with the correct term: positive, negative, 0, undefined.

31. The product of two negative integers is _____.

32. The quotient of two negative integers is _____.

33. The quotient of two positive integers is _____.

34. The product of two positive integers is _____.

35. The quotient of a positive integer and a negative integer is _____.

36. The product of three negative integers is _____.

37. If x is an integer, then $0 \cdot x =$ _____.

38. If x is an integer and $x \neq 0$, then $\dfrac{0}{x} =$ _____.

39. If x is an integer, then $\dfrac{x}{0} =$ _____.

40. If x is a negative integer, then x^2 is _____.

8.7 NEGATIVE RATIONAL NUMBERS

So far in Chapter 8, we have worked only with integers. Whenever we have added, subtracted, multiplied, or divided, the result has always been an integer. We can also have negative decimal numbers and negative rational numbers as well as other kinds of negative radical numbers, such as $-\sqrt{2}$ and $-\sqrt{3}$, that will be discussed in Chapter 10.

We have already learned how to divide integers such as $\dfrac{-10}{-5} = +2$.

But the quotient can be a rational number that is not an integer. For example,

$$\frac{-10}{-12} = \frac{5}{6} \quad \text{and} \quad \frac{-18}{+10} = -\frac{9}{5}$$

In general, if a and b are positive integers,

$$\frac{-a}{+b} = \frac{+a}{-b} = -\frac{a}{b} \quad \text{and} \quad \frac{-a}{-b} = \frac{+a}{+b} = +\frac{a}{b}$$

The rules for **order of operations** apply to positive and negative rational numbers just as they do to whole numbers as discussed in Section 2.2. The rules are restated here for your convenience.

RULES FOR ORDER OF OPERATIONS

1. First, simplify expressions within parentheses.

2. Second, find any powers indicated by exponents.

3. Third, moving from left to right, perform any multiplications or divisions in the order they appear.

4. Fourth, moving from left to right, perform any additions or subtractions in the order they appear.

EXAMPLES

Simplify the following expressions using the rules for order of operations and your knowledge of negative numbers.

1. $14.6 + 3(-2.5)$

$$14.6 + 3(-2.5) = 14.6 + (-7.5)$$
$$= 7.1$$

2. $\dfrac{3}{4} - \dfrac{7}{8}$

$$\frac{3}{4} - \frac{7}{8} = \frac{6}{8} + \left(\frac{-7}{8}\right)$$

$$= \frac{6-7}{8} = \frac{-1}{8} = -\frac{1}{8}$$

3. $(-2.3)^2 + (-1.7)^2$

$$(-2.3)^2 + (-1.7)^2 = 5.29 + 2.89 = 8.18$$

4. $\left(2\dfrac{1}{2}\right) \div \left(-3\dfrac{3}{4}\right) + \left(-1\dfrac{5}{8}\right)$

$$\left(2\frac{1}{2}\right) \div \left(-3\frac{3}{4}\right) + \left(-1\frac{5}{8}\right) = \left(\frac{5}{2}\right) \div \left(-\frac{15}{4}\right) + \left(-\frac{13}{8}\right)$$

$$= \frac{5}{2}\left(-\frac{4}{15}\right) + \left(\frac{-13}{8}\right)$$

$$= \frac{-2}{3} + \frac{-13}{8}$$

$$= \frac{-16}{24} + \frac{-39}{24} = -\frac{55}{24} = -2\frac{7}{24}$$

EXERCISES 8.7

Simplify the following expressions using the rules for order of operations and your knowledge of negative numbers.

1. $21.3 + (7)(-2.8)$

2. $-8.14 + (15)(-3.1)$

3. $4 - (0.017)(-5.3)$

4. $0.0423 - 8.8$

5. $(12.3)(-0.15)$

6. $(-7.3)(-0.26)$

7. $(3.366) \div (-0.612)$

8. $(-79.53) \div (-24.1)$

9. $\dfrac{-27.09}{-0.0301}$

10. $\dfrac{-0.329}{0.94}$

11. $(2.1)(5) - (-7)(-1.3)$

12. $\left(\dfrac{-2}{3}\right)\left(\dfrac{7}{-9}\right)$

13. $\dfrac{-5}{6} + \dfrac{-3}{8}$

14. $\dfrac{1}{6} - \dfrac{3}{8}$

15. $-\dfrac{7}{8} - \dfrac{5}{6}$

16. $\left(\dfrac{-1}{2} + \dfrac{3}{4}\right)\left(\dfrac{4}{7}\right)$

17. $\left(-11\dfrac{2}{3}\right)\left(4\dfrac{3}{5}\right)$

18. $\left(7\dfrac{3}{4}\right)\left(-13\dfrac{2}{3}\right)$

19. $\left(2\dfrac{3}{4}\right) \div \left(-7\dfrac{1}{2}\right)$

20. $\left(-6\dfrac{4}{5}\right) \div \left(-10\dfrac{2}{3}\right)$

21. $(-9)\left(8 - 1\dfrac{2}{3}\right)$

22. $(7.2 - 4.2)\left(-3\dfrac{1}{2}\right)$

23. $\left(7.5 - 7\dfrac{1}{2}\right)(6.1)$

24. $(2.3 - 5.03)\left(\dfrac{3}{4} - 0.75\right)$

Evaluate the following expressions (remember the rules for the order of operations).

25. $4 \cdot 3 - 5 \cdot 7$

26. $18 \cdot 2 + \dfrac{6}{-2}$

27. $\dfrac{15}{-3} + 4(-8)$

28. $\dfrac{-18}{-9} + \dfrac{-35}{+7} + \dfrac{-10}{-2}$

29. $16 \div (-4) \cdot 2$

30. $(+12)(-6) \div 3 \cdot 2$

31. $-5(6 + 3) - \dfrac{12}{-6}$

32. $(16 - 25)(32 - 21)$

33. $8 \div 4 \cdot 3 - 2 - 16$

34. $15 \div (-3) - 2(-8)$

35. $(-5)^2 + (-5)^3$

36. $\dfrac{8}{4} \cdot 2 - 8 \cdot \dfrac{4}{2}$

37. $6^2 - 4 \cdot 2 + 5(7 + 3^2)$

38. $\dfrac{5^2}{5 \cdot 5}$

39. $(6 \cdot 10 - 5) \div 11 \cdot 5 + 7 - 3^2$

40. $8 + (-4)(4) \div 8 - 16 \div 4 \div 2^2$

SUMMARY: CHAPTER 8

A shaded dot on a number line is the **graph** of the number corresponding to that point, and the number is the **coordinate** of the point.

The whole numbers and their opposites form the set of **integers.**

Integers	$\ldots, -3, -2, -1, 0, 1, 2, 3, \ldots$
Positive integers	$1, 2, 3, 4, 5, \ldots$
Negative integers	$\ldots, -3, -2, -1$

The number 0 is neither positive nor negative.

The **absolute value,** symbolized $| \ |$, of a number is its distance from 0 on a number line, regardless of the direction.

RULES FOR ADDING INTEGERS

1. To add two integers with like signs, add their absolute values and use the common sign.

2. To add two integers with unlike signs, subtract their absolute values (smaller from larger) and use the sign of the number with the larger absolute value.

DEFINITION

The **difference** of two integers a and b is the sum of a and the opposite of b. Symbolically, $(a) - (b) = (a) + (-b)$.

If a and b are positive integers,

1. $a \cdot b = ab$

2. $a(-b) = -ab$

3. $(-a)(-b) = ab$

Or, in words,

---

1. The product of two positive integers is positive.

2. The product of a positive integer and a negative integer is negative.

3. The product of two negative integers is positive.

DEFINITION

If a and b are integers and $b \neq 0$,

$$\frac{a}{b} = x \quad \text{means} \quad a = b \cdot x$$

If a and b are positive integers,

1. $\dfrac{+a}{+b} = +\dfrac{a}{b}$

2. $\dfrac{-a}{+b} = -\dfrac{a}{b}$

3. $\dfrac{+a}{-b} = -\dfrac{a}{b}$

4. $\dfrac{-a}{-b} = +\dfrac{a}{b}$

If a and b are positive integers,

$$\dfrac{-a}{+b} = \dfrac{+a}{-b} = -\dfrac{a}{b} \quad \text{and} \quad \dfrac{-a}{-b} = \dfrac{+a}{+b} = +\dfrac{a}{b}$$

The rules for **order of operations** apply to positive and negative rational numbers just as they do to whole numbers.

REVIEW QUESTIONS: CHAPTER 8

Draw a number line and graph each set of integers.

1. $\{-5, -4, 0\}$

2. $\{-1, 1, 3, 4\}$

Perform the indicated operations and simplify.

3. $10 + (-3)$

4. $(14) + (5)$

5. $(-51) + (-5)$

6. $(-2) + (+8)$

7. $(6) + (7) + (-4)$

8. $(17) - (-2)$

9. $(4) - (5)$

10. $(-18) - (-15)$

11. $(-4) - (-4)$

12. $(72) + (-72)$

13. $(-41) - (41)$

14. $7 + 4$

15. $16 - 5$

16. $36 - 2 + 3$

17. $-15 - 8$

18. $-9 - 7 + 16$

19. $35 - 10 - 8 - 11$

20. $76 + 4 + 30 - 100$

Find the value of each of the following absolute values.

21. $|-6|$

22. $|-4 + 7 - 8|$

23. $|5 - 2 - 8 + 7|$

Add.

24.
$$\begin{array}{r} -50 \\ +65 \\ -13 \\ \hline \end{array}$$

25.
$$\begin{array}{r} -98 \\ -46 \\ -5 \\ +14 \\ \hline \end{array}$$

26.
$$\begin{array}{r} 250 \\ -136 \\ -192 \\ \hline \end{array}$$

Simplify each of the following expressions.

27. $7(-9)$ **28.** $(-4)(-10)$ **29.** $(-3)(-1)(-1)$

30. $(-5)(-6)(-2)$ **31.** $(-4)^2$ **32.** $(-5)^3$

33. $(-2)(-3)(-4)(-5)$ **34.** $(|-3|)^3$ **35.** $\dfrac{51}{-17}$

36. $\dfrac{-62}{-31}$ **37.** $\dfrac{-18}{3}$ **38.** $\dfrac{75}{-15}$

39. $(-1.4)^2 + (-1.3)^2$ **40.** $\left(3\dfrac{1}{2}\right) \div \left(-2\dfrac{1}{4}\right) + 1\dfrac{1}{8}$

41. $(7.4 - 9.3)\left(\dfrac{1}{4} - 0.25\right)$ **42.** $5^2 - 10 \cdot 3 + 4(8 - 3^3)$

Determine whether each of the following is true or false. If a statement is false, give an example to show why.

43. The sum of two positive numbers must be a positive number.

44. The sum of two negative numbers must be a negative number.

45. The difference of two positive numbers must be a positive number.

46. The difference of two negative numbers must be a negative number.

47. The sum of a positive and a negative number must be positive.

48. The sum of a positive and a negative number must be negative.

49. The difference of a positive number and its opposite must be 0.

50. The sum of a positive number and its opposite must be 0.

51. The product of a positive number and its opposite must be negative.

52. The quotient of two negative numbers must be negative.

CHAPTER TEST: CHAPTER 8

1. Graph the set of integers $\{-3, -2, 0, 1\}$ on a number line.

Simplify each of the following expressions.

2. $8 + (-2)$

3. $(16) + (9)$

4. $(-26) + (-3)$

5. $(-4) + (+12)$

6. $(8) + (9) + (-7)$

7. $(20) - (-1)$

8. $(6) - (7)$

9. $(-17) - (-21)$

10. $(-32) - (-32)$

11. $(-10) - (10)$

12. $13 - 14$

13. $96 - 5 + 2$

14. $22 - 15 - 4$

15. $-18 - 3 - 25$

16. $-80 - 90 - 50$

17. $8(-10)$

18. $(-13)(-5)$

19. $(-1)^3(-5)$

20. $(-3)^2(-7)(4)$

21. $\dfrac{56}{-4}$

22. $\dfrac{-85}{5}$

23. $\dfrac{-36}{-18}$

24. $(-2.1)^2 + (-1.2)^2$

25. $\left(-9\dfrac{1}{2}\right) \div \left(2\dfrac{1}{4}\right) - 1\dfrac{5}{8}$

26. $(5.8 - 6.4)\left(\dfrac{3}{4} - 0.5\right)$

27. $(-8)^2 - 5 \cdot 4 \div 2 + 3(5 - 4^2)$

28. $|-20 + 6 - 9 + 13|$

29. True or false: The sum of a positive number and a negative number can be 0. Explain your answer.

30. True or false: The sum of two positive numbers can be 0. Explain your answer.

9
SOLVING EQUATIONS

9.1 COMBINING LIKE TERMS AND EVALUATING EXPRESSIONS

Any one number is called a **constant.** A **variable** is a symbol or letter that can represent more than one number. A number written next to a letter (as in $5x$) or two variables written next to each other (as in ab) indicates multiplication. In $5x$, the constant 5 is called the **coefficient** of the variable x.

An expression that involves only multiplications and/or divisions with constants and/or variables is called a **term.** Examples of terms are

$$5x, \quad -6y, \quad 13x^2, \quad \frac{-15x}{3a}, \quad \text{and} \quad -42$$

Like terms (or **similar terms**) can be constants, or they can be terms that contain variables that are of the same power in each term. Thus,

7, 32, -5, and -13 are like terms
$3x$, $5x$, $-4x$, and $12x$ are like terms
$-8y^2$, $2y^2$, $17y^2$, and $-10y^2$ are like terms
25, $-7x$, and $4x^2$ are **not** like terms
$-9x^2$ and $13y^2$ are **not** like terms

To simplify expressions that contain like terms, we want to **combine like terms.** For example,

$$5x + 3x = 8x$$
$$-10x - 2x = -12x$$

and
$$4a - 5a + 7a = 6a$$

An explanation of how to combine like terms involves the **distributive property** as applied to integers. (The distributive property was introduced in Section 1.6.)

DISTRIBUTIVE PROPERTY OF MULTIPLICATION OVER ADDITION

If a, b, and c are integers, then

$$a(b + c) = ab + ac$$

The form of the distributive property most useful for our purposes is

$$ba + ca = (b + c)a$$

Using this form, we combine like terms as follows:

$$5x + 3x = (5 + 3)x = 8x$$
$$-10x - 2x = (-10 - 2)x = -12x$$
and $$4a - 5a + 7a = (4 - 5 + 7)a = 6a$$

EXAMPLES

In the following expressions, combine like terms whenever possible.

1. $2y - 3y$
   $2y - 3y = (2 - 3)y = -1y$ or $-y$ [NOTE: In the term $-y$, the coefficient is understood to be -1.]

2. $-4x + 5x + 3x$
   $-4x + 5x + 3x = (-4 + 5 + 3)x = 4x$

3. $7a + a + 2 + 3$
   $7a + a + 2 + 3 = (7 + 1)a + 5$ [NOTE: In the term a, the coefficient is understood to be $+1$.]
   $= 8a + 5$

4. $-6a + 4b - 8$
   $-6a + 4b - 8$ has no like terms

With some practice, you should be able to combine like terms mentally without writing down the step involving the distributive property.

After simplifying an expression, we may want to evaluate that expression for one or more values of the variables. This involves applying the rules for order of operations as stated in Section 2.2. The rules are restated here for convenience.

RULES FOR ORDER OF OPERATIONS

1. First, simplify expressions within parentheses.

2. Second, find any powers indicated by exponents.

3. Third, moving from left to right, perform any multiplications or divisions in the order they appear.

4. Fourth, moving from left to right, perform any additions or subtractions in the order they appear.

To evaluate an expression, first combine like terms, then substitute the desired value(s) for the variable(s) and follow the rules for order of operations.

EXAMPLES

Evaluate each of the following expressions for the given values of the variables.

1. $x - 5$ for $x = 4$ and $x = -3$

Solution:
For $x = 4$, $x - 5 = 4 - 5 = -1$
For $x = -3$, $x - 5 = -3 - 5 = -8$

2. $3x + 2$ for $x = 4$ and $x = -3$

Solution:
For $x = 4$, $3x + 2 = 3 \cdot 4 + 2 = 12 + 2 = 14$
For $x = -3$, $3x + 2 = 3(-3) + 2 = -9 + 2 = -7$

3. $-2y - 6y - 1$ for $y = -2$ and $y = 0$

Solution:
Simplify first:

$$-2y - 6y - 1 = -8y - 1$$

For $y = -2$, $-8y - 1 = -8(-2) - 1 = 16 - 1 = 15$
For $y = 0$, $-8y - 1 = -8(0) - 1 = 0 - 1 = -1$

PRACTICE QUIZ	Simplify each expression by combining like terms.	ANSWERS
	1. $5x - 6x$	**1.** $-1x$ or $-x$
	2. $3y + 4y - 2y + 6 - 8$	**2.** $5y - 2$
	Evaluate the following expression for $x = -3$.	
	3. $-7x - 14$	**3.** 7

EXERCISES 9.1

Simplify the following expressions by combining like terms whenever possible.

1. $6x + 2x$	**2.** $4x - 3x$	**3.** $5x + x$
4. $7x - 3x$	**5.** $-10a + 3a$	**6.** $-11y + 4y$
7. $-18y + 6y$	**8.** $-2x - 5x$	**9.** $-5x - 4x$
10. $-x - 2x$	**11.** $-7x - x$	**12.** $2x - 2x$
13. $5x - 5x$	**14.** $16p - 17p$	**15.** $9c - 10c$
16. $3x - 5x + 12x$	**17.** $2a + 14a - 25a$	

18. $6c - 13c + 5c$ **19.** $40p - 30p - 10p$

20. $16x - 15x - 3x$ **21.** $2x + 3x - 7$

22. $5x - 6x + 2$ **23.** $7x - 8x + 5$

24. $-5x - 7x - 4$ **25.** $-8a - 3a - 2$

26. $-4x + x + 1 - 3$ **27.** $-2x + 5x + 6 - 5$

28. $4x + 7 - 8 + 3x$ **29.** $-5x - 1 + 8 + 9x$

30. $10y + 3 - 4 - 6y$

Evaluate each of the following expressions for $x = -3$, $y = 2$, $z = 3$, $a = -1$, and $c = -2$.

31. $x - 2$ **32.** $y - 2$ **33.** $z - 3$ **34.** $2x + z$

35. $3y - x$ **36.** $x - 4z$ **37.** $20 - 2a$ **38.** $10 + 2c$

39. $3c - 5$ **40.** $2x + 3x - 7$ **41.** $7a - a + 3$

42. $-3y - 4y + 6 - 2$ **43.** $-2x - 3x + 1 - 4$

44. $5y - 2y - 3y + 4$ **45.** $2x - 3x + x - 8$

9.2   WRITING EXPRESSIONS

If a word problem is stated in English, we need to be able to translate the English phrases into algebraic expressions. Then the problem can be solved by relating the expressions to each other according to the relationships stated in the problem and applying the rules of algebra.

Observing certain key words in a phrase helps in translating that phrase into its algebraic equivalent. The following examples illustrate how to translate some of these key words into algebraic symbols. The key words are in boldface type.

EXAMPLES

English Phrase	Algebraic Expression

1. 7 **multiplied by** the variable x
  the **product** of 7 and x
  7 **times** x
	$7x$

2. 5 **added to** the unknown y
  the **sum** of 5 and y
  5 **plus** y
	$5 + y$

3. 8 **subtracted from** a number **times** 6 ⎫
   the **difference** between $6x$ and 8   ⎬ $6x - 8$
   $6x$ **minus** 8   ⎭

4. **twice** a number **plus** 3 ⎫
   3 **added to** 2 **times** a number ⎬ $2x + 3$
   $2x$ **increased by** 3 ⎭

5. three **times** the **quantity** found by ⎫
      adding 1 to a number   ⎬
   the **product** of 3 with the **quantity** $x + 1$ ⎬ $3(x + 1)$
   three **times** the **sum** of a number and 1 ⎭

Always read word problems carefully and look for the key words. Some of these important words and their meanings are provided in the following list.

ADDITION	SUBTRACTION	MULTIPLICATION	DIVISION
add	subtract	multiply	divide
sum	difference	product	quotient
plus	minus	times	
more than	less than	twice	
increased by	decreased by	of	

EXERCISES 9.2

Write the following English phrases as algebraic expressions. Use any letter as the unknown number.

1. 5 more than a number   2. 6 added to a number

3. a number plus 10   4. a number increased by 1

5. 8 less than a number   6. a number decreased by 4

7. 14 subtracted from a number   8. a number minus 3

9. the sum of a number and 11

10. the difference between 6 and a number

11. the product of 2 and a number   12. 3 multiplied by a number

13. the quotient of a number and -7

14. -18 divided by a number

15. 4 times a number minus 3   16. -2 times a number plus 17

17. 4 times the difference between a number and 3

18. 5 times the quantity $x - 7$

19. -2 times the quantity $x + 5$

20. 10 minus twice a certain number

21. 13 plus twice a certain number

22. a certain number increased by 3 times that number

23. 6 times the sum of a number and 1

24. the difference of 9 and twice a number

25. the sum of -3 and 3 times a number

26. 5 more than 8 times a number

27. the product of 9 with the quantity $x - 3$

28. -4 times the quantity found by decreasing a number by 2

29. 3 less than the product of a number and 7

30. 16 increased by twice a number

Translate each of the following expressions into an English phrase. (There may be more than one correct answer.)

31. $x + 6$ **32.** $x - 7$ **33.** $5x$ **34.** $\dfrac{x}{2}$

35. $\dfrac{5}{x}$ **36.** $12 + x$ **37.** $20 - x$ **38.** $2x + 5$

39. $2x - 3$ **40.** $3 + 4x$ **41.** $5 - 3x$ **42.** $10(x + 3)$

43. $-4(x - 7)$ **44.** $5(x - 8)$ **45.** $3(x - 11)$

Answer each of the following questions by writing an algebraic expression using the given variable.

46. If you pay $17 for an item and it goes up d dollars in price, what is the new price?

47. If Jim is 4 years older than Sue, and Sue is y years old, how old is Jim?

48. If Jim is 4 years older than Sue, and Jim is x years old, how old is Sue?

49. A person bowled a total of T points in three games. What was the person's average score?

50. How many minutes are in h hours?

51. The length of a rectangle is 7 meters greater than its width. If w is the width, what is the length?

52. If d dollars are invested at 16% for 6 months, what is the interest earned?

53. If an item originally sold for p dollars, what was the selling price after a 25% discount?

54. If Virginia drove her car d kilometers in 3.2 hours, what was her average speed?

55. How far can you drive in 2.5 hours at an average rate of r kilometers per hour?

9.3 SOLVING EQUATIONS I

The equation $6 + 7 = 13$ is **true,** and the equation $5 + 10 = 12$ is **false.** If an equation contains a variable, such as $2x + 4 = 10$, we want to find the value (or values) for the variable that will give a true statement when substituted for the variable. This procedure is called **solving the equation,** and the value (or values) found is called the **solution** (or solutions) of the equation.

Two equations that have exactly the same solutions are said to be **equivalent.** Thus,

$$2x + 4 = 10$$
and $$3x - 1 = 8$$

are equivalent because they both have the solution $x = 3$. Substituting $x = 3$ into both equations,

$$
\begin{array}{ll}
2x + 4 = 10 & 3x - 1 = 8 \\
2 \cdot 3 + 4 \overset{?}{=} 8 & 3 \cdot 3 - 1 \overset{?}{=} 8 \\
6 + 4 \overset{?}{=} 10 & 9 - 1 \overset{?}{=} 8 \\
10 = 10 & 8 = 8
\end{array}
$$

This verifies that $x = 3$ is indeed the solution for both equations.

Now we need to know how to find the solution, such as $x = 3$, to an equation. Solving an equation in a step-by-step manner involves the following two basic ideas.

1. Whatever is done to one side of the equation must be done to the other side. (This does not include simplifying expressions.)

2. The object is to find a very simple equation, such as $x = 3$, that is equivalent to the original equation.

The following examples illustrate how an equation can be solved in a step-by-step manner. Note that **equivalent equations are written one under the other.** Do not write equations side by side. Study the examples carefully so that you can follow the same procedure in the exercises. [NOTE: While some equations studied in algebra have more than one solution, the equations studied in this chapter will have only one solution.]

EXAMPLES

1. $x + 4 = 10$ Write the equation.

    $x + 4 - 4 = 10 - 4$ Add -4 to both sides. (-4 is the opposite of $+4$.)

        $x = 6$ Simplify.

2. $x - 15 = 3$ Write the equation.

    $x - 15 + 15 = 3 + 15$ Add 15 to both sides. (15 is the opposite of -15.)

        $x = 18$ Simplify.

3. $6y = -42$ Write the equation.

   $\dfrac{6y}{6} = \dfrac{-42}{6}$ Divide both sides by 6, the coefficient of the variable.

    $y = -7$ Simplify.

4. $-7 = x + 13$ Write the equation.

  $-7 - 13 = x + 13 - 13$ Add -13 to both sides.

    $-20 = x$ Simplify. Note that the variable can be on the right side.

Each solution may be checked by substituting the solution into the original equation. If there are no errors, each substitution will give a true statement.

Check:

1. $x + 4 = 10$ 2. $x - 15 = 3$
   $6 + 4 \overset{?}{=} 10$ $18 - 15 \overset{?}{=} 3$
     $10 = 10$ $3 = 3$

3. $6y = -42$ 4. $-7 = x + 13$
 $6(-7) \overset{?}{=} -42$ $-7 \overset{?}{=} -20 + 13$
   $-42 = -42$ $-7 = -7$

The equations in this section are easily solved in one step. Equations that require several steps to find their solutions will be discussed in Section 9.4.

In summary,

1. If a constant is added to a variable, add its opposite to both sides of the equation.

2. If a constant is multiplied by a variable, divide both sides by that constant.

3. Remember that the object is to isolate the variable on one side of the equation, right side or left side.

PRACTICE QUIZ

Solve the following equations.

1. $x + 10 = 3$

2. $y - 4 = 20$

3. $-5x = -45$

ANSWERS

1. $x = -7$

2. $y = 24$

3. $x = 9$

EXERCISES 9.3

Solve the following equations.

1. $x + 4 = 10$	**2.** $x + 13 = 20$	**3.** $y - 5 = 17$
4. $y - 12 = 4$	**5.** $y + 8 = 3$	**6.** $x + 10 = 7$
7. $x - 5 = -7$	**8.** $x - 14 = -10$	**9.** $y - 8 = -6$
10. $x - 12 = -5$	**11.** $5x = 30$	**12.** $3y = 15$
13. $10y = -40$	**14.** $6x = -48$	**15.** $-2x = 12$
16. $-4x = 24$	**17.** $-8y = -40$	**18.** $-12y = -36$
19. $16 = x + 3$	**20.** $25 = x + 14$	**21.** $10 = x - 13$
22. $4 = x - 15$	**23.** $-18 = y + 4$	**24.** $-20 = y + 5$
25. $-13 = y - 20$	**26.** $-11 = y - 18$	**27.** $10 = -2x$
28. $40 = -8x$	**29.** $-50 = -10y$	**30.** $-60 = -6y$

9.4   SOLVING EQUATIONS II

Solving equations can involve several steps instead of just one step as was shown in Section 9.3. We may need to simplify one side or the other by combining like terms; there may be variables on both sides of the equation; we may need to use the distributive property.

The following examples illustrate how to solve equations involving these and other situations. In all cases, after both sides have been simplified, **the object of the procedures is to get the variable terms on one side and the constant terms on the other side.** Be sure to write equivalent equations one under the other.

EXAMPLES

1.

$$4x + 3 = 11$$ Write the equation.

$$4x + 3 - 3 = 11 - 3$$ Add -3 to both sides.

$$4x = 8$$ Simplify.

$$\frac{4x}{4} = \frac{8}{4}$$ Divide both sides by 4, the coefficient of x.

$$x = 2$$ Simplify.

2. $$2x - 4 + 3x = 26$$ Write the equation.

$$5x - 4 = 26$$ Combine like terms on the left side.

$$5x - 4 + 4 = 26 + 4$$ Add $+4$ to both sides.

$$5x = 30$$ Simplify.

$$\frac{5x}{5} = \frac{30}{5}$$ Divide both sides by 5.

$$x = 6$$ Simplify.

3.

$$5x + 2 = 3x - 8$$ Write the equation.

$$5x + 2 - 2 = 3x - 8 - 2$$ Add -2 to both sides.

$$5x = 3x - 10$$ Simplify.

$$5x - 3x = 3x - 10 - 3x$$ Add $-3x$ to both sides.

$$2x = -10$$ Simplify; now one side has the term with variable, and the other side has the term with constant.

$$\frac{2x}{2} = \frac{-10}{2}$$ Divide both sides by 2.

$$x = -5$$ Simplify.

4.

$$3(x + 2) = 18 - x$$ Write the equation.

$$3x + 6 = 18 - x$$ Use the distributive property.

$$3x + 6 - 6 = 18 - x - 6$$ Add -6 to both sides.

$$3x = 12 - x$$ Simplify.

$$3x + x = 12 - x + x$$ Add $+x$ to both sides.

$$4x = 12$$ Simplify.

$$\frac{4x}{4} = \frac{12}{4}$$ Divide both sides by 4.

$$x = 3$$ Simplify.

Each of the examples can be checked by substituting the solution into the original equation. However, checking can be time-consuming and need not be done for every problem. **Particularly on an exam, checking should be done only after you have finished the entire exam.**

Check:

2.
$$2x - 4 + 3x = 26$$
$$2(6) - 4 + 3(6) \stackrel{?}{=} 26$$
$$12 - 4 + 18 \stackrel{?}{=} 26$$
$$26 = 26$$

3.
$$5x + 2 = 3x - 8$$
$$5(-5) + 2 \stackrel{?}{=} 3(-5) - 8$$
$$-25 + 2 \stackrel{?}{=} -15 - 8$$
$$-23 = -23$$

If there are fractions in an equation, multiply each term on both sides of the equation by the LCM (least common multiple) of all the denominators. Then solve the equation just as before. The following example illustrates how to proceed with fractions.

EXAMPLE

$$\frac{1}{2}x + \frac{3}{4}x = \frac{1}{6}x - 26$$

$$12\left(\frac{1}{2}x\right) + 12\left(\frac{3}{4}x\right) = 12\left(\frac{1}{6}x\right) - 12(26) \qquad \text{(12 is the LCM of 2, 4, 6)}$$

$$6x + 9x = 2x - 312$$
$$15x - 2x = 2x - 312 - 2x$$
$$13x = -312$$
$$\frac{13x}{13} = \frac{-312}{13}$$
$$x = -24$$

PRACTICE QUIZ	Solve the following equations.	ANSWERS
	1. $10x + 4 = 14$	1. $x = 1$
	2. $x + 5 - 2x = 7 + x$	2. $x = -1$
	3. $2(x - 7) = 5(x + 2)$	3. $x = -8$
	4. $\frac{2}{3}x = \frac{1}{2}x + 1$	4. $x = 6$

EXERCISES 9.4

Solve the following equations.

1. $2x + 3 = 5$ 2. $3x - 4 = 8$ 3. $4y + 1 = 9$

4. $3x - 10 = 11$

5. $6x + 4 = -14$

6. $7y - 8 = -1$

7. $3 + 6y = 15$

8. $6 + 5y = 21$

9. $2x + 3 = -9$

10. $3x - 1 = -4$

11. $5y + 12 = -3$

12. $10y + 3 = -17$

13. $15 = 2x - 3$

14. $20 = 3x - 1$

15. $-17 = 5y - 2$

16. $30 = 4y + 6$

17. $4 = 5x + 9$

18. $28 = 10x - 2$

19. $-24 = 7x - 3$

20. $96 = 25y - 4$

21. $3x = x - 10$

22. $5y = 2y + 12$

23. $7y = 6y + 5$

24. $6x = 2x + 20$

25. $5x = 2x$

26. $4x = 3x$

27. $4x + 3 = 2x + 9$

28. $5y - 2 = 4y - 6$

29. $7x + 14 = 10x + 5$

30. $5x + 20 = 8x - 4$

31. $5(x - 2) = 3(x - 8)$

32. $2(y + 1) = 3y + 3$

33. $4(x - 1) = 2x + 6$

34. $6y - 3 = 3(y + 2)$

35. $7y - 6y + 12 = 4y$

36. $6x + 5 + 3x = 3x - 13$

37. $5x - 2x + 4 = 3x + x - 1$

38. $7x + x - 6 = 2(x + 9)$

39. $x - 5 + 4x = 4(x - 3)$

40. $3(x + 6) = 3x + 2(x + 1)$

41. $\frac{1}{2}x + \frac{3}{4}x = -15$

42. $\frac{2}{3}y - 5 = \frac{1}{3}y + 20$

43. $\frac{3}{5}x - 4 = \frac{1}{5}x - 5$

44. $\frac{5}{8}y - \frac{1}{4} = \frac{2}{5}y + \frac{1}{3}$

45. $\frac{x}{8} + \frac{1}{6} = \frac{x}{10} + 2$

9.5 SOLVING WORD PROBLEMS

In this section, we will restrict the discussion to solving word problems that relate to number concepts and use the phrases discussed in Section 9.2. Word problems that relate to such fields as physics, chemistry, engineering, geometry, art, business, computer science, and so on will be part of later courses in mathematics.

Three number concepts of interest are **consecutive integers, consecutive even integers,** and **consecutive odd integers.** The numbers 17, 18, and 19 are consecutive integers because, when written in order, each integer is **one more** than the previous integer.

Symbolically, for three consecutive integers,

if $\qquad n = $ first integer

then $\qquad n + 1 = $ second integer

and $\qquad n + 2 = $ third integer

For example,

if	$n = 17$
then	$n + 1 = 18$
and	$n + 2 = 19$

Similarly,

if	$n = -10$
then	$n + 1 = -9$
and	$n + 2 = -8$

The numbers 30, 32, 34, and 36 are **consecutive even integers** because all are even numbers and when they are written in order, each integer is **two more** than the previous integer. Thus, four consecutive even integers can be represented as n, $n + 2$, $n + 4$, and $n + 6$.
So,

if	$n = 30$
then	$n + 2 = 32$
	$n + 4 = 34$
and	$n + 6 = 36$

The numbers, 7, 9, 11, and 13 are **consecutive odd integers** because all are odd and when they are written in order, each integer is **two more** than the previous integer. Thus, four consecutive odd integers can be represented in exactly the same way as consecutive even integers: n, $n + 2$, $n + 4$, and $n + 6$.
So,

if	$n = 7$
then	$n + 2 = 9$
	$n + 4 = 11$
and	$n + 6 = 13$

[NOTE: You do not know whether n, $n + 2$, $n + 4$, and $n + 6$ are four consecutive even integers or four consecutive odd integers until you know the value of n.]

Follow the steps listed here to solve any type of word problem.

STEPS IN SOLVING WORD PROBLEMS

1. Read the problem carefully. Read the problem a second time.

2. Decide what is unknown and represent it with a letter.

3. Translate the English phrases into mathematical phrases and form an equation indicated by the sentence.

4. Solve the equation.

5. Check to see that the solution of the equation makes sense in the problem.

EXAMPLES

1. Five times a number is increased by 3, and the result is 38. What is the number?

Solution:
Let n = the number
Translate "five times a number is increased by 3" to "$5n + 3$."
Translate "the result is" to "$=$."
The equation to be solved is

$$5n + 3 = 38$$
$$5n + 3 - 3 = 38 - 3$$
$$5n = 35$$
$$\frac{5n}{5} = \frac{35}{5}$$
$$n = 7$$

Check:
5 times 7 increased by 3 is $5 \cdot 7 + 3$, and $5 \cdot 7 + 3 = 35 + 3 = 38$.

2. The sum of three consecutive odd integers is -57. What are the integers?

Solution:
Let n = first odd integer
$n + 2$ = second consecutive odd integer
$n + 4$ = third consecutive odd integer
The equation to be solved is

$$n + (n + 2) + (n + 4) = -57$$
$$n + n + 2 + n + 4 = -57$$
$$3n + 6 = -57$$
$$3n + 6 - 6 = -57 - 6$$

$$3n = -63$$

$$\frac{3n}{3} = \frac{-63}{3}$$

$$n = -21$$

$$n + 2 = -19$$

$$n + 4 = -17$$

Check: $(-21) + (-19) + (-17) = -57$

3.   Seven less than four times a number is equal to twice the number increased by five. Find the number.

Solution:
Let n = the number
Translate "seven less than four times a number" to $4n - 7$.
Translate "twice the number increased by five" to $2n + 5$.
The equation to be solved is

$$4n - 7 = 2n + 5$$

$$4n - 7 - 2n = 2n + 5 - 2n$$

$$2n - 7 = +5$$

$$2n - 7 + 7 = 5 + 7$$

$$2n = 12$$

$$\frac{2n}{2} = \frac{12}{2}$$

$$n = 6$$

Check: $4(6) - 7 \overset{?}{=} 2(6) + 5$

$$24 - 7 \overset{?}{=} 12 + 5$$

$$17 = 17$$

EXERCISES 9.5

Solve each of the following problems.

1.   Find a number whose product with 3 is 57.

2.   Find a number that when multiplied by 7 gives 84.

3.   The sum of a number and 32 is 86. What is the number?

4.   The difference of a number and 16 is -48. What is the number?

5.   If the product of a number and 4 is decreased by 10, the result is 50. Find the number.

6. If the product of a number and 5 is added to 12, the result is 7. Find the number.

7. If the product of a number and 8 is increased by 24, the result is twice the number. What is the number?

8. The sum of a number and 2 is equal to three times the number. What is the number?

9. If twice a number is decreased by 4, the result is the number. Find the number.

10. Three times the sum of a number and 4 is -60. What is the number?

11. Twice a number plus 5 is equal to 20 more than the number. What is the number?

12. If 7 is subtracted from a number, the result is 8 times the number. Find the number.

13. Twenty plus a number is equal to twice the number plus three times the same number. What is the number?

14. The sum of two consecutive integers is 37. Find the two integers.

15. The sum of three consecutive integers is -42. Find the three integers.

16. The sum of three consecutive odd integers is 27. Find the three integers.

17. Find three consecutive odd integers whose sum is 81.

18. Find three consecutive even integers whose sum is 30.

19. Find four consecutive even integers whose sum is 54 more than the smallest one.

20. If the product of 4 and the sum of a number and 3 is diminished by 6, the result is 26. Find the number.

21. What is the number whose sum with 18 is ten times the number?

22. If 7 times a number decreased by four times the number is equal to the sum of the number and 10, what is the number?

23. What is the number whose product with 6 is equal to twice the number decreased by 12?

24. The difference of a number and 3 is equal to the difference of five times the number and 15. What is the number?

25. If the sum of two consecutive integers is multiplied by 3, the result is -15. What are the two integers?

26. If the sum of two consecutive even integers is multiplied by 5, the result is 90. What are the two integers?

27. If twice a number is increased by three times that number, the result is 4 more than the number. What is the number?

28. The sum of a number, twice the number, and six times the number is equal to four times the number increased by 40. Find the number.

29. The product of a number and 5 is equal to the product of 4 with the difference of the number and 7. Find the number.

30. Six times the sum of a number and 5 is equal to four times the quantity $x + 9$ (where x represents the number). What is the number?

SUMMARY: CHAPTER 9

A **variable** is a symbol or letter that can represent more than one number. An expression that involves only multiplications and/or divisions with constants and/or variables is called a **term.**

Like terms (or **similar terms**) can be constants, or they can be terms that contain variables that are of the same power in each term. For example, $-3x$ and $7x$ are like terms, but $-3x^2$ and $7x$ are **not** like terms. Also, $5y^2$ and $4x^2$ are not like terms.

DISTRIBUTIVE PROPERTY OF MULTIPLICATION OVER ADDITION

If a, b, and c are integers, then

$$a(b + c) = ab + ac$$

KEY WORDS

ADDITION	SUBTRACTION	MULTIPLICATION	DIVISION
add	subtract	multiply	divide
sum	difference	product	quotient
plus	minus	times	
more than	less than	twice	
increased by	decreased by	of	

Two equations that have exactly the same solutions are said to be **equivalent.**

TWO BASIC IDEAS IN SOLVING EQUATIONS

1. Whatever is done to one side of the equation must be done to the other side. (This does not include simplifying expressions.)

2. The object is to find a very simple equation, such as $x = 3$, that is equivalent to the original equation.

In solving an equation, the object of the procedures is to get the variable terms on one side and the constant terms on the other side.

$n, n + 1, n + 2$ represent **consecutive integers.**

$n, n + 2, n + 4$ represent **consecutive odd integers** if n is odd.

$n, n + 2, n + 4$ represent **consecutive even integers** if n is even.

REVIEW QUESTIONS: CHAPTER 9

Simplify each expression by combining like terms.

1. $8x + 7x$ **2.** $9y + 3y$ **3.** $4x - x$

4. $-5x - x$ **5.** $13w + w$ **6.** $10y - 10y$

7. $8y - 11y + y$ **8.** $-30p + 2p - 6p$ **9.** $-7a - 2a + a + 6$

10. $5x + 13 - 7x + 1$ **11.** $18y - 10 + 3 - 4y$

12. $-2x - 8x - 2 - 8$

Evaluate each expression for $x = 5$ and $y = -2$.

13. $20 - 2y + x$ **14.** $-10 + 6x + 2y$ **15.** $7x - 3y + x - 5$

Write the following English phrases as algebraic expressions. Choose any letter you like for the unknown number.

16. 8 more than a number

17. 3 less than five times a number

18. the quotient of a number and 9

19. -3 times the quantity $x + 2$

20. the sum of 10 and four times a number

21. 18 decreased by twice a number

Translate each of the following expressions into an English phrase.

22. $-4x$ **23.** $5 + 7x$ **24.** $-2(x + 6)$

Write an algebraic expression that will answer each of the following questions.

25. How many hours are in m minutes?

26. The width of a rectangle is 3 meters less than its length. If l is the length, what is the width?

27. How far can you drive if you average r miles per hour for t hours?

Solve the following equations.

28. $5x = 40$ **29.** $13x = -39$ **30.** $-70 = -7y$

31. $y - 8 = 19$ **32.** $x + 14 = 21$ **33.** $y - 25 = -11$

34. $2(x - 1) = x + 1$ **35.** $5(x + 3) = 4x + 13$ **36.** $y = 5y - 20$

37. $8x - 3x = 14 - 9x$ **38.** $2(x - 10) = 4(16 - x)$

39. $2x + 16 = 3x + 24$ **40.** $\dfrac{1}{5}x + \dfrac{1}{3} = \dfrac{2}{3} + \dfrac{4}{15}x$

41. $\dfrac{1}{2}x - 2 = \dfrac{2}{3}x$ **42.** $\dfrac{x}{4} + \dfrac{x}{5} = 2x + \dfrac{31}{20}$

43. If the product of a number and 10 is decreased by 15, the result is -35. Find the number.

44. If 14 is added to twice a number, the result is that number minus 13. Find the number.

45. The sum of three consecutive integers is 78. Find the three integers.

46. Find three consecutive odd integers with the property that $\dfrac{1}{3}$ of the first number is equal to the sum of the other two minus 41.

47. The product of 7 and a number is equal to the product of 5 with the sum of the number and 2. What is the number?

48. What is the number whose sum with -8 is equal to 6 less than four times the number?

CHAPTER TEST: CHAPTER 9

Simplify each expression by combining like terms.

1. $9x - 10x$ **2.** $14y + 2y$ **3.** $-8x - 7x$

4. $20p - 3 - 6p$ **5.** $12x - 4 - 8x - 4x$ **6.** $-5 - 6x - 2 + 3x$

Evaluate each expression for $x = -1$ and $y = 4$.

7. $5x - 10 + 3 - x$ **8.** $-3y - 2x + 14 + 5x$

Write the following English phrases as algebraic expressions. The unknown number can be any letter.

9. 5 less than $\frac{3}{4}$ of a number

10. 6 more than the product of a number and 20

11. -3 times the quantity $x - 7$

12. 1 plus twice the quantity of a number minus 9

13. Translate the expression $6 - 4x$ into a phrase in English.

Write an algebraic expression that will answer each of the following questions.

14. The width of a rectangle is w. The length is 3 less than four times the width. What is the length?

15. The cost of a calculator has gone up 15% since you bought yours for d dollars. What is the new price?

Solve the following equations.

16. $x - 20 = 20$ **17.** $-7x = -42$

18. $3x + 14 = -10$ **19.** $81 = 21 - 5y$

20. $4(x - 7) = x - 37$ **21.** $-x - 13 = 2(-4 - x)$

22. $\frac{1}{2}x + 2 = \frac{3}{4}x - 3$ **23.** $\frac{1}{3}y = \frac{3}{5}y + 4$

24. Twice the sum of a number and 3 is equal to the number minus 10. Find the number.

25. The sum of three consecutive even integers is equal to 252. Find the integers.

26. If $\frac{1}{3}$ of a number is added to $\frac{1}{2}$ of the same number, the result is equal to 6 less than the number. Find the number.

10

REAL NUMBERS AND GRAPHING LINEAR EQUATIONS

10.1 SQUARE ROOTS

A number is squared when it is multiplied by itself. If a whole number is squared, the result is called a **perfect square.** For example, squaring 7 gives $7^2 = 49$, and 49 is a perfect square. Table 10.1 shows the perfect square numbers found by squaring the whole numbers from 1 to 20. A more complete table is located on the inside back cover of this text.

TABLE 10.1 SQUARES OF WHOLE NUMBERS FROM 1 TO 20

$1^2 = 1$	$6^2 = 36$	$11^2 = 121$	$16^2 = 256$
$2^2 = 4$	$7^2 = 49$	$12^2 = 144$	$17^2 = 289$
$3^2 = 9$	$8^2 = 64$	$13^2 = 169$	$18^2 = 324$
$4^2 = 16$	$9^2 = 81$	$14^2 = 196$	$19^2 = 361$
$5^2 = 25$	$10^2 = 100$	$15^2 = 225$	$20^2 = 400$

Since $5^2 = 25$, the **square root** of 25 is 5. The symbol for square root is $\sqrt{}$, called a **square root sign** or **radical sign.** Thus,

$$\sqrt{25} = 5 \quad \text{since} \quad 5^2 = 25$$
$$\sqrt{49} = 7 \quad \text{since} \quad 7^2 = 49$$
$$\sqrt{64} = 8 \quad \text{since} \quad 8^2 = 64$$

Table 10.2 contains the square roots of the perfect square numbers from 1 to 400. Both Table 10.1 and Table 10.2 should be memorized.

TABLE 10.2 SQUARE ROOTS OF PERFECT SQUARES FROM 1 TO 400

$\sqrt{1} = 1$	$\sqrt{36} = 6$	$\sqrt{121} = 11$	$\sqrt{256} = 16$
$\sqrt{4} = 2$	$\sqrt{49} = 7$	$\sqrt{144} = 12$	$\sqrt{289} = 17$
$\sqrt{9} = 3$	$\sqrt{64} = 8$	$\sqrt{169} = 13$	$\sqrt{324} = 18$
$\sqrt{16} = 4$	$\sqrt{81} = 9$	$\sqrt{196} = 14$	$\sqrt{361} = 19$
$\sqrt{25} = 5$	$\sqrt{100} = 10$	$\sqrt{225} = 15$	$\sqrt{400} = 20$

The square roots of some numbers are not as easily found as those in the tables. In fact, most square roots are **irrational numbers (nonrepeating infinite decimals);** that is, most square roots can only be approximated with decimals. One method for approximating square roots (other than using a calculator) is explained in Appendix IV.

Decimal approximations to $\sqrt{2}$ are shown here so that you will understand that there is no finite decimal number whose square is 2.

1.4	1.414	1.41421	1.41422
1.4	1.414	1.41421	1.41422
56	5656	141421	282844
1 4	1414	282842	282844
1.96	5656	565684	565688
	1 414	141421	141422
	1.999396	565684	565688
		1 41421	1 41422
		1.9999899241	2.0000182084

So, $\sqrt{2}$ is between 1.41421 and 1.41422.

Many numbers that are square roots ($\sqrt{}$), cube roots ($\sqrt[3]{}$), fourth roots ($\sqrt[4]{}$), and so on are infinite nonrepeating decimals and are classified as **irrational numbers.** All the numbers we have studied—rational numbers (including whole numbers, integers, and positive and negative fractions) and irrational numbers—come under the classification of **real numbers.**

Real numbers are the basis for the discussions in the remainder of Chapter 10. Real numbers and their properties, such as the commutative properties of addition and multiplication, are an important part of the courses in algebra that follow this course and will not be discussed here.

The symbol $\sqrt{}$ is called a radical sign, and the number under the radical sign is called the **radicand.** In $\sqrt{18}$, 18 is the radicand. Also, $\sqrt{18}$ is an irrational number, since 18 is not a perfect square. Expressions with radical signs, such as $\sqrt{18}$, are called **radicals.**

To simplify radicals in general, we need the following property.

If a and b are positive real numbers, then

$$\sqrt{ab} = \sqrt{a}\sqrt{b}$$

With this property, we can look for a perfect square factor and write

$$\sqrt{18} = \sqrt{9 \cdot 2} = \sqrt{9} \cdot \sqrt{2} = 3\sqrt{2} \qquad \text{(9 is a square factor.)}$$

The expression $3\sqrt{2}$ is simplified because the radicand, 2, has no perfect square factor.

EXAMPLES

Simplify the following radicals.

1. $\sqrt{50}$

$$\sqrt{50} = \sqrt{25 \cdot 2} = \sqrt{25} \cdot \sqrt{2} = 5\sqrt{2} \qquad \text{(25 is a square factor.)}$$

2. $\sqrt{75}$

   $\sqrt{75} = \sqrt{25 \cdot 3} = \sqrt{25} \cdot \sqrt{3} = 5\sqrt{3}$

3. $\sqrt{450}$

   $\sqrt{450} = \sqrt{9 \cdot 50} = \sqrt{9 \cdot 25 \cdot 2} = \sqrt{9} \cdot \sqrt{25} \cdot \sqrt{2} = 3 \cdot 5\sqrt{2} = 15\sqrt{2}$

   or

   $\sqrt{450} = \sqrt{225 \cdot 2} = \sqrt{225} \cdot \sqrt{2} = 15\sqrt{2}$

PRACTICE QUIZ	Simplify the following radicals.	ANSWERS
	1. $\sqrt{8}$	**1.** $2\sqrt{2}$
	2. $\sqrt{20}$	**2.** $2\sqrt{5}$
	3. $\sqrt{135}$	**3.** $3\sqrt{15}$

EXERCISES 10.1

In the following exercises, you may use Tables 10.1 and 10.2 and the table located on the inside back cover as aids.

State whether or not each number is a perfect square.

1. 144	**2.** 169	**3.** 81	**4.** 16	**5.** 400
6. 225	**7.** 242	**8.** 48	**9.** 45	**10.** 40

11. Show by squaring that $\sqrt{3}$ is between 1.732 and 1733.

12. Show by squaring that $\sqrt{5}$ is between 2.236 and 2.237.

Answer the following questions using the set $A = \left\{ -10, \dfrac{3}{4}, 7.2, \sqrt{3} \right\}$.

13. Which of the numbers in set A is an integer?

14. Which of the numbers in set A is an irrational number?

15. Which of the numbers in set A is a real number?

Simplify the following radicals.

16. $\sqrt{12}$	**17.** $\sqrt{28}$	**18.** $\sqrt{24}$	**19.** $\sqrt{32}$	**20.** $\sqrt{48}$
21. $\sqrt{288}$	**22.** $\sqrt{363}$	**23.** $\sqrt{242}$	**24.** $\sqrt{500}$	**25.** $\sqrt{300}$
26. $\sqrt{128}$	**27.** $\sqrt{125}$	**28.** $\sqrt{72}$	**29.** $\sqrt{98}$	**30.** $\sqrt{605}$
31. $\sqrt{150}$	**32.** $\sqrt{169}$	**33.** $\sqrt{196}$	**34.** $\sqrt{800}$	**35.** $\sqrt{80}$
36. $\sqrt{90}$	**37.** $\sqrt{40}$	**38.** $\sqrt{256}$	**39.** $\sqrt{361}$	**40.** $\sqrt{108}$

10.2   INTERVALS OF REAL NUMBERS AND INEQUALITIES

In Section 8.2, we discussed number lines and graphing sets of integers on number lines. The graph of the set $A = \{-2, -1, 1, 3\}$ is shown in Figure 10.1.

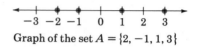

Graph of the set $A = \{2, -1, 1, 3\}$

Figure 10.1

On each number line there are many points (an infinite number) between the integers. **Each point on a number line corresponds to one real number, and each real number corresponds to one point on a number line.** In fact, number lines are called **real number lines.** The locations of some real numbers, other than integers, are shown in Figure 10.2.

Figure 10.2

If we want to compare two numbers, we can use the following symbols of equality and inequality (reading from left to right as shown):

$a = b$ a is equal to b
$a < b$ a is less than b
$a \leq b$ a is less than or equal to b
$a > b$ a is greater than b
$a \geq b$ a is greater than or equal to b

(A slash, /, through a symbol negates that symbol. For example, $\neq$ is read "is not equal to" and $\not<$ is read "is not less than.") We can also read the symbols from right to left. For example, we can read $a < b$ as "b is greater than a."

On a number line, smaller numbers are to the left of larger numbers. Thus, referring to Figure 10.3, we have

$$-1 < 2 \qquad -2 < -1 \qquad c > a \qquad b > 0$$

Figure 10.3

Suppose that a and b are two real numbers and that $a < b$. We may be interested not in a or b but in the numbers between a and b. The real numbers between two real numbers are in an **interval of real numbers.** Intervals are classified and graphed in the following manner. **(x is understood to represent real numbers.)**

Open Interval $a < x < b$

x is any real number greater than a and less than b.

Closed Interval $a \leq x \leq b$

x is any real number greater than or equal to a and less than or equal to b.

Half-Open Interval $a \leq x < b$

$a < x \leq b$

One of the end points is not included.

Open Interval $x > a$

$x < a$

Half-Open Interval $x \geq b$

$x \leq b$

EXAMPLES

1. Graph the closed interval $4 \leq x \leq 6$.

   Solution:

   Note that all real numbers between 4 and 6 and including 4 and 6 are in this interval. For example, $4\frac{1}{2}$ and 5.99 are both in this interval because $4 \leq 4\frac{1}{2} \leq 6$ and $4 \leq 5.99 \leq 6$.

2. Represent the following graph using interval notation and tell what kind of interval it is.

   Solution:

   $-2 < x < 0$ is an open interval.

Equations of the form $ax + b = c$ are called **linear equations** or **first degree equations** since the variable x is first degree. Similarly, inequalities of the form

$$ax + b < c, \qquad ax + b > c$$
$$ax + b \le c, \qquad ax + b \ge c$$

and
$$c < ax + b < d, \qquad c \le ax + b \le d$$

are called **linear inequalities** or **first degree inequalities.**

Intervals of real numbers are the solutions to linear inequalities, and we can solve linear inequalities just as we can solve linear equations. The rules are the same with one important exception: multiplying both sides of an inequality by a negative number "reverses the sense" of the inequality. Consider the following examples.

We know that $4 < 10$:

<table>
<tr><td>Add 3</td><td>Add -5</td></tr>
<tr><td>$4 < 10$</td><td>$4 < 10$</td></tr>
<tr><td>$4 + 3 < 10 + 3$</td><td>$4 - 5 < 10 - 5$</td></tr>
<tr><td>$7 < 13$</td><td>$-1 < 5$</td></tr>
</table>

<table>
<tr><td>Multiply by 2</td><td>Multiply by -2</td><td></td></tr>
<tr><td>$4 < 10$</td><td>$4 < 10$</td><td></td></tr>
<tr><td>$2 \cdot 4 < 2 \cdot 10$</td><td>$-2(4) > -2(10)$</td><td>(The sense is reversed</td></tr>
<tr><td>$8 < 20$</td><td>$-8 > -20$</td><td>from $<$ to $>$.)</td></tr>
</table>

Figure 10.4 illustrates the effect of multiplying both sides of the inequality $-1 < 2$ by -3.

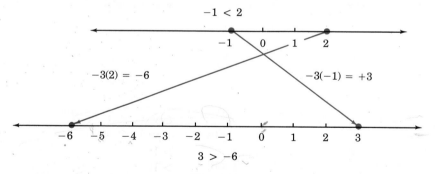

Figure 10.4

RULES FOR SOLVING INEQUALITIES

1. The same number may be added to both sides, and the sense of the inequality will remain the same.

2. Both sides may be multiplied by (or divided by) the same positive number, and the sense of the inequality will remain the same.

3. Both sides may be multiplied by (or divided by) the same negative number, but the sense of the inequality must be reversed.

As with solving equations, the object of solving inequalites is to find equivalent inequalities that are simpler than the original and to isolate the variable on one side of the inequality. The following examples illustrate the techniques.

$-3 + 3 \quad 0$

EXAMPLES

Solve the following inequalites and graph each solution on a number line.

1. $x - 3 < 2$

$$x - 3 < 2$$

$$x + 3 + 3 < 2 + 3 \qquad \text{Add 3 to both sides.}$$

$$x < 5$$

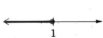

2. $-2x + 6 \geq 4$

$$-2x + 6 \geq 4$$

$$-2x + 6 - 6 \geq 4 - 6 \qquad \text{Add } -6 \text{ to both sides.}$$

$$-2x \geq -2$$

$$\frac{-2x}{-2} \leq \frac{-2}{-2} \qquad \begin{array}{l}\text{Divide both sides by } -2 \text{ and} \\ \textbf{reverse the sense.}\end{array}$$

$$x \leq 1$$

3. $7y - 8 > y + 10$

$$7y + 8 > y + 10$$

$$7y - 8 + y > y + 10 - y \qquad \text{Add } -y \text{ to both sides.}$$

$$6y - 8 > 10$$

$$6y + 8 + 8 > 10 + 8 \qquad \text{Add 8 to both sides.}$$

$$6y > 18$$

$$\frac{6y}{6} > \frac{18}{6} \qquad \begin{array}{l}\text{Divide both sides by 6.} \\ \text{The sense of the inequality} \\ \text{is unchanged.}\end{array}$$

$$y > 3$$

4. Find the values of x that satisfy both $5 < 2x + 3$ and $2x + 3 < 10$ and graph the solution.

Solution:

We can write these two inequalities in one expression and solve both at the same time since both are to be satisfied by the same values of x.

$$5 < 2x + 3 < 10$$
$$5 - 3 < 2x + 3 - 3 < 10 - 3 \qquad \text{Add } -3 \text{ to each member.}$$
$$2 < 2x < 7$$

$$\frac{2}{2} < \frac{2x}{2} < \frac{7}{2} \qquad \text{Divide each member by 2.}$$

$$1 < x < \frac{7}{2}$$

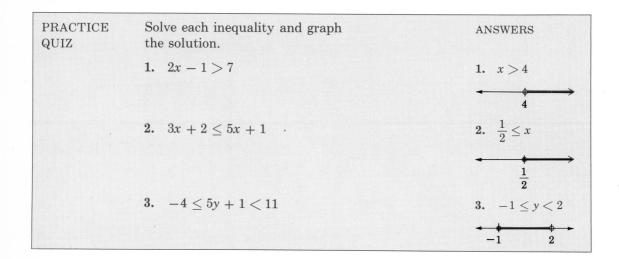

PRACTICE QUIZ	Solve each inequality and graph the solution.	ANSWERS
	1. $2x - 1 > 7$	**1.** $x > 4$
	2. $3x + 2 \le 5x + 1$	**2.** $\frac{1}{2} \le x$
	3. $-4 \le 5y + 1 < 11$	**3.** $-1 \le y < 2$

EXERCISES 10.2

State whether each of the following inequalities is true or false.

1. $3 \ne -3$ **2.** $-5 < -2$ **3.** $-13 > -1$ **4.** $|-7| \ne |+7|$

5. $|-7| < |+7|$ **6.** $|-4| > |+3|$ **7.** $-\frac{1}{2} < -\frac{3}{4}$ **8.** $\sqrt{2} > \sqrt{3}$

9. $-4 < -6$ **10.** $|-6| > 0$

Represent each of the following graphs with interval notation and tell what kind of interval it is.

11.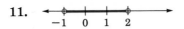

12.

13.

14.

15.

Graph each of the following intervals and tell what kind of interval it is.

16. $7 < x < 10$ **17.** $-2 < x < 0$ **18.** $1 \le y \le 4$

19. $-3 \le y \le 5$ **20.** $-1 < y \le 2$ **21.** $4 \le x < 7$

22. $68 \le x \le 72$ **23.** $-12 < z < -7$ **24.** $x > -3$

25. $x \ge 0$ **26.** $x < -\sqrt{3}$ **27.** $x \le \dfrac{2}{3}$

Solve and graph the solution for each of the following inequalities.

28. $x + 4 < 7$ **29.** $x - 6 > -2$ **30.** $y - 3 \ge -1$

31. $y + 5 \le 2$ **32.** $2y \le 3$ **33.** $5y > -6$

34. $6y + 1 > 5$ **35.** $7x - 2 < 9$ **36.** $x + 2 < 3x + 2$

37. $x - 4 > 2x + 1$ **38.** $2x - 5 \ge x + 2$ **39.** $3x - 8 \le x + 2$

40. $\dfrac{1}{2}x - 4 < 2$ **41.** $\dfrac{1}{3}x + 1 > -1$

42. $-x + 3 < -2$ **43.** $-x - 5 \ge -4$

44. $7y - 1 \ge 5y + 1$ **45.** $6x + 3 > x - 2$

46. $-4x - 4 \ge -4 + x$ **47.** $3x + 15 < x + 5$

48. $4 \le x + 7 \le 5$ **49.** $-2 \le x - 3 \le 1$

50. $-1 \le 2y + 1 \le 0$ **51.** $0 \le 3x - 1 \le 5$

52. $-3 < 4x + 1 \le 5$ **53.** $-5 \le y - 2 \le -1$

54. $7 \le -2x - 3 < 9$ **55.** $14 < 5x - 1 \le 24$

10.3 ORDERED PAIRS OF REAL NUMBERS

Equations such as $d = 40t$, $I = .18P$, and $y = 2x - 5$ represent relationships between pairs of variables. The first equation, $d = 40t$, can be interpreted as follows: The distance (d) traveled in time (t) at a rate of 40 miles per hour is found by multiplying 40 by t (assuming t is measured in hours). Thus, if $t = 3$ hours, then $d = 40 \cdot 3 = 120$ miles. The pair (3, 120) is called an **ordered pair** and is in the form (t, d). Similarly, (5, 200) represents $t = 5$ and $d = 200$. The ordered pairs (5, 200) and (200, 5) are not the same because the numbers are not in the same order.

We say that the ordered pair (3, 120) is **a solution of** or **satisfies** the equation $d = 40t$. In the same way, (100, 18) satisfies $I = .18P$ where $P = 100$ and $I = .18(100) = 18$. Also, (3, 1) satisfies $y = 2x - 5$ where $x = 3$ and $y = 2 \cdot 3 - 5 = 1$.

In an ordered pair such as (x, y), x is called the **first component** (or **first coordinate**), and y is called the **second component** (or **second coordinate**). To find ordered pairs that satisfy an equation in two variables, we can **choose any value** for one variable and find the corresponding value for the other variable by substituting into the equation. For example, for the equation $y = 2x - 5$,

choose $x = 2$, then $y = 2 \cdot 2 - 5 = -1$ ordered pair $(2, -1)$
choose $x = -1$, then $y = 2(-1) - 5 = -7$ ordered pair $(-1, -7)$
choose $x = 0$, then $y = 3 \cdot 0 - 5 = -5$ ordered pair $(0, -5)$
choose $x = 5$, then $y = 3 \cdot 5 - 5 = 10$ ordered pair $(5, 10)$

All the ordered pairs, $(2, -1)$, $(-1, -7)$, $(0, -5)$, and $(5, 10)$, satisfy the equation $y = 2x - 5$.

Since the value of y "depends" on the choice of x, the first component (x) is called the **independent variable** and the second component (y) is called the **dependent variable.** The ordered pairs can be written in table form. Remember that the choices for the independent variable are arbitrary; other values could have been chosen.

$d = 40t$			$I = .18P$			$y = 2x - 5$	
t	d		P	I		x	y
1	40		100	18		-2	-9
2	80		200	36		0	-5
5	200		1000	180		1	-3
7	280		5000	900		4	3
						5	5

We can graph ordered pairs of real numbers as points on a plane using the **Cartesian*** **coordinate system.** In this system, the plane is separated into four **quadrants** by two number lines that are perpendicular to each other. The lines intersect at a point called the **origin,** represented by the ordered pair (0, 0). The horizontal number line is called the **horizontal**

---

*The system is named after the famous mathematician René Descartes (1596–1650).

axis or **x-axis.** The vertical number line is called the **vertical axis** or **y-axis.** (See Figure 10.5.)

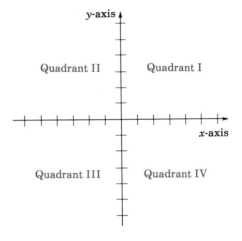

Figure 10.5

On the x-axis, positive numbers are marked to the right and negative numbers to the left. On the y-axis, positive numbers are marked up and negative numbers down. **Each point in a plane corresponds to one ordered pair of real numbers, and each ordered pair of real numbers corresponds to one point in a plane.** (See Figure 10.6.)

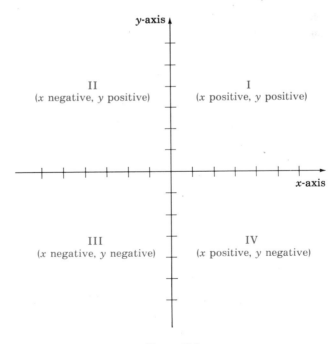

Figure 10.6

The graphs of the points $A(3, 1)$, $B(-2, 3)$, $C(-3, -1)$, $D(1, -2)$, and $E(2, 0)$ are shown in Figure 10.7. The point $E(2, 0)$ is on an axis and not in any quadrant. Each ordered pair is called the **coordinates** of the corresponding point.

Point	Quadrant
$A(3, 1)$	I
$B(-2, 3)$	II
$C(-3, -1)$	III
$D(1, -2)$	IV
$E(2, 0)$	x-axis

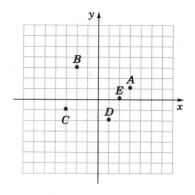

Figure 10.7

EXAMPLES

1. Graph the set of ordered pairs $\{(-2, 1), (0, 3), (1, 2), (2, -2)\}$.

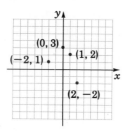

2. Graph the set of ordered pairs $\{(-3, -5), (-2, -3), (-1, -1), (0, 1), (1, 3)\}$.

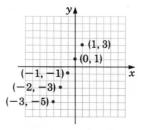

3. The graph of a set of ordered pairs is given. List the ordered pairs in the graph.

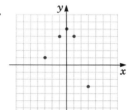

Solution: $\{(-4, 1), (-3, -1), (-1, 0), (0, 2), (1, 4), (1, -2), (3, 0)\}$

EXERCISES 10.3

List the ordered pairs that correspond to the points in the graph.

1. 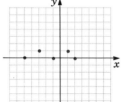   **2.** 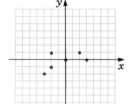   **3.**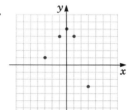

4. 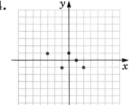   **5.** 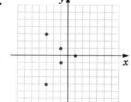   **6.**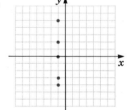

7. 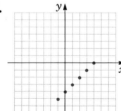   **8.** 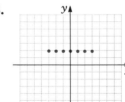   **9.**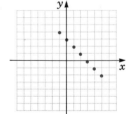

10.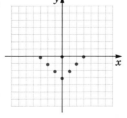

Graph each of the following sets of ordered pairs.

11. $\{(-2, 4), (-1, 3), (0, 1), (1, -2), (1, 3)\}$

12. $\{(-5, 1), (-3, 2), (-2, -1), (0, 2), (2, -1)\}$

13. $\{(-1, 2), (1, 3), (2, -2), (3, 4), (4, -2)\}$

14. $\{(-2, 3), (-1, 0), (0, -3), (2, 3), (4, -1)\}$

15. $\{(0, -3), (1, -1), (2, 1), (3, 3), (4, 5)\}$

16. $\{(-3, 3), (-2, 2), (-1, 1), (0, 0), (1, -1)\}$

17. $\{(-2, -1), (0, -1), (2, -1), (4, -1), (6, -1)\}$

18. $\{(-3, 1), (-2, 1), (-1, 1), (0, 1), (1, 1)\}$

19. $\{(-3, 3), (-1, 1), (0, 0), (1, 1), (3, 3)\}$

20. $\{(-2, -2), (-1, -1), (0, 0), (1, -1), (2, -2)\}$

21. $\{(-3, 9), (-2, 4), (-1, 1), (1, 1), (2, 4), (3, 9)\}$

22. $\{(-3, -9), (-2, -4), (-1, -1), (1, -1), (2, -4), (3, -9)\}$

23. $\{(-4, 0), (-2, 0), (0, 0), (2, 0), (4, 0)\}$

24. $\{(0, -3), (0, -1), (0, 0), (0, 1), (0, 3)\}$

25. $\{(-2, 1), (1, 4), (2, 5), (3, 6), (4, 7)\}$

26. $\{(-1, -5), (0, -2), (1, 1), (2, 4), (3, 7)\}$

27. $\{(-2, -7), (-1, -5), (2, 1), (3, 3), (4, 5)\}$

28. $\{(0, 1), (1, -1), (2, -3), (3, -5), (5, -9)\}$

29. $\{(-3, 11), (-2, 8), (0, 2), (2, -4), (3, -7)\}$

30. $\{(-2, -1), (-1, 1), (1, 5), (2, 7), (3, 9)\}$

10.4   GRAPHING LINEAR EQUATIONS

There are an infinite number of ordered pairs of real numbers that satisfy the equation $y = 3x + 1$. In Section 10.3, we substituted only integers for x and calculated integer values for y. We can also substitute fractions and radicals. For example,

$$\text{if } x = \frac{1}{3}, \quad \text{then } y = 3 \cdot \frac{1}{3} + 1 = 1 + 1 = 2$$

$$\text{if } x = -\frac{3}{4}, \text{ then } y = 3\left(-\frac{3}{4}\right) + 1 = -\frac{9}{4} + \frac{4}{4} = -\frac{5}{4}$$

$$\text{if } x = \sqrt{2}, \quad \text{then } y = 3\sqrt{2} + 1$$

Further studies of these kinds of substitutions will be reserved for later courses. The important idea here is that we do not have enough time or paper to substitute all real values for x. But, graphing a few points will show the trend that is of interest here. (See Figure 10.8.)

x	$y = 3x + 1$
-2	$y = 3(-2) + 1 = -5$
-1	$y = 3(-1) + 1 = -2$
0	$y = 3(0) + 1 = 1$
$\dfrac{2}{3}$	$y = 3\left(\dfrac{2}{3}\right) + 1 = 3$
2	$y = 3(2) + 1 = 7$

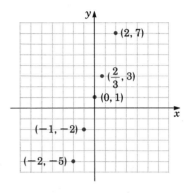

Figure 10.8

It appears that the points in Figure 10.8 lie on a straight line. In fact, they do lie on a straight line. We can draw a straight line through all the points, as shown in Figure 10.9, and **any point that lies on the line will satisfy the equation.**

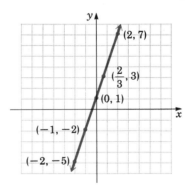

Figure 10.9

The points (ordered pairs) that satisfy any equation of the form

$$Ax + By = C \quad (A \text{ and } B \text{ not both } 0)$$

will lie on a straight line. The equation is called a **linear equation** and is in the **standard form** for the equation of a line. The linear equation $y = 3x + 1$ can be written in the standard form $-3x + y = 1$. Both forms are acceptable and correct.

Since we know that the graph of a linear equation is a straight line, we need only graph two points (two points determine a line) and draw the line through these two points. Choose any two values of x or any two values of y. (As a check against possible error, it is a good idea to locate three points instead of just two.)

EXAMPLE

Draw the graph of the linear equation $x + 2y = 6$.

$$
\begin{array}{ccc}
x = -2 & x = 0 & x = 2 \\
-2 + 2y = 6 & 0 + 2y = 6 & 2 + 2y = 6 \\
2y = 8 & 2y = 6 & 2y = 4 \\
y = 4 & y = 3 & y = 2
\end{array}
$$

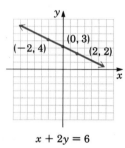

$x + 2y = 6$

(Locating three points helps in avoiding errors. Avoid choosing points close together.)

Letting $x = 0$ will locate the point where the line crosses the y-axis. This point is called the **y-intercept.** Letting $y = 0$ will locate the point where the line crosses the x-axis. This point is called the **x-intercept.** These two points are generally easy to locate and are frequently used for drawing the graph of a linear equation.

EXAMPLES

1. Draw the graph of the linear equation $x - 2y = 8$ by locating the y-intercept and the x-intercept.

$$
\begin{array}{cc}
x = 0 & y = 0 \\
0 - 2y = 8 & x - 2 \cdot 0 = 8 \\
y = -4 & x = 8
\end{array}
$$

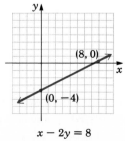

$x - 2y = 8$

2. Locate the y-intercept and the x-intercept and draw the graph of the linear equation $3x - y = 3$.

$$x = 0 \qquad\qquad y = 0$$
$$3 \cdot 0 - y = 3 \qquad 3x - 0 = 3$$
$$y = -3 \qquad\qquad x = 1$$

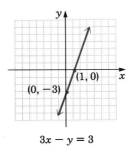

$$3x - y = 3$$

If $A = 0$ in the standard form $Ax + By = C$, the equation takes the form $By = C$ or $y = \dfrac{C}{B}$. For example, we can write $0x + 3y = 6$ as $y = 2$. Thus, no matter what value x has, the value of y is 2. The graph of the equation $y = 2$ is a horizontal line, as shown in Figure 10.10.

The y-coordinate is 2 for all points on the line $y = 2$.

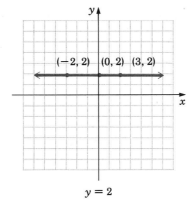

$$y = 2$$

Figure 10.10

If $B = 0$ in the standard form $AX + By = C$, the equation takes the form $Ax = C$ or $x = \dfrac{C}{A}$. For example, we can write $5x + 0y = -5$ as $x = -1$. Thus, no matter what value y has, the value of x is -1. The graph of the equation $x = -1$ is a vertical line, as shown in Figure 10.11.

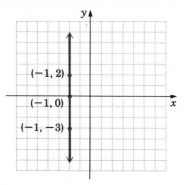

The x-coordinate is -1 for all points on the line $x = -1$.

Figure 10.11

EXAMPLE

Graph the horizontal line $y = -1$ and the vertical line $x = 3$ using the same axes.

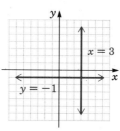

EXERCISES 10.4

Graph the following linear equations.

1. $y = x + 1$ 2. $y = x + 2$ 3. $y = x - 4$

4. $y = x - 6$ 5. $y = 2x$ 6. $y = 3x$

7. $y = -x$ 8. $y = -4x$ 9. $y = x$

10. $y = 2 - x$ 11. $y = 3 - x$ 12. $y = 5 - x$

13. $y = 2x + 1$ 14. $y = 2x - 1$ 15. $y = 2x - 3$

16. $y = 2x + 5$ 17. $y = -2x + 1$ 18. $y = -2x - 2$

19. $y = -3x + 2$ 20. $y = -3x - 4$ 21. $x - 2y = 4$

22. $x - 3y = 6$ 23. $-2x + 3y = 6$ 24. $2x - 5y = 10$

25. $-2x + y = 4$ 26. $2x + 3y = 6$ 27. $3x + 5y = 15$

28. $4x + y = 8$ 29. $x + 4y = 8$ 30. $3x - 4y = 12$

31. $x = 4$ 32. $y = 5$ 33. $y = -5$

34. $x = -2$ 35. $x - 5 = 0$ 36. $y + 3 = 0$

37. $2y - 3 = 0$ 38. $3x = -1$ 39. $4x - 5 = 0$

40. $3y + 7 = 0$

10.5 THE PYTHAGOREAN THEOREM (DISTANCE BETWEEN TWO POINTS)

If two sides of a triangle are perpendicular to each other, the angle has a measure of 90° and is called a **right angle.** If a triangle has a right angle, it is called a **right triangle.** The longest side (opposite the 90° angle) is called the **hypotenuse.** (See Figure 10.12.)

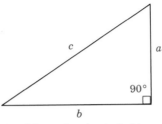

a and *b* are the shortest sides.
c is the hypotenuse.

Figure 10.12

Pythagoras, a famous Greek mathematician, is given credit for discovering the following theorem. He discovered that if he squared each side of a right triangle, the sum of the squares of the two sides was equal to the square of the hypotenuse.

PYTHAGOREAN THEOREM

In a right triangle the square of the hypotenuse is equal to the sum of the squares of the two sides:

$$c^2 = a^2 + b^2$$

In effect, Pythagoras and his Society had discovered a new kind of number, irrational numbers. But they were living in an age when numbers were considered mystical, and the Society was so unpopular that its members had to repress their knowledge of the new numbers or risk punishment and ridicule.

The following examples illustrate the Pythagorean Theorem.

EXAMPLES   1.

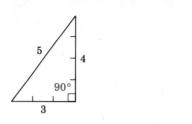

$5^2 = 3^2 + 4^2$
$25 = 9 + 16$

2.

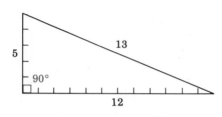

$13^2 = 5^2 + 12^2$
$169 = 25 + 144$

In most right triangles, all three sides do not have integer values. Examples 3 and 4 illustrate irrational numbers.

3.

$c^2 = 1^2 + 1^2$
$c^2 = 1 + 1 = 2$
$c = \sqrt{2}$

That is, the length of the hypotenuse is $\sqrt{2}$.

4.

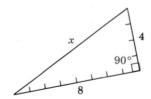

$x^2 = 8^2 + 4^2$
$x^2 = 64 + 16 = 80$
$x = \sqrt{80} = \sqrt{16 \cdot 5} = \sqrt{16} \cdot \sqrt{5} = 4\sqrt{5}$

The length of the hypotenuse is $4\sqrt{5}$.

Subscript notation is very useful in working with graphs and points. A **subscript** is a small number or letter written below and to the right of another letter. For example, P_1 (read "P sub 1") can represent one point, P_2 (read "P sub 2") can represent a second point, P_3 a third point, and so on. Similarly, the coordinates of P_1 might be (x_1, y_1), the coordinates of P_2 might be (x_2, y_2), and so on.

The Pythagorean Theorem can be used to find the distance between two points. Suppose P_1 has coordinates $(2, 3)$ and P_2 has coordinates $(6, 8)$, as shown in Figure 10.13.

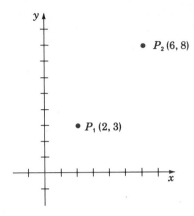

Figure 10.13

Form a right triangle by drawing a line through P_1 parallel to the x-axis and a line through P_2 parallel to the y-axis. The line segment joining P_1 and P_2 is the hypotenuse of the triangle. The point P_3 has coordinates $(6, 3)$ since it is on a vertical line through P_2 and is on a horizontal line through P_1. (See Figure 10.14.)

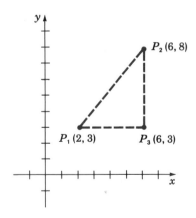

Figure 10.14

The distance between P_1 and P_3 is the difference in the x-coordinates, $a = |6 - 2| = 4$. The distance between P_2 and P_3 is the difference in the y-coordinates, $b = |8 - 3| = 5$. By letting d be the distance between P_1 and P_2 (the hypotenuse) and applying the Pythagorean Theorem, we get

$$d^2 = a^2 + b^2 = 4^2 + 5^2 = 16 + 25 = 41$$

So, $d = \sqrt{41}$. (See Figure 10.15.)

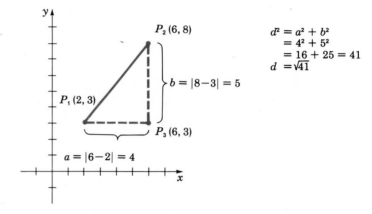

$$d^2 = a^2 + b^2$$
$$= 4^2 + 5^2$$
$$= 16 + 25 = 41$$
$$d = \sqrt{41}$$

Figure 10.15

This technique leads to the **distance formula**

$$d = \sqrt{(x_2 - x_1)^2 + (y_2 - y_1)^2}$$

In the calculation, be sure to add the squares before taking the square root.

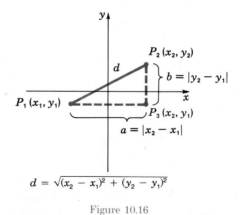

$$d = \sqrt{(x_2 - x_1)^2 + (y_2 - y_1)^2}$$

Figure 10.16

EXAMPLES   Draw a right triangle similar to the one in Figure 10.14 and label the points.

Then use the distance formula to find the distance between the given points as illustrated in Figures 10.15 and 10.16.

1. $P_1(-2, -1)$ and $P_2(5, 3)$

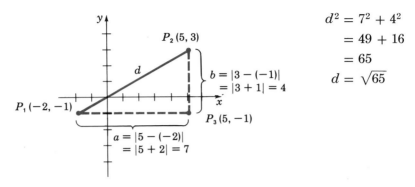

$$d^2 = 7^2 + 4^2$$
$$= 49 + 16$$
$$= 65$$
$$d = \sqrt{65}$$

[NOTE: We subtract negative numbers to find a and b.]

2. $P_1(-2, 6)$ and $P_2(4, -2)$

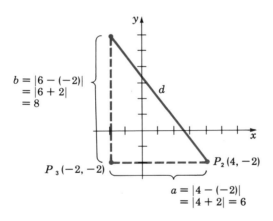

$$d^2 = 6^2 + 8^2$$
$$= 36 + 64$$
$$= 100$$
$$d = \sqrt{100} = 10$$

EXERCISES 10.5

For each problem, draw a right triangle similar to the one in Figure 10.14 and label the points. Then use the distance formula to find the distance between the given points.

1. $P_1(2, 0)$ and $P_2(3, 5)$ 2. $P_1(3, 0)$ and $P_2(5, 2)$

3. $P_1(0, 3)$ and $P_2(5, 7)$ 4. $P_1(0, 2)$ and $P_2(6, 7)$

5. $P_1(-4, 0)$ and $P_2(0, 3)$ 6. $P_1(-3, 0)$ and $P_2(0, 4)$

7. $P_1(0, -12)$ and $P_2(5, 0)$ 8. $P_1(-5, 0)$ and $P_2(0, 12)$

9. $P_1(-3, 2)$ and $P_2(3, -6)$ 10. $P_1(1, -3)$ and $P_2(9, 3)$

11. $P_1(1, 5)$ and $P_2(4, 2)$ 12. $P_1(1, 4)$ and $P_2(4, 1)$

13. $P_1(-6, 1)$ and $P_2(-1, -4)$ 14. $P_1(-7, -2)$ and $P_2(0, -5)$

15. $P_1(-2, 5)$ and $P_2(2, -5)$ 16. $P_1(-3, 4)$ and $P_2(4, -3)$

17. $P_1(-1, 3)$ and $P_2(6, -2)$ 18. $P_1(1, -7)$ and $P_2(8, -1)$

19. $P_1(-3, 7)$ and $P_2(9, 2)$ 20. $P_1(6, 4)$ and $P_2(8, -2)$

Use the distance formula to find the length of each side and use the Pythagorean Theorem to determine whether or not each triangle is a right triangle. Graph the points and draw the triangle.

21. $A(1, 1), B(7, 4), C(5, 8)$ 22. $A(-5, -1), B(2, 1), C(-1, 6)$

SUMMARY: CHAPTER 10

The symbol for **square root** is $\sqrt{}$, called a **square root sign** or **radical sign**. A **radicand** is the number under the radical sign.

Irrational numbers are infinite nonrepeating decimals. (There is no pattern to their decimal representation.) Most radicals are irrational numbers.

The **real numbers** include all rational numbers (whole numbers, integers, and positive and negative fractions) and all irrational numbers.

If a and b are positive real numbers, then

$$\sqrt{ab} = \sqrt{a}\sqrt{b}$$

Each point on a number line corresponds to one real number, and each real number corresponds to one point on a number line.

Intervals are classified as

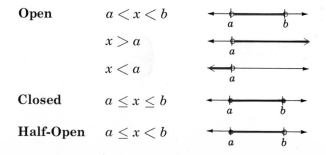

Open	$a < x < b$	
	$x > a$	
	$x < a$	
Closed	$a \le x \le b$	
Half-Open	$a \le x < b$	

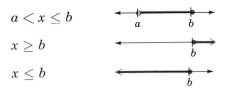

$$a < x \leq b$$

$$x \geq b$$

$$x \leq b$$

Linear inequalities are of the form $ax + b < c$. ($>$, $\leq$, or $\geq$ might appear instead of $<$.)

RULES FOR SOLVING INEQUALITIES

1. The same number may be added to both sides, and the sense of the inequality will remain the same.

2. Both sides may be multiplied by (or divided by) the same positive number, and the sense of the inequality will remain the same.

3. Both sides may be multiplied by (or divided by) the same negative number, but the sense of the inequality must be reversed.

In an **ordered pair** such as (x, y), x is called the **first component** (or **first coordinate**), and y is called the **second component** (or **second coordinate**). x is also called the **independent variable** and y is called the **dependent variable.**

In the **Cartesian coordinate system,** two perpendicular number lines (called **axes**) intersect at a point (called the **origin**) and separate a plane into four **quadrants.** The horizontal number line is called the **horizontal axis** or **x-axis.** The vertical number line is called the **vertical axis** or **y-axis.**

Each point in a plane corresponds to one ordered pair of real numbers, and each ordered pair of real numbers corresponds to one point in a plane.

$Ax + By = C$ (A and B not both 0) is the **standard form** of a **linear equation.** The graph of a linear equation is a straight line.

PYTHAGOREAN THEOREM

In a right triangle, the square of the hypotenuse is equal to the sum of the squares of the two sides:

$$c^2 = a^2 + b^2$$

The distance between two points (x_1, y_1) and (x_2, y_2) is given by

$$d = \sqrt{(x_2 - x_1)^2 + (y_2 - y_1)^2}$$

REVIEW QUESTIONS: CHAPTER 10

1. Show by squaring that $\sqrt{7}$ is between 2.645 and 2.646.

Answer the following questions using the set $B = \left\{ -\sqrt{5}, -\frac{1}{2}, 0, 6.13 \right\}$.

2. Which of the numbers in set B is an integer?

3. Which of the numbers in set B is a rational number?

4. Which of the numbers in set B is a real number?

Simplify the following radicals.

5. $\sqrt{169}$ **6.** $\sqrt{225}$ **7.** $\sqrt{128}$ **8.** $\sqrt{63}$

9. $\sqrt{54}$ **10.** $\sqrt{242}$ **11.** $\sqrt{300}$ **12.** $\sqrt{250}$

Graph each of the following intervals and tell what kind of interval it is.

13. $-\frac{1}{2} < x < \frac{3}{4}$ **14.** $0 \leq x < 5$ **15.** $x \geq \sqrt{2}$

16. $-3 \leq x \leq 3.1$ **17.** $y < \frac{1}{3}$ **18.** $14 < y \leq 15$

Represent each of the following graphs using interval notation and tell what kind of interval it is.

19.

20.

Solve and graph the solution for each of the following inequalities.

21. $x + 3 < -1$ **22.** $y - 5 \leq 2$ **23.** $3y \geq 8$

24. $x - 1 > 3x + 5$ **25.** $-4x + 6 < 16 + x$ **26.** $9 \leq 5x - 1 \leq 14$

List the ordered pairs that correspond to the points in the graph.

27. 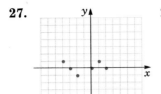   **28.**   **29.**

Graph each of the following sets of ordered pairs.

30. $\{(-3, 1), (-2, 1), (-1, 2), (0, 2), (1, 3)\}$

31. $\{(-4, 5), (-3, 2), (0, -4), (1, 1), (3, 1)\}$

32. $\{(1, 4), (1, 3), (1, 1), (1, 0)\}$

Graph the following linear equations.

33. $y = 2x - 1$ **34.** $y = x + 5$ **35.** $y = -x - 2$

36. $3x + y = 6$ **37.** $2x - 3y = 12$ **38.** $y = -3$

Use the distance formula to find the distance between the given points.

39. $P_1(0, 3)$ and $P_2(4, 6)$ **40.** $P_1(-2, 1)$ and $P_2(5, -1)$

41. Graph the three points $A(-1, 2)$, $B(3, 4)$, and $C(3, -4)$ and draw the triangle. Use the Pythagorean Theorem to determine whether or not the triangle is a right triangle.

CHAPTER TEST: CHAPTER 10

1. Show by squaring that $\sqrt{8}$ is between 2.828 and 2.829.

Answer the following questions using the set $A = \left\{ -7, -3.14, \frac{1}{4}, \sqrt{6}, 2\frac{1}{2} \right\}$.

2. Which of the numbers in set A is an integer?

3. Which of the numbers in set A is a rational number?

4. Which of the numbers in set A is an irrational number?

5. Which of the numbers in set A is a real number?

Simplify the following radicals.

6. $\sqrt{256}$ **7.** $\sqrt{98}$ **8.** $\sqrt{500}$ **9.** $\sqrt{112}$

Graph each of the following intervals and tell what kind of interval it is.

10. $\frac{2}{3} \le x \le 4$ **11.** $0 < x < 2.1$ **12.** $x < -\sqrt{3}$

Represent each of the following graphs using interval notation and tell what kind of interval it is.

13. ⟵ ○———○ ⟶
        1   3.6

14. ⟵ ●———● ⟶
      $-\frac{2}{3}$ $1\frac{2}{5}$

Solve and graph the solutions for each of the following inequalities.

15. $x - 12 \ge -8$ **16.** $-4y < -4$

17. $2x - 5 \le 5x - 2$ **18.** $-3 < 7x + 4 < 11$

19. List the ordered pairs that correspond to the points in the graph.

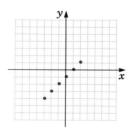

20. Graph the set of ordered pairs $\{(-2, 5), (-1, 3), (0, 1), (1, -1)\}$.

Graph the following linear equations.

21. $y = -2x + 1$ **22.** $x = 2$ **23.** $2x + 3y = 8$

24. Use the distance formula to find the distance between $P_1(-3, 4)$ and $P_2(1, 8)$.

25. Use the Pythagorean Theorem to show that the points $A(-3, -1)$, $B(2, -1)$, and $C(2, 5)$ do form a right triangle.

Appendix I
ANCIENT NUMERATION SYSTEMS

I.1   EGYPTIAN, MAYAN, ATTIC GREEK, AND ROMAN SYSTEMS

The number systems used by ancient peoples are interesting from a historical point of view, but from a mathematical point of view they are difficult to work with. One of the many things that determine the progress of any civilization is its system of numeration. Humankind has made its most rapid progress since the invention of the zero and the place value system (which we will discuss in the next section) by the Hindu-Arabic peoples about A.D. 800.

Egyptian Numerals (Hieroglyphics)

The ancient Egyptians used a set of symbols called hieroglyphics as early as 3500 B.C. (See Table I.1.)

TABLE I.1   EGYPTIAN HIEROGLYPHIC NUMERALS

SYMBOL	NAME		VALUE
I	Staff (vertical stroke)	1	one
∩	Heel bone (arch)	10	ten
୧	Coil of rope (scroll)	100	one hundred
⚱	Lotus flower	1000	one thousand
⌐	Pointing finger	10,000	ten thousand
∾	Bourbot (tadpole)	100,000	one hundred thousand
⛑	Astonished man	1,000,000	one million

To write the numeral for a number, the Egyptians wrote the symbols next to each other from left to right, and the number represented was the sum of the values of the symbols. The most times any symbol was used was nine. Instead of using a symbol ten times, they used the symbol for the next higher number. They also grouped the symbols in threes or fours.

EXAMPLE

⚱ ୧ ୧ ୧ ∩ ∩ ∣∣∣∣   represents the number one thousand
   ୧ ୧ ୧   ∣∣∣   six hundred twenty-seven, or
                         $1000 + 600 + 20 + 7 = 1627$

Mayan System

The Mayans used a system of dots and bars (for numbers from 1 to 19) combined with a place value system. A dot represented one and a bar represented five. They had a symbol, ⬭ , for zero and based their system, with one exception, on twenty. (See Table I.2.) The symbols were arranged vertically, smaller values starting at the bottom. The value of the third place up was 360 (18 times the value of the second place), but all other places were 20 times the value of the previous place.

TABLE I.2 MAYAN NUMERALS

SYMBOL		VALUE	
·		1	one
——		5	five
⬭		0	zero

EXAMPLES

1. ··· $(3 + 5 = 8)$

2. ⁞⁞⁞ $(3 \cdot 5 + 4 = 19)$

3. ···   3 20's
    ⬭   0 units
    $(3 \cdot 20 + 0 = 60)$

4. ··   2 7200's
    ⬭   0 360's
    ·   6 20's
    ··   7 units
    $(2 \cdot 7200 + 0 \cdot 360 + 6 \cdot 20 + 7 = 14{,}527)$

[NOTE: ⬭ is used as a place holder.]

Attic Greek System

The Greeks used two numeration systems, the Attic (see Table I.3) and the Alexandrian (see Section I.2 for information on the Alexandrian system). In the Attic system, no numeral was used more than four times. When a symbol was needed five or more times, the symbol for five was used, as shown in the examples.

TABLE I.3 ATTIC GREEK NUMERALS

SYMBOL	VALUE	
I	1	one
Γ	5	five
Δ	10	ten
H	100	one hundred
X	1000	one thousand
M	10,000	ten thousand

EXAMPLES

1. X X ⼛ H H ⼌ I I I I $(2 \cdot 1000 + 7 \cdot 100 + 5 \cdot 10 + 4 = 2754)$

2. ⼓ H H H H Δ Δ Γ $(5 \cdot 1000 + 4 \cdot 100 + 2 \cdot 10 + 5 = 5425)$

Roman System

The Romans used a system (Table I.4) that we still see in evidence as hours on clocks and dates on buildings.

TABLE I.4 ROMAN NUMERALS

SYMBOL	VALUE	
I	1	one
V	5	five
X	10	ten
L	50	fifty
C	100	one hundred
D	500	five hundred
M	1000	one thousand

The symbols were written largest to smallest, from left to right. The value of the numeral was the sum of the values of the individual symbols. Each symbol was used as many times as necessary, with the following exceptions: When the Romans got to 4, 9, 40, 90, 400, or 900, they used a system of subtraction.

$IV = 5 - 1 = 4$ $XL = 50 - 10 = 40$ $CD = 500 - 100 = 400$

$IX = 10 - 1 = 9$ $XC = 100 - 10 = 90$ $CM = 1000 - 100 = 900$

EXAMPLES

1. VII   represents 7

2. DXLIV   represents 544

3. MCCCXXVIII   represents 1328

EXERCISES I.1

Find the values of the following ancient numbers.

1. ＄＄ ∩∩∩ / ∩∩ / |||

2.

3.

4. ....

5. ⊖

6. ☰

7. Ɓ I I I

8. Ꞁ H H Δ Δ Δ Ɓ I

9. X X H H H Ɓ Δ I I

10. XCVII

11. DCCXLIV

12. MMMCDLXV

13. CMLXXVIII

14. Write 64
    (a) as an Egyptian numeral   (b) as a Mayan numeral
    (c) as an Attic Greek numeral   (d) as a Roman numeral

15. Follow the instructions for Problem 14, using 532 in place of 64.

16. Follow the same instructions, using 1969.

17. Follow the same instructions, using 846.

I.2 BABYLONIAN, ALEXANDRIAN GREEK, AND CHINESE-JAPANESE SYSTEMS

Babylonian System (Cuneiform Numerals)

The Babylonians (about 3500 B.C.) used a place value system based on the number sixty, called a sexagesimal system. They had only two symbols, $\vee$ and $<$. (See Table I.5.) These wedge shapes are called cuneiform numerals, since **cuneus** means **wedge** in Latin.

TABLE I.5 CUNEIFORM NUMERALS

SYMBOL	VALUE	
$\vee$	1	one
$<$	10	ten

The symbol for one was used as many as nine times, and the symbol for ten as many as five times; however, since there was no symbol for zero, many Babylonian numbers could be read several ways. For our purposes, we will group the symbols to avoid some of the ambiguities inherent in the system.

EXAMPLES

1. $\vee\vee\vee$ $<<\vee\vee\vee \atop \vee\vee$ $<<<\vee\vee$

$$(3 \cdot 60^2) + (25 \cdot 60^1) + (32 \cdot 1) = (3 \cdot 3600) + (25 \cdot 60) + 32$$
$$= 10{,}800 + 1500 + 32$$
$$= 12{,}332$$

2.

$$V \quad <<<<\begin{matrix}VVV\\VVV\end{matrix} \quad <\begin{matrix}VVVV\\VVV\end{matrix}$$

$$(1 \cdot 60^2) + (46 \cdot 60^1) + (17 \cdot 1) = 3600 + 2760 + 17 = 6377$$

Alexandrian Greek System

The Greeks used two numeration systems, the Attic and the Alexandrian. We discussed the Attic Greek system in Section I.1.

In the Alexandrian system (Table I.6), the letters were written next to each other, largest to smallest, from left to right. Since the numerals were also part of the Greek alphabet, an accent mark or bar was sometimes used above a letter to indicate that it represented a number. Multiples of 1000 were indicated by strikes in front of the unit symbols, and multiples of 10,000 were indicated by placing the unit symbols above the symbol M.

TABLE I.6   ALEXANDRIAN GREEK SYMBOLS

SYMBOL	NAME	VALUE		SYMBOL	NAME	VALUE	
A	Alpha	1	one	Ξ	Xi	60	sixty
B	Beta	2	two	O	Omicron	70	seventy
Γ	Gamma	3	three	Π	Pi	80	eighty
Δ	Delta	4	four	Ϙ	Koppa	90	ninety
E	Epsilon	5	five	P	Rho	100	one hundred
F	Digamma (or Vau)	6	six	Σ	Sigma	200	two hundred
Z	Zeta	7	seven	T	Tau	300	three hundred
H	Eta	8	eight	Υ	Upsilon	400	four hundred
Θ	Theta	9	nine	Φ	Phi	500	five hundred
I	Iota	10	ten	X	Chi	600	six hundred
K	Kappa	20	twenty	Ψ	Psi	700	seven hundred
Λ	Lambda	30	thirty	Ω	Omega	800	eight hundred
M	Mu	40	forty	ϡ	Sampi	900	nine hundred
N	Nu	50	fifty				

EXAMPLES

1. $\overline{\Phi \; \Xi \; Z}$ $(500 + 60 + 7 = 567)$

2. $\overset{\rule{1em}{0.4pt}}{B}$
   M T N Δ $(20,000 + 300 + 50 + 4 = 20,354)$

Chinese-Japanese System

The Chinese-Japanese system (Table I.7) uses a different numeral for each of the digits up to ten, then a symbol for each power of ten. A digit written above a power of ten is to be multiplied by that power, and all such results are to be added to find the value of the numeral.

TABLE I.7 CHINESE-JAPANESE NUMERALS

SYMBOL	VALUE	SYMBOL	VALUE
一	1 one	七	7 seven
二	2 two	八	8 eight
三	3 three	九	9 nine
四	4 four	十	10 ten
五	5 five	百	100 one hundred
六	6 six	千	1000 one thousand

EXAMPLES

1.
$$\left.\begin{array}{c} 三 \\ 十 \end{array}\right\} 30$$

九 9

(30 + 9 = 39)

2.
$$\left.\begin{array}{c} 五 \\ 千 \end{array}\right\} 5000$$

$$\left.\begin{array}{c} 四 \\ 百 \end{array}\right\} 400$$

$$\left.\begin{array}{c} 八 \\ 十 \end{array}\right\} 80$$

$$二 \Big\} \; 2$$

(5000 + 400 + 80 + 2 = 5482)

EXERCISES I.2

Find the value of each of the following ancient numerals.

1. V <<VVV VVV   2. <<<VV   3. <<< VV

4. Υ N E   5. $\overline{\Delta}$ 6. $\overline{\Sigma \; K \; B}$

M /Z π

7.
四
十
六

8.
五
千
一
百
八

9.
九
百
九
十
九

Write the following numbers as (a) Babylonian numerals, (b) Alexandrian Greek numerals, and (c) Chinese-Japanese numerals.

10. 472 **11.** 596 **12.** 5047 **13.** 3665 **14.** 7293 **15.** 10,852

Appendix II
BASE TWO AND BASE FIVE

II.1 THE BINARY SYSTEM (BASE TWO)

In the decimal system, ten is the base. You might ask if another number could be chosen as the base in a place value system. And, if so, would the system be any better or more useful than the decimal system? The fact is that computers do operate under a place value system whose base is two. In the **binary system** (or base two system), only two digits are needed, 0 and 1. These two digits correspond to the two possible conditions of an electric current, either **on** or **off.**

Any number can be represented in base two or in base ten. However, base ten has a definite advantage when large numbers are involved, as you will see. The advantage of base two over base ten is that for base two only two digits are needed, while ten digits are needed for base ten.

If the base of a place value system were not ten but two, then the beginning point would be not a decimal point but a **binary point.** The value of each place would be a power of two, as shown in Figure II.1.

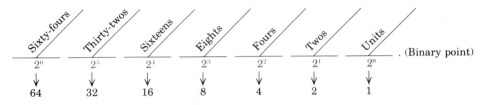

Figure II.1

TO WRITE NUMBERS IN THE BASE TWO SYSTEM,
REMEMBER THREE THINGS

1. $\{0, 1\}$ is the set of digits we can use.

2. The value of each place from the binary point is in powers of two.

3. The symbol 2 does not exist in the binary system, just as there is no digit for ten in the decimal system.

To avoid confusion with the base ten numerals, we will write $_{(2)}$ to the lower right of each base two numeral. We could write $_{(10)}$ to the lower right of each base ten numeral, but this would not be practical since most of the numerals we work with are in base ten. Therefore, **if no base is indicated, the numeral will be understood to be in base ten.**

EXAMPLES

1. Find the value of the numeral $1101._{(2)}$.
   Writing the value of each place under the digit gives

$$\frac{1}{2^3} \quad \frac{1}{2^2} \quad \frac{0}{2^1} \quad \frac{1}{2^0} \quad \cdot_{(2)}$$

In expanded notation,

$$
\begin{aligned}
1101._{(2)} &= 1(2^3) + 1(2^2) + 0(2^1) + 1(2^0) \\
&= 1(8) + 1(4) + 0(2) + 1(1) \\
&= 8 + 4 + 0 + 1 \\
&= 13
\end{aligned}
$$

Thus, to a computer, the symbol $1101._{(2)}$ means "thirteen."

2. $\underline{1}._{(2)} = 1$

3. $\underline{1}\ \underline{0}._{(2)} = 1(2) + 0 = 2$

4. $\underline{1}\ \underline{1}._{(2)} = 1(2) + 1 = 2 + 1 = 3$

5. $\underline{1}\ \underline{0}\ \underline{0}._{(2)} = 1(2^2) + 0(2) + 0(1) = 4 + 0 + 0 = 4$

6. $\underline{1}\ \underline{0}\ \underline{1}._{(2)} = 1(2^2) + 0(2) + 1(1) = 4 + 0 + 1 = 5$

7. $\underline{1}\ \underline{1}\ \underline{0}._{(2)} = 1(2^2) + 1(2) + 0(1) = 4 + 2 + 0 = 6$

8. $\underline{1}\ \underline{1}\ \underline{1}._{(2)} = 1(2^2) + 1(2) + 1(1) = 4 + 2 + 1 = 7$

9. $\underline{1}\ \underline{0}\ \underline{0}\ \underline{0}._{(2)} = 1(2^3) + 0(2^2) + 0(2) + 0(1) = 8 + 0 + 0 + 0 = 8$

10. $\underline{1}\ \underline{0}\ \underline{0}\ \underline{1}._{(2)} = 1(2^3) + 0(2^2) + 0(2) + 1(1) = 8 + 0 + 0 + 1 = 9$

11. $\underline{1}\ \underline{0}\ \underline{1}\ \underline{0}._{(2)} = 1(2^3) + 0(2^2) + 1(2) + 0(1) = 8 + 0 + 2 + 0 = 10$

Do **not** read $100_{(2)}$ as "one hundred" because the 1 is not in the hundreds place. The 1 is in the fours place. So, $100_{(2)}$ is read "four" or "one, zero, zero—base two." Similarly, $111_{(2)}$ is read "seven" or "one, one, one—base two."

EXERCISES II.1

Write the following base ten numerals in expanded form using components.

Example: $273 = 2(10^2) + 7(10^1) + 3(10^0)$

1. 35   **2.** 761   **3.** 8469   **4.** 500   **5.** 62,322

Write the following base two numerals in expanded form and find the value of each numeral.

Example: $110_{(2)} = 1(2^2) + 1(2^1) + 0(2^0)$
$= 1(4) + 1(2) + 0(1)$
$= 4 + 2 + 0$
$= 6$

6. $11_{(2)}$ **7.** $101_{(2)}$ **8.** $111_{(2)}$ **9.** $1011_{(2)}$

10. $1101_{(2)}$ **11.** $110111_{(2)}$ **12.** $11110_{(2)}$ **13.** $101011_{(2)}$

14. $11010_{(2)}$ **15.** $1000_{(2)}$ **16.** $1000010_{(2)}$ **17.** $11101_{(2)}$

18. $10110_{(2)}$ **19.** $111111_{(2)}$ **20.** $1111_{(2)}$

21. A computer is directed to place some information in memory space number $1101111_{(2)}$. What is the number of this memory space in base ten?

II.2 THE QUINARY SYSTEM (BASE FIVE)

Many numbers may be used as bases for place value systems. To illustrate this point and to emphasize the concept of place value, we will discuss one more base system, base five. Interested students may want to try writing numerals in base three or base eight or base eleven.

Again, the system relies on powers of the base and a set of digits. In the **quinary system** (base five system), the powers of 5 are $5^0, 5^1, 5^2, 5^3, 5^4$, and so on, and the digits to be used are $\{0, 1, 2, 3, 4\}$. The **quinary point** is the beginning point, as shown in Figure II.2.

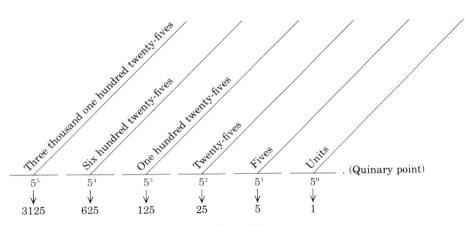

Figure II.2

EXAMPLES

1. $\underline{1}_{\cdot(5)} = 1$

2. $\underline{2}_{\cdot(5)} = 2$

3. $\underline{3} \cdot_{(5)} = 3$

4. $\underline{4} \cdot_{(5)} = 4$

5. $\underline{1}\ \underline{0} \cdot_{(5)} = 1(5) + 0(1) = 5 + 0 = 5$

6. $\underline{1}\ \underline{1} \cdot_{(5)} = 1(5) + 1(1) = 5 + 1 = 6$

7. $\underline{1}\ \underline{2} \cdot_{(5)} = 1(5) + 2(1) = 5 + 2 = 7$

8. $\underline{1}\ \underline{3} \cdot_{(5)} = 1(5) + 3(1) = 5 + 3 = 8$

9. $\underline{1}\ \underline{4} \cdot_{(5)} = 1(5) + 4(1) = 5 + 4 = 9$

10. $\underline{2}\ \underline{0} \cdot_{(5)} = 2(5) + 0(1) = 10 + 0 = 10$

11. $\underline{2}\ \underline{1} \cdot_{(5)} = 2(5) + 1(1) = 10 + 1 = 11$

12. $\underline{2}\ \underline{2} \cdot_{(5)} = 2(5) + 2(1) = 10 + 2 = 12$

13. $\underline{3}\ \underline{2}\ \underline{4} \cdot_{(5)} = 3(5^2) + 2(5^1) + 4(5^0)$
$$= 3(25) + 2(5) + 4(1)$$
$$= 75 + 10 + 4$$
$$= 89$$

EXERCISES II.2

Write the following base five numerals in expanded form and find the value of each.

1. $24_{(5)}$ **2.** $13_{(5)}$ **3.** $10_{(5)}$ **4.** $43_{(5)}$ **5.** $104_{(5)}$

6. $312_{(5)}$ **7.** $32_{(5)}$ **8.** $230_{(5)}$ **9.** $423_{(5)}$ **10.** $444_{(5)}$

11. $1034_{(5)}$ **12.** $4124_{(5)}$ **13.** $244_{(5)}$ **14.** $3204_{(5)}$ **15.** $13042_{(5)}$

16. Do the numerals $101_{(2)}$ and $10_{(5)}$ represent the same number? If so, what is the number?

17. Answer the questions in Problem 16 about the numerals $1101_{(2)}$ and $23_{(5)}$.

18. Answer the questions in Problem 16 about the numerals $11,100_{(2)}$ and $103_{(5)}$.

19. What set of digits do you think would be used in a base eight system?

20. What set of digits do you think would be used in a base twelve system? [HINT: New symbols for some new digits must be introduced.]

II.3 ADDITION AND MULTIPLICATION IN BASE TWO AND BASE FIVE

Now that we have two new numeration systems, base two and base five, a natural question to ask is, How are addition and multiplication* performed in these systems? The basic techniques are the same as for base ten, since place value is involved. However, because different bases are involved, the numerals will be different. For example, to add five plus seven in base ten, we write $5 + 7 = 12$. In base two, this same sum is written $101_{(2)} + 111_{(2)} = 1100_{(2)}$.

Writing the numerals vertically (one under the other) gives

$$
\begin{array}{cc}
5 & 101_{(2)} \\
7 & 111_{(2)} \\
\hline
12 & 1100_{(2)}
\end{array}
$$

Now a step-by-step analysis of the sum in base two will be provided.

EXAMPLES

1. (a) $101_{(2)}$
$111_{(2)}$

The numerals are written so that the digits of the same place value line up.

(b) $\begin{array}{r} 1 \\ 101_{(2)} \\ 111_{(2)} \\ \hline 0_{(2)} \end{array}$

Adding $1 + 1$ in the units column gives "two," which is written $10_{(2)}$. 0 is written in the units column, and 1 is "carried" to the "twos" column.

(c) $\begin{array}{r} 1\,1 \\ 101_{(2)} \\ 111_{(2)} \\ \hline 00_{(2)} \end{array}$

Now, in the twos column, $1 + 0 + 1$ is again "two," or $10_{(2)}$. Again 0 is written, and 1 is "carried" to the next column, the "fours" column.

(d) $\begin{array}{r} 1\,1 \\ 101_{(2)} \\ 111_{(2)} \\ \hline 1100_{(2)} \end{array}$

In the fours column (or third column), $1 + 1 + 1$ is "three," or $11_{(2)}$. Since there are no digits in the "eights" column (or fourth column), 11 is written, and the sum is $1100_{(2)}$.

Checking in Base Ten:

$$
\begin{array}{lllr}
101_{(2)} = & 1(2^2) + 0(2^1) + 1(2^0) = & 4 + 0 + 1 = 5 & 5 \\
111_{(2)} = & 1(2^2) + 1(2^1) + 1(2^0) = & 4 + 2 + 1 = 7 & 7 \\
\hline
1100_{(2)} & 1(2^3) + 1(2^2) + 0(2^1) + 0(2^0) = 8 + 4 + 0 + 0 = 12 & & 12
\end{array}
$$

---

*Subtraction and division may also be performed in base two and base five, but will not be discussed here for reasons of time. Some students may want to investigate these operations on their own.

Addition in base five is similar. Although the thinking is done in base ten, which is familiar to us, the numerals written must be in base five.

2.   (a) $143_{(5)}$
$34_{(5)}$

The numerals are written so that the digits of the same place value line up.

1
(b) $143_{(5)}$
$34_{(5)}$
$2_{(5)}$

Adding $3 + 4$ in the units column gives "seven," which is written $12_{(5)}$. 2 is written in the units column, and 1 is carried to the fives column.

1 1
(c) $143_{(5)}$
$34_{(5)}$
$32_{(5)}$

In the fives column, $1 + 4 + 3$ gives "eight," which is $13_{(5)}$. The 3 is written, and 1 is carried to the next column (the twenty-fives column).

1 1
(d) $143_{(5)}$
$34_{(5)}$
$232_{(5)}$

In the third column, $1 + 1$ gives 2. The sum is $232_{(5)}$.

Checking in Base Ten:

$$143_{(5)} = 1(5^2) + 4(5^1) + 3(5^0) = 25 + 20 + 3 = 48 \qquad 48$$
$$\phantom{143_{(5)} =} \underline{ 34_{(5)} = 3(5^1) + 4(5^0) = 15 + 4 = 19 \qquad 19}$$
$$232_{(5)} = 2(5^2) + 3(5^1) + 2(5^0) = 50 + 15 + 2 = 67 \qquad \overline{67}$$

Multiplication in each base is performed and checked in the same manner as addition. Of course, the difference is that you multiply instead of add. When you multiply, be sure to write the correct symbol for the number in the base being used. Also remember to add in the correct base.

3. $101_{(2)}$
$111_{(2)}$
${}_1101$
${}_1101$
101
$100011_{(2)}$

Multiplication in base two is easy, since we are multiplying by only 1's or 0's. The adding must be done in base two.

Checking gives:

$101_{(2)} = 5$
$\underline{111_{(2)} = 7}$
$101 \qquad 35$
101
$\underline{101}$
$100011_{(2)}$

Remember to **multiply** the checking numbers.

$$100,011_{(2)} = 1(2^5) + 0(2^4) + 0(2^3) + 0(2^2) + 1(2) + 1(1)$$
$$= 32 + 2 + 1 = 35$$

4.

$$\begin{array}{r} {}^{1}34_{(5)} \\ 23_{(5)} \\ \hline 212 \\ 123 \\ \hline 1442_{(5)} \end{array}$$

with carry 2 above.

Multiplying 3×4 gives "twelve," which is $22_{(5)}$. Write 2 and carry 2 just as in regular multiplication. Then, 3×3 is "nine," and "nine" plus 2 is "eleven"; but in base five, "eleven" is $21_{(5)}$. Similarly, 2×4 is "eight," or $13_{(5)}$. Write the 3, carry the 1. 2×3 is "six," and "six" plus 1 is "seven," or $12_{(5)}$.

Checking gives:

$$\begin{array}{r} 34_{(5)} = 19 \\ 23_{(5)} = 13 \\ \hline 212 \qquad 57 \\ 123 \qquad 19 \\ \hline 1442_{(5)} \quad 247 \end{array}$$

Remember to **multiply** the checking numbers.
$$1442_{(5)} = 1(5^3) + 4(5^2) + 4(5) + 2(1)$$
$$= 125 + 100 + 20 + 2 = 247$$

EXERCISES II.3

Add in the base indicated and check your work in base ten.

1. $\begin{array}{r} 101_{(2)} \\ 11_{(2)} \end{array}$ **2.** $\begin{array}{r} 43_{(5)} \\ 213_{(5)} \end{array}$ **3.** $\begin{array}{r} 1101_{(2)} \\ 1011_{(2)} \end{array}$ **4.** $\begin{array}{r} 111_{(2)} \\ 1010_{(2)} \end{array}$ **5.** $\begin{array}{r} 134_{(5)} \\ 243_{(5)} \end{array}$

6. $\begin{array}{r} 11_{(2)} \\ 10_{(2)} \\ 11_{(2)} \end{array}$ **7.** $\begin{array}{r} 11_{(2)} \\ 11_{(2)} \\ 101_{(2)} \end{array}$ **8.** $\begin{array}{r} 214_{(5)} \\ 343_{(5)} \end{array}$ **9.** $\begin{array}{r} 14_{(5)} \\ 321_{(5)} \\ 43_{(5)} \end{array}$ **10.** $\begin{array}{r} 431_{(5)} \\ 214_{(5)} \\ 102_{(5)} \end{array}$

11. $\begin{array}{r} 11_{(2)} \\ 101_{(2)} \\ 111_{(2)} \\ 101_{(2)} \end{array}$ **12.** $\begin{array}{r} 111_{(2)} \\ 11_{(2)} \\ 110_{(2)} \\ 111_{(2)} \end{array}$ **13.** $\begin{array}{r} 101_{(2)} \\ 101_{(2)} \\ 101_{(2)} \\ 101_{(2)} \end{array}$ **14.** $\begin{array}{r} 23_{(5)} \\ 103_{(5)} \\ 214_{(5)} \\ 322_{(5)} \end{array}$ **15.** $\begin{array}{r} 414_{(5)} \\ 211_{(5)} \\ 334_{(5)} \\ 222_{(5)} \end{array}$

Multiply in the base indicated and check your work in base ten.

16. $\begin{array}{r} 1101_{(2)} \\ 111_{(2)} \end{array}$ **17.** $\begin{array}{r} 1011_{(2)} \\ 101_{(2)} \end{array}$ **18.** $\begin{array}{r} 423_{(5)} \\ 30_{(5)} \end{array}$ **19.** $\begin{array}{r} 104_{(5)} \\ 23_{(5)} \end{array}$

20. $\begin{array}{r} 223_{(5)} \\ 44_{(5)} \end{array}$ **21.** $\begin{array}{r} 423_{(5)} \\ 32_{(5)} \end{array}$ **22.** $\begin{array}{r} 1111_{(2)} \\ 111_{(2)} \end{array}$ **23.** $\begin{array}{r} 111_{(2)} \\ 111_{(2)} \end{array}$

24. $\begin{array}{r} 2212_{(5)} \\ 43_{(5)} \end{array}$ **25.** $\begin{array}{r} 10111_{(2)} \\ 110_{(2)} \end{array}$

Appendix III
GREATEST COMMON DIVISOR (GCD)

Consider the two numbers 12 and 18. Is there a number (or numbers) that will divide into **both** 12 and 18? To help answer this question, the divisors for 12 and 18 are listed below.

Set of divisors for 12: {1, 2, 3, 4, 6, 12}
Set of divisors for 18: {1, 2, 3, 6, 9, 18}

The **common divisors** for 12 and 18 are 1, 2, 3, and 6. The **greatest common divisor (GCD)** for 12 and 18 is 6; that is, of all the common divisors of 12 and 18, 6 is the largest divisor.

EXAMPLE   List the divisors of each number in the set {36, 24, 48} and find the greatest common divisor (GCD).

Set of divisors for 36: {**1, 2, 3, 4, 6,** 9, **12,** 18, 36}
Set of divisors for 24: {**1, 2, 3, 4, 6,** 8, **12,** 24}
Set of divisors for 48: {**1, 2, 3, 4, 6,** 8, **12,** 16, 24, 48}

The common divisors are **1, 2, 3, 4, 6,** and **12. GCD = 12.**

DEFINITION   The **Greatest Common Divisor (GCD)*** of a set of natural numbers is the largest natural number that will divide into all the numbers in the set.

As the above example illustrates, listing all the divisors of each number before finding the GCD can be tedious and difficult. **The use of prime factorizations leads to a simple technique for finding the GCD.**

TECHNIQUE FOR FINDING THE GCD OF A SET OF NATURAL NUMBERS

1. Find the prime factorization of each number.

2. Find the prime factors common to all factorizations.

3. Form the product of these primes, using each prime the number of times it is common to **all** factorizations.

4. This product is the GCD. If there are no primes common to all factorizations, the GCD is 1.

---

*The largest common divisor is, of course, the largest common factor, and the GCD could be called the **greatest common factor** and be abbreviated **GCF.**

EXAMPLES

1. Find the GCD for $\{36, 24, 48\}$.

$$\left.\begin{array}{l} 36 = 2 \cdot 2 \cdot 3 \cdot 3 \\ 24 = 2 \cdot 2 \cdot 2 \cdot 3 \\ 48 = 2 \cdot 2 \cdot 2 \cdot 2 \cdot 3 \end{array}\right\} \text{GCD} = 2 \cdot 2 \cdot 3 = 12$$

The factor 2 appears twice, and the factor 3 appears once in **all** the prime factorizations.

2. Find the GCD for $\{360, 75, 30\}$.

$$\left.\begin{array}{l} 360 = 36 \cdot 10 = 4 \cdot 9 \cdot 2 \cdot 5 = 2 \cdot 2 \cdot 2 \cdot 3 \cdot 3 \cdot 5 \\ 75 = 3 \cdot 25 = 3 \cdot 5 \cdot 5 \\ 30 = 6 \cdot 5 = 2 \cdot 3 \cdot 5 \end{array}\right\} \begin{array}{l} \text{GCD} = 3 \cdot 5 \\ = 15 \end{array}$$

Each of the factors 3 and 5 appears only once in **all** the prime factorizations.

3. Find the GCD for $\{168, 420, 504\}$.

$$\left.\begin{array}{l} 168 = 8 \cdot 21 = 2 \cdot 2 \cdot 2 \cdot 3 \cdot 7 \\ 420 = 10 \cdot 42 = 2 \cdot 5 \cdot 6 \cdot 7 \\ \quad\quad = 2 \cdot 2 \cdot 3 \cdot 5 \cdot 7 \\ 504 = 4 \cdot 126 = 2 \cdot 2 \cdot 6 \cdot 21 \\ \quad\quad = 2 \cdot 2 \cdot 2 \cdot 3 \cdot 3 \cdot 7 \end{array}\right\} \text{GCD} = 2 \cdot 2 \cdot 3 \cdot 7 = 84$$

In **all** the prime factorizations, 2 appears twice, 3 once, and 7 once.

If the GCD of two numbers is 1 (that is, they have no common prime factors), then the two numbers are said to be **relatively prime.** The numbers themselves may be prime or they may be composite.

EXAMPLES

1. Find the GCD for $\{15, 8\}$.

$$\left.\begin{array}{l} 15 = 3 \cdot 5 \\ 8 = 2 \cdot 2 \cdot 2 \end{array}\right\} \text{GCD} = 1 \qquad \text{8 and 15 are relatively prime.}$$

2. Find the GCD for $\{20, 21\}$.

$$\left.\begin{array}{l} 20 = 2 \cdot 2 \cdot 5 \\ 21 = 3 \cdot 7 \end{array}\right\} \text{GCD} = 1 \qquad \text{20 and 21 are relatively prime.}$$

EXERCISES III

Find the GCD for each of the following sets of numbers.

1. $\{12, 8\}$ 2. $\{16, 28\}$ 3. $\{85, 51\}$

4. $\{20, 75\}$ 5. $\{20, 30\}$ 6. $\{42, 48\}$

7. $\{15, 21\}$ 8. $\{27, 18\}$ 9. $\{18, 24\}$

10. $\{77, 66\}$ 11. $\{182, 184\}$ 12. $\{110, 66\}$

13. $\{8, 16, 64\}$ 14. $\{121, 44\}$ 15. $\{28, 52, 56\}$

16. $\{98, 147\}$ 17. $\{60, 24, 96\}$ 18. $\{33, 55, 77\}$

19. $\{25, 50, 75\}$ 20. $\{30, 78, 60\}$ 21. $\{17, 15, 21\}$

22. $\{520, 220\}$ 23. $\{14, 55\}$ 24. $\{210, 231, 84\}$

25. $\{140, 245, 420\}$

Which of the following pairs of numbers are relatively prime?

26. $\{35, 24\}$ 27. $\{11, 23\}$ 28. $\{14, 36\}$ 29. $\{72, 35\}$

30. $\{42, 77\}$ 31. $\{16, 51\}$ 32. $\{20, 21\}$ 33. $\{8, 15\}$

34. $\{66, 22\}$ 35. $\{10, 27\}$

Appendix IV

THE DIVIDE-AND-AVERAGE METHOD FOR FINDING SQUARE ROOTS

> TO FIND THE SQUARE ROOT OF A NUMBER BY THE DIVIDE-AND-AVERAGE METHOD
>
> 1. Estimate the square root.
>
> 2. Divide the number by the estimated square root.
>
> 3. Carry the quotient out to one more place than the estimate. If the number in this place is even, leave it; if the number in this place is odd, replace it by the next larger even number.
>
> 4. Then average the estimate and the quotient.
>
> This average is the new estimate of the square root. The process of dividing and averaging is continued until the digits in the quotient and in the average are the same out to the desired decimal place.

An interesting fact about this process is that regardless of what number is used as the first estimate, eventually the same square root will result. This method is particularly adaptable for high-speed computers.

EXAMPLE

Find $\sqrt{300}$ correct to the nearest hundredth by the divide-and-average method. Let the first estimate be 17 since $17^2 = 289$.

$$\frac{17.6 + 17}{2} = \frac{34.6}{2} = 17.3$$

$$
\begin{array}{r}
17.6 \\
17\overline{)300.0} \\
17 \\
\hline
130 \\
119 \\
\hline
11\,0 \\
10\,2 \\
\hline
8
\end{array}
$$

$$\frac{17.34 + 17.3}{2} = \frac{34.64}{2} = 17.32$$

$$
\begin{array}{r}
17.34 \\
17.3.\overline{)300.0.00} \\
173 \\
\hline
127\,0 \\
121\,1 \\
\hline
5\,9\,0 \\
5\,1\,9 \\
\hline
7\,10 \\
6\,92 \\
\hline
18
\end{array}
$$

$$\frac{17.322 + 17.32}{2} = \frac{34.642}{2} = 17.321$$

$$
\begin{array}{r}
17.321 \text{ use } 17.322 \\
17.32.\,\overline{)300.00.0000} \\
173\ 2 \\
\hline
126\ 80 \\
121\ 24 \\
\hline
5\ 56\ 0 \\
5\ 19\ 6 \\
\hline
36\ 40 \\
34\ 64 \\
\hline
1\ 760 \\
1\ 732 \\
\hline
28
\end{array}
$$

$\sqrt{300} \approx 17.32$, to the nearest hundredth.

EXERCISES IV

Use the divide-and-average method to find the following square roots correct to the nearest hundredth.

1. $\sqrt{500}$ 2. $\sqrt{370}$ 3. $\sqrt{2}$ 4. $\sqrt{8}$ 5. $\sqrt{28,900}$

6. $\sqrt{22,500}$ 7. $\sqrt{250}$ 8. $\sqrt{425}$ 9. $\sqrt{95}$ 10. $\sqrt{70}$

11. $\sqrt{86}$ 12. $\sqrt{280}$ 13. $\sqrt{630}$ 14. $\sqrt{18}$ 15. $\sqrt{2500}$

16. If a number is multiplied by 9, and the square root of the new number is found, how will this square root compare with the square root of the original number? Do Problems 10 and 13 and Problems 3 and 14 confirm your answer?

ANSWER KEY

CHAPTER 1

EXERCISES 1.1 PAGE 4

1. $3(10) + 7(1)$.
   thirty-seven

2. $8(10) + 4(1)$
   eighty-four

3. $9(10) + 8(1)$
   ninety-eight

5. $1(100) + 2(10) + 2(1)$
   one hundred twenty-two

6. $4(100) + 9(10) + 3(1)$
   four hundred ninety-three

7. $8(100) + 2(10) + 1(1)$
   eight hundred twenty-one

9. $1(1000) + 8(100) + 9(10) + 2(1)$
   one thousand, eight hundred ninety-two

10. $5(1000) + 4(100) + 9(10) + 6(1)$
    five thousand, four hundred ninety-six

11. $1(10,000) + 2(1000) + 5(100) + 1(10) + 7(1)$
    twelve thousand, five hundred seventeen

13. $2(100,000) + 4(10,000) + 3(1000) + 4(100) + 0(10) + 0(1)$
    two hundred forty-three thousand, four hundred

14. $8(100,000) + 9(10,000) + 1(1000) + 5(100) + 4(10) + 0(1)$
    eight hundred ninety-one thousand, five hundred forty

15. $4(10,000) + 3(1000) + 6(100) + 5(10) + 5(1)$
    forty-three thousand, six hundred fifty-five

17. $8(1,000,000) + 4(100,000) + 0(10,000) + 0(1000) + 8(100) + 1(10) + 0(1)$
    eight million, four hundred thousand, eight hundred ten

18. $5(1,000,000) + 6(100,000) + 6(10,000) + 3(1000) + 7(100) + 0(10) + 1(1)$
    five million, six hundred sixty-three thousand, seven hundred one

19. $1(10,000,000) + 6(1,000,000) + 3(100,000) + 0(10,000) + 2(1000) + 5(100) + 9(10) + 0(1)$
    sixteen million, three hundred two thousand, five hundred ninety

21. $8(10,000,000) + 3(1,000,000) + 0(100,000) + 0(10,000) + 0(1000) + 6(100) + 0(10) + 5(1)$
    eighty-three million, six hundred five

22. $1(100,000,000) + 5(10,000,000) + 2(1,000,000) + 4(100,000) + 0(10,000) + 3(1000) + 6(100) + 7(10) + 2(1)$
    one hundred fifty-two million, four hundred three thousand, six hundred seventy-two

23. $6(100,000,000) + 7(10,000,000) + 9(1,000,000) + 0(100,000) + 7(10,000) + 8(1000) + 1(100) + 0(10) + 0(1)$
    six hundred seventy-nine million, seventy-eight thousand, one hundred

25. $8(1,000,000,000) + 5(100,000,000) + 7(10,000,000) + 2(1,000,000) + 0(100,000) + 0(10,000) + 3(1000) + 4(100) + 2(10) + 5(1)$
    eight billion, five hundred seventy-two million, three thousand, four hundred twenty-five

26. 76

27. 132

29. 3842

30. 2005		**31.** 192,151		**33.** 21,400	
34. 33,333		**35.** 5,045,000		**37.** 10,639,582	
38. 281,300,501		**39.** 530,000,700		**41.** 90,090,090	
42. 82,700,000		**43.** 175,000,002		**45.** 757	

EXERCISES 1.2   PAGE 8

1. 16	**2.** 15	**3.** 13	**5.** 21	**6.** 20	
7. 19	**9.** 12	**10.** 17	**11.** 18	**13.** 12	
14. 21	**15.** 17	**17.** 27	**18.** 18	**19.** 21	
21. commutative		**22.** commutative		**23.** associative	
25. associative		**26.** associative		**27.** identity	
29. identity		**30.** associative		**31.** 162	
33. 239	**34.** 835	**35.** 1298	**37.** 1236	**38.** 4168	
39. 6869		**41.** 1,603,426		**42.** 1,463,930	
43. 2,610,667		**45.** 2762 miles		**46.** $6,313,323	
47. $18,463		**49.** 1518 students		**50.** 33,830 appliances	

EXERCISES 1.3   PAGE 12

1. 3	**2.** 13	**3.** 0	**5.** 9	**6.** 17	**7.** 9
9. 5	**10.** 6	**11.** 0	**13.** 13	**14.** 20	**15.** 20
17. 5	**18.** 45	**19.** 13	**21.** 94	**22.** 126	**23.** 218
25. 475		**26.** 376		**27.** 593	**29.** 188
30. 478		**31.** 1569		**33.** 1568	**34.** 1531
35. 0		**37.** 694		**38.** 5871	**39.** 2517
41. 2,806,644		**42.** 3,800,559		**43.** 1,006,958	**45.** 5,671,011
46. 222		**47.** 140		**49.** 32 years	**50.** 44 points
51. $250,404		**53.** $934		**54.** $235,456	**55.** $39,100
57. $3700		**58.** $868			

EXERCISES 1.4   PAGE 16

1. 72	**2.** 42	**3.** 48	**5.** 30	**6.** 45	**7.** 27
9. 24	**10.** 28	**11.** 56	**13.** 0	**14.** 0	**15.** 9
17. 0	**18.** 0	**19.** 6	**21.** 42	**22.** 60	**23.** 60
25. 30	**26.** 24	**27.** 40	**29.** 96	**30.** 105	**31.** 72
33. 0	**34.** 0	**35.** 0			

45.

GIVEN NO.	ADD 5	TRIPLE	SUBTRACT 15
2	7	21	6
1	6	18	3
7	12	36	21
8	13	39	24
5	10	30	15
6	11	33	18

EXERCISES 1.5   PAGE 19

1. 250	**2.** 7600	**3.** 47,000	**5.** 720	**6.** 13	
7. 3000	**9.** 400	**10.** 3600	**11.** 1200	**13.** 6300	
14. 9000	**15.** 4000	**17.** 15,000	**18.** 3600	**19.** 5200	
21. 16,000	**22.** 180,000	**23.** 10,000		**25.** 60,000	
26. 25,000	**27.** 12,000	**29.** 240,000		**30.** 800,000	
31. 48,000	**33.** 600,000	**34.** 16,000,000		**35.** 630,000	

EXERCISES 1.6   PAGE 22

1. 224	**2.** 162	**3.** 432	**5.** 344	**6.** 432	
7. 455	**9.** 252	**10.** 760	**11.** 2352	**13.** 330	
14. 1189	**15.** 2412	**17.** 960	**18.** 2790	**19.** 7055	
21. 544	**22.** 2548	**23.** 880	**25.** 375	**26.** 4371	
27. 2064	**29.** 156	**30.** 2916	**31.** 5166	**33.** 2850	
34. 7632	**35.** 29,601	**37.** 9800		**38.** 174,045	
39. 125,178	**41.** 31,200	**42.** 66,960		**43.** 380,000	
45. 496,400	**46.** 1,504,000	**47.** 217,300		**49.** 249,600	
50. 897,000	**51.** 6821	**53.** $32,760; $93,600			
54. $24,480; $42,600		**55.** 275 mi; 200 mi			

EXERCISES 1.7   PAGE 27

1. 30	**2.** 10	**3.** 21	**5.** 12
6. 30	**7.** 5	**9.** 6 R4	**10.** 7 R2
11. 24	**13.** 10 R11	**14.** 14	**15.** 11 R7
17. 41	**18.** 45 R6	**19.** 47 R8	**21.** 5 R2
22. 2 R3	**23.** 6 R1	**25.** 6 **26.** 12	**27.** 9
29. 32 R2	**30.** 3 R10	**31.** 9	**33.** 8
34. 15 R5	**35.** 11 R8	**37.** 42 R3	**38.** 50
39. 20	**41.** 30	**42.** 400 R3	**43.** 300 R13
45. 301 R4	**46.** 2 R2	**47.** 3 R3	**49.** 61 R15
50. 54 R3	**51.** 2	**53.** 22 R74	**54.** 4 R192
55. 7 R358	**57.** 196 R370	**58.** 221 R308	**59.** 107 R215
61. Yes; Yes; 23		**62.** Yes; Yes; 62	**63.** 30
65. 22; Yes		**66.** 203; Yes	**67.** $64
69. $10,035		**70.** 102 chairs	

EXERCISES 1.8   PAGE 30

1. 103	**2.** 54	**3.** 6	**5.** 485	**6.** 502	
7. 3000	**9.** 586	**10.** 706	**11.** 82	**13.** 85	
14. $42	**15.** $57	**17.** $932	**18.** 7 hr	**19.** $10,432	

REVIEW QUESTIONS: CHAPTER 1   PAGE 33

1. $4(100) + 9(10) + 5(1)$
four hundred ninety-five

2. $1(1000) + 9(100) + 7(10) + 5(1)$
    one thousand, nine hundred seventy-five
3. $6(10,000) + 0(1000) + 3(100) + 0(10) + 8(1)$
    sixty thousand, three hundred eight
4. 4856   5. 15,032,197   6. 672,340,083
7. commutative prop. of add.   8. associative prop. of mult.
9. associative prop. of add.   10. commutative prop. of mult.
11. $32 \div 8 = 4;\ 2 \div 2 = 1$; division is not associative
12. 10,541   13. 1674   14. 2400   15. 0
16. 480,000   17. 508   18. 2384   19. 2102
20. 5952   21. 3,822,498   22. 14,388,000   23. 292 R2
24. 135 R81   25. 703 R7   26. 1059   27. 35
28. 9   29. $1485; $99   30. 83
31. If a is a whole number, then there is a unique whole number 0 with
    the property that $a + 0 = a$. If a is a whole number, then there is a
    unique whole number 1 with the property that $a \cdot 1 = a$.
32. 70

33.

GIVEN NO.	ADD 100	DOUBLE	SUBTRACT 200
3	103	206	6
20	120	240	40
15	115	230	30
8	108	216	16

34. 20 times

CHAPTER TEST: CHAPTER 1   PAGE 35

1. $8(1000) + 9(100) + 5(10) + 2(1)$
    eight thousand, nine hundred fifty-two
2. identity   3. $7 \cdot 9 = 9 \cdot 7$
4. 12,009   5. 1735   6. 13,781,661   7. 488
8. 1229   9. 5707   10. 2584   11. 220,405
12. 210,938   13. 403   14. 172 R388   15. 2005
16. 74   17. 54   18. $306; $51; $224

CHAPTER 2

EXERCISES 2.1   PAGE 38

	Exponent	Base	Power		Exponent	Base	Power
1.	3	2	8	2.	5	2	32
3.	2	5	25	5.	0	7	1
6.	2	11	121	7.	4	1	1
9.	0	4	1	10.	6	3	729

11. 2	3	9		**13.** 0	5	1
14. 50	1	1		**15.** 1	62	62
17. 2	10	100		**18.** 3	10	1000
19. 2	4	16		**21.** 4	10	10,000
22. 3	5	125		**23.** 3	6	216
25. 0	9	1				

26. 2^2	**27.** 5^2	**29.** 3^3	**30.** 2^5
31. 11^2	**33.** 2^3	**34.** 3^2	**35.** 6^2
37. 9^2 or 3^4	**38.** 8^2 or 4^3 or 2^6	**39.** 10^2	**41.** 10^4
42. 6^3	**43.** 12^2	**45.** 3^5	**46.** 25^2 or 5^4
47. 15^2	**49.** 7^3	**50.** 10^5	**51.** 6^5
53. $2^2 \cdot 7^2$	**54.** $5^2 \cdot 9^3$	**55.** $2^2 \cdot 3^3$	**57.** $7^2 \cdot 13$
58. 11^3	**59.** $2 \cdot 3^2 \cdot 11^2$		

EXERCISES 2.2   PAGE 41

1. 3	**2.** 15	**3.** 7	**5.** 22	**6.** 31
7. 3	**9.** 5	**10.** 10	**11.** 5	**13.** 5
14. 0	**15.** 3	**17.** 7	**18.** 0	**19.** 0
21. 6	**22.** 3	**23.** 3	**25.** 27	**26.** 0
27. 26	**29.** 0	**30.** 5	**31.** 69	**33.** 68
34. 80	**35.** 140	**37.** 5	**38.** 9	**39.** 0
41. 9	**42.** 93	**43.** 12	**45.** 24	**46.** 118
47. 70	**49.** 230	**50.** 34		

EXERCISES 2.3   PAGE 45

1. 2, 3, 4, 9	**2.** 3, 9	**3.** 3, 5	**5.** 2, 3, 5, 10
6. 3	**7.** 2, 4	**9.** 2, 3, 4	**10.** 3, 5
11. none	**13.** 3, 9	**14.** 2, 3, 4, 5, 10	**15.** none
17. none	**18.** 2	**9.** 3	**21.** 3, 5
22. 3	**23.** 2, 3, 4, 5, 9, 10	**25.** 2, 3	**26.** 2
27. none	**29.** 2	**30.** 2, 3	**31.** 3, 5
33. 3, 9	**34.** 3, 5	**35.** 2, 3, 9	**37.** 2, 4, 5, 10
38. 2, 3, 4, 9	**39.** 3, 9	**41.** 2, 3, 4, 5, 10	**42.** none
43. 2, 4	**45.** 2, 3, 4, 5, 10	**46.** 2, 3, 5, 10	**47.** 3, 5
49. none	**50.** 2	**51.** 2, 3, 9	**53.** 2, 3, 4
54. 5	**55.** 2, 3, 4, 9	**57.** 3, 9	**58.** 2
59. 2, 3, 4	**61.** yes; 18, 36, 54, 72, 90	**62.** no; 9, 18, 36, 45, 63	

EXERCISES 2.4   PAGE 50

1. 5, 10, 15, 20, . . .	**2.** 7, 14, 21, 28, . . .	**3.** 11, 22, 33, 44, . . .
5. 12, 24, 36, 48, . . .	**6.** 9, 18, 27, 36, . . .	**7.** 20, 40, 60, 80, . . .
9. 16, 32, 48, 64, . . .	**10.** 25, 50, 75, 100, . . .	

11.

1	(2)	(3)	4	(5)	6	(7)	8	9	10
(11)	12	(13)	14	15	16	(17)	18	(19)	20
21	22	(23)	24	25	26	27	28	(29)	30
(31)	32	33	34	35	36	(37)	38	39	40
(41)	42	(43)	44	45	46	(47)	48	49	50
51	52	(53)	54	55	56	57	58	(59)	60
(61)	62	63	64	65	66	(67)	68	69	70
(71)	72	(73)	74	75	76	77	78	(79)	80
81	82	(83)	84	85	86	87	88	(89)	90
91	92	93	94	95	96	(97)	98	99	100

13. prime   **14.** composite; 4, 7   **15.** composite; 4, 8
17. prime   **18.** composite; 2, 8   **19.** composite; 7, 9
21. composite; 3, 17   **22.** prime   **23.** prime
25. prime   **26.** composite; 4, 13   **27.** composite; 3, 19
29. composite; 2, 43   **30.** prime   **31.** prime
33. 3, 4   **34.** 8, 2   **35.** 12, 1   **37.** 25, 2   **38.** 5, 4
39. 8, 3   **41.** 12, 3   **42.** 7, 1   **43.** 21, 3   **45.** 5, 5
46. 4, 4   **47.** 12, 5   **49.** 9, 3   **50.** 18, 4   **51.** 2
53. $1 \cdot 52$ $\dfrac{52}{1\overline{)52}}$ $\dfrac{1}{52\overline{)52}}$ $\dfrac{26}{2\overline{)52}}$ $\dfrac{2}{26\overline{)52}}$ $\dfrac{13}{4\overline{)52}}$ $\dfrac{4}{13\overline{)52}}$
     $2 \cdot 26$
     $4 \cdot 13$

EXERCISES 2.5   PAGE 54

1. $2^3 \cdot 3$ **2.** $2^2 \cdot 7$ **3.** 3^3 **5.** $2^2 \cdot 3^2$
6. $2^2 \cdot 3 \cdot 5$ **7.** $2^3 \cdot 3^2$ **9.** 3^4 **10.** $3 \cdot 5 \cdot 7$
11. 5^3 **13.** $3 \cdot 5^2$ **14.** $2 \cdot 3 \cdot 5^2$ **15.** $2 \cdot 3 \cdot 5 \cdot 7$
17. $2 \cdot 5^3$ **18.** $3 \cdot 31$ **19.** $2^3 \cdot 3 \cdot 7$ **21.** $2 \cdot 3^2 \cdot 7$
22. $2^4 \cdot 3$ **23.** 17   **25.** $3 \cdot 17$ **26.** $2^4 \cdot 3^2$
27. 11^2 **29.** $3^2 \cdot 5^2$ **30.** $2^2 \cdot 13$ **31.** 2^5
33. $2^2 \cdot 3^3$ **34.** 103   **35.** 101   **37.** $2 \cdot 3 \cdot 13$
38. $2^2 \cdot 5^3$ **39.** $2^4 \cdot 5^4$ **41.** 1, 2, 3, 4, 6, 12
42. 1, 2, 3, 6, 9, 18   **43.** 1, 2, 4, 7, 14, 28
45. 1, 11, 121   **46.** 1, 3, 5, 9, 15, 45
47. 1, 3, 5, 7, 15, 21, 35, 105   **49.** 1, 97
50. 1, 2, 3, 4, 6, 8, 9, 12, 16, 18, 24, 36, 48, 72, 144

EXERCISES 2.6   PAGE 58

1. 24   **2.** 105   **3.** 36   **5.** 110   **6.** 252

7.	60	**9.**	200	**10.**	600	**11.**	196	**13.** 240
14.	112	**15.**	100	**17.**	700	**18.**	432	**19.** 252
21.	8	**22.**	210	**23.**	1560	**25.**	60	**26.** 120
27.	240	**29.**	2250	**30.**	918	**31.**	726	**33.** 2610
34.	324	**35.**	675	**37.**	120	**38.**	600	**39.** 1680

41. 120: $8 \cdot 15$ **42.** 30: $6 \cdot 5$ **43.** 120: $10 \cdot 12$
$10 \cdot 12$ $15 \cdot 2$ $15 \cdot 8$
$15 \cdot 8$ $30 \cdot 1$ $24 \cdot 5$

45. 270: $6 \cdot 45$ **46.** 1140: $12 \cdot 95$ **47.** 4410: $45 \cdot 98$
$18 \cdot 15$ $95 \cdot 12$ $63 \cdot 70$
$27 \cdot 10$ $228 \cdot 5$ $98 \cdot 45$
$45 \cdot 6$

49. 14,157: $99 \cdot 143$ **50.** 3375: $125 \cdot 27$ **51.** 70 min;
$143 \cdot 99$ $135 \cdot 25$ 7 times,
$363 \cdot 39$ $225 \cdot 19$ 5 times

53. 48 hr; 4 orbits and 3 orbits, respectively
54. Once every 30 days; once every 30 days
55. 180 days; 18 trips, 15 trips, 12 trips, 10 trips

REVIEW QUESTIONS: CHAPTER 2   PAGE 61

1. base, exponent, power   **2.** different factors   **3.** prime
4. composite   **5.** 2^7 **6.** 13^2
7. 15   **8.** 46   **9.** 35
10. 13   **11.** 2   **12.** 2
13. 3, 5, 9   **14.** 2, 3, 4, 9   **15.** none
16. 2, 3, 4, 5, 9, 10   **17.** 2, 4   **18.** 3
19. 3, 6, 9, 12, . . . ; yes, 3 is prime.   **20.** yes
21. 2, 3, 5, 7, 11, 13, 17, 19, 23, 29, 31, 37, 41, 43, 47, 53, 59
22. 6 and 4   **23.** 5 and 12   **24.** $2 \cdot 3 \cdot 5^2$ **25.** $5 \cdot 13$
26. $2^2 \cdot 3 \cdot 7$ **27.** $2^2 \cdot 23$ **28.** 168   **29.** 1800
30. 270   **31.** 1638: $18 \cdot 91$, $39 \cdot 42$, $63 \cdot 26$
32. 210 sec; 14 laps, 12 laps

CHAPTER TEST: CHAPTER 2   PAGE 63

1. exponent   **2.** 5^3 **3.** $35 = 5 \cdot 7$
4. The last two digits are divisible by 4 so the number is divisible by 4.
The last digit is not 0 or 5 so the number is not divisible by 5.
5. 6, 12, 18, 24, . . . ; no   **6.** 57   **7.** 5
8. 1   **9.** $2^2 \cdot 3 \cdot 5^2$ **10.** $2^2 \cdot 13$
11. 2^6 **12.** $2^2 \cdot 3^2 \cdot 5$ **13.** 75: $15 \cdot 5$, $25 \cdot 3$, $75 \cdot 1$
14. 1470: $42 \cdot 35$, $105 \cdot 14$, $147 \cdot 10$ **15.** 300 sec; 5 laps, 4 laps

CHAPTER 3

EXERCISES 3.1   PAGE 69

1. (a) 0, (b) 0, (c) 0, (d) 0   **2.** (a) undefined, (b) undefined,
(c) undefined, (d) undefined

3. $\dfrac{1}{3}$ **5.** $\dfrac{1}{4}$ **6.** $\dfrac{3}{16}$ **7.** $\dfrac{1}{4}$

9.

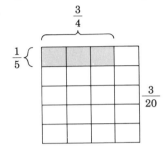

10.

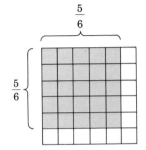

11.

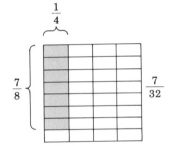

13.

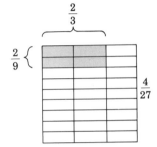

14. $\dfrac{3}{25}$ **15.** $\dfrac{1}{16}$ **17.** $\dfrac{12}{35}$ **18.** $\dfrac{12}{35}$

19. $\dfrac{1}{9}$ **21.** $\dfrac{3}{32}$ **22.** $\dfrac{4}{25}$ **23.** $\dfrac{9}{49}$

25. $\dfrac{15}{32}$ **26.** $\dfrac{4}{81}$ **27.** 0   **29.** $\dfrac{35}{12}$

30. $\dfrac{12}{1} = 12$ **31.** $\dfrac{10}{1} = 10$ **33.** $\dfrac{45}{2}$ **34.** $\dfrac{42}{5}$

35. $\dfrac{32}{15}$ **37.** $\dfrac{99}{20}$ **38.** $\dfrac{6}{385}$ **39.** $\dfrac{48}{455}$

41. $\dfrac{1}{360}$ **42.** $\dfrac{27}{100}$ **43.** $\dfrac{840}{1} = 840$ **45.** 0

46. 0   **47.** 0   **49.** $\dfrac{728}{45}$ **50.** $\dfrac{9}{500}$

EXERCISES 3.2   PAGE 74

1. 21   **2.** 4   **3.** 10   **5.** 45   **6.** 20   **7.** 16

9. 42	**10.** 60	**11.** 54	**13.** 40	**14.** 15	**15.** 44
17. 10	**18.** 32	**19.** 45	**21.** 6	**22.** 9	**23.** 9
25. 6	**26.** 6	**27.** 1	**29.** 11	**30.** 8	**31.** 30

33. $\dfrac{1}{3}$ **34.** $\dfrac{2}{3}$ **35.** $\dfrac{3}{4}$ **37.** $\dfrac{2}{5}$ **38.** $\dfrac{4}{5}$ **39.** $\dfrac{7}{18}$

41. $\dfrac{0}{25} = 0$ **42.** $\dfrac{3}{4}$ **43.** $\dfrac{2}{5}$ **45.** $\dfrac{5}{6}$ **46.** $\dfrac{1}{4}$

47. $\dfrac{2}{3}$ **49.** $\dfrac{2}{3}$ **50.** $\dfrac{3}{4}$ **51.** $\dfrac{6}{25}$ **53.** $\dfrac{2}{3}$

54. $\dfrac{2}{3}$ **55.** $\dfrac{12}{35}$ **57.** $\dfrac{2}{9}$ **58.** $\dfrac{3}{7}$ **59.** $\dfrac{25}{76}$

61. $\dfrac{1}{2}$ **62.** $\dfrac{4}{1} = 4$ **63.** $\dfrac{6}{1} = 6$ **65.** $\dfrac{2}{17}$ **66.** $\dfrac{3}{8}$

67. $\dfrac{9}{20}$ **69.** $\dfrac{10}{9}$ **70.** $\dfrac{11}{15}$ **71.** $\dfrac{5}{4}$ **73.** $\dfrac{5}{18}$

74. $\dfrac{1}{6}$ **75.** $\dfrac{1}{4}$ **77.** $\dfrac{21}{4}$ **78.** $\dfrac{189}{52}$ **79.** $\dfrac{77}{4}$

81. $\dfrac{8}{5}$ **82.** $\dfrac{3}{10}$ **83.** $\dfrac{2}{7}$

EXERCISES 3.3 PAGE 80

1.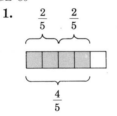

2.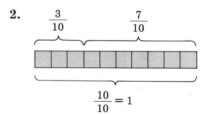

3.

5. $\dfrac{5}{16} + \dfrac{7}{16} = \dfrac{1}{16} \cdot 5 + \dfrac{1}{16} \cdot 7 = \dfrac{1}{16}(5 + 7) = \dfrac{1}{16} \cdot 12 = \dfrac{12}{16}$

6. $\dfrac{10}{10} = 1$ **7.** $\dfrac{5}{14}$ **9.** $\dfrac{3}{2}$ **10.** $\dfrac{3}{2}$

11. $\dfrac{10}{5} = 2$ **13.** $\dfrac{5}{3}$ **14.** $\dfrac{3}{5}$ **15.** $\dfrac{13}{18}$

17. $\dfrac{11}{16}$ **18.** $\dfrac{23}{20}$ **19.** $\dfrac{3}{5}$ **21.** $\dfrac{12}{12} = 1$

22. $\dfrac{11}{16}$ 23. $\dfrac{17}{20}$ 25. $\dfrac{17}{21}$ 26. $\dfrac{3}{4}$

27. $\dfrac{9}{13}$ 29. $\dfrac{23}{54}$ 30. $\dfrac{151}{140}$ 31. $\dfrac{23}{72}$

33. $\dfrac{9}{4}$ 34. $\dfrac{3}{5}$ 35. $\dfrac{1}{2}$ 37. $\dfrac{5}{8}$

38. $\dfrac{49}{60}$ 39. $\dfrac{5}{6}$ 41. $\dfrac{343}{432}$ 42. $\dfrac{31}{96}$

43. $\dfrac{13}{9}$ 45. $\dfrac{1}{5}$ 46. $\dfrac{7}{6}$ 47. $\dfrac{317}{1000}$

49. $\dfrac{271}{10,000}$ 50. $\dfrac{631}{100}$ 51. $\dfrac{8191}{1000}$ 53. $\dfrac{753}{1000}$

54. $\dfrac{63}{50}$ 55. $\dfrac{89}{200}$ 57. $\dfrac{613}{1000}$ 58. $\dfrac{27,683}{10,000}$

59. $\dfrac{5134}{1000} = \dfrac{2567}{500}$

EXERCISES 3.4   PAGE 83

1. (a)   (b)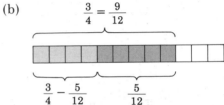

(c)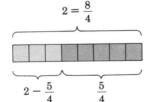

2. $\dfrac{3}{4}$ by $\dfrac{1}{12}$ 3. $\dfrac{7}{8}$ by $\dfrac{1}{24}$ 5. $\dfrac{4}{10}$ by $\dfrac{1}{40}$ 6. $\dfrac{13}{20}$ by $\dfrac{1}{40}$

7. $\dfrac{21}{25}$ by $\dfrac{11}{400}$ 9. $\dfrac{7}{24}$ by $\dfrac{1}{72}$ 10. $\dfrac{11}{48}$ by $\dfrac{1}{60}$ 11. $\dfrac{37}{100}$ by $\dfrac{1}{20}$

13. $\dfrac{8}{9}, \dfrac{9}{10}, \dfrac{11}{12}$ 14. $\dfrac{11}{12}, \dfrac{19}{20}, \dfrac{7}{6}$ 15. $\dfrac{40}{36}, \dfrac{31}{24}, \dfrac{17}{12}$ 17. $\dfrac{13}{18}, \dfrac{31}{36}, \dfrac{7}{8}$

18. $\dfrac{20}{10,000}, \dfrac{3}{1000}, \dfrac{1}{100}$ 19. $\dfrac{298}{1000}, \dfrac{3}{10}, \dfrac{32}{100}, \dfrac{3,333}{10,000}$

21. $\dfrac{3}{7}$ 22. $\dfrac{6}{10} = \dfrac{3}{5}$ 23. $\dfrac{4}{8} = \dfrac{1}{2}$ 25. $\dfrac{9}{15} = \dfrac{3}{5}$ 26. $\dfrac{3}{6} = \dfrac{1}{2}$

27. $\dfrac{13}{30}$ **29.** $\dfrac{9}{32}$ **30.** $\dfrac{1}{40}$ **31.** $\dfrac{7}{54}$ **33.** $\dfrac{8}{15}$

34. $\dfrac{19}{180}$ **35.** $\dfrac{31}{288}$ **37.** $\dfrac{27}{8}$ **38.** $\dfrac{23}{16}$ **39.** $\dfrac{3}{16}$

41. $\dfrac{87}{100}$ **42.** $\dfrac{59}{1000}$ **43.** $\dfrac{3759}{5000}$ **45.** $\dfrac{11}{48}$ **46.** $\dfrac{17}{18}$

47. $\dfrac{3}{32}$

EXERCISES 3.5   PAGE 88

1. $\dfrac{4}{3}$ **2.** $\dfrac{5}{2}$ **3.** $\dfrac{4}{3}$ **5.** $\dfrac{3}{2}$ **6.** $\dfrac{3}{2}$

7. $\dfrac{7}{5}$ **9.** $\dfrac{5}{4}$ **10.** $\dfrac{5}{4}$ **11.** $4\dfrac{1}{6}$ **13.** $1\dfrac{1}{3}$

14. $1\dfrac{1}{4}$ **15.** $1\dfrac{1}{2}$ **17.** $6\dfrac{1}{7}$ **18.** $2\dfrac{1}{8}$ **19.** $7\dfrac{1}{2}$

21. $3\dfrac{1}{9}$ **22.** $2\dfrac{1}{15}$ **23.** 3 **25.** $4\dfrac{1}{2}$ **26.** $2\dfrac{1}{17}$

27. $\dfrac{3}{1}=3$ **29.** $1\dfrac{3}{4}$ **30.** $1\dfrac{17}{20}$ **31.** $\dfrac{37}{8}$ **33.** $\dfrac{76}{15}$

34. $\dfrac{8}{5}$ **35.** $\dfrac{46}{11}$ **37.** $\dfrac{7}{3}$ **38.** $\dfrac{34}{7}$ **39.** $\dfrac{32}{3}$

41. $\dfrac{34}{5}$ **42.** $\dfrac{71}{5}$ **43.** $\dfrac{50}{3}$ **45.** $\dfrac{101}{5}$ **46.** $\dfrac{47}{5}$

47. $\dfrac{92}{7}$ **49.** $\dfrac{17}{1}=17$ **50.** $\dfrac{151}{50}$

EXERCISES 3.6   PAGE 92

1. $7\dfrac{2}{3}$ **2.** $10\dfrac{3}{8}$ **3.** $42\dfrac{7}{20}$ **5.** $10\dfrac{3}{4}$ **6.** $7\dfrac{13}{35}$

7. $12\dfrac{7}{27}$ **9.** $18\dfrac{23}{45}$ **10.** $18\dfrac{1}{15}$ **11.** $28\dfrac{1}{2}$ **13.** $9\dfrac{29}{40}$

14. $12\dfrac{17}{24}$ **15.** $12\dfrac{47}{60}$ **17.** $63\dfrac{11}{24}$ **18.** $26\dfrac{35}{48}$ **19.** $53\dfrac{83}{192}$

21. $3\dfrac{1}{2}$ **22.** $4\dfrac{3}{5}$ **23.** $3\dfrac{5}{8}$ **25.** $3\dfrac{7}{8}$ **26.** $10\dfrac{4}{5}$

27. $3\dfrac{29}{32}$ **29.** $3\dfrac{1}{60}$ **30.** $1\dfrac{7}{16}$ **31.** $3\dfrac{9}{10}$ **33.** $7\dfrac{4}{5}$

34. $\dfrac{5}{8}$ **35.** $57\dfrac{1}{6}$ **37.** $\dfrac{13}{15}$ **38.** $1\dfrac{13}{16}$ **39.** 17

41. $8\dfrac{7}{12}$ hr, or 8 hr 35 min   **42.** $35\dfrac{7}{10}$ km   **43.** $\dfrac{3}{5}$ hr; 36 min

45. $\dfrac{5}{6}$ hr   **46.** $219\dfrac{13}{16}$ lb   **47.** $89\dfrac{1}{8}$ ft

49. $6\frac{3}{5}$ min   **50.** $4\frac{5}{8}$ dollars

EXERCISES 3.7   PAGE 96

1. $7\frac{7}{12}$ **2.** $1\frac{13}{35}$ **3.** $10\frac{1}{2}$ **5.** $22\frac{1}{2}$ **6.** 12

7. $31\frac{1}{6}$ **9.** 8   **10.** 12   **11.** 30   **13.** $21\frac{3}{4}$

14. 1   **15.** 1   **17.** $27\frac{1}{2}$ **18.** $10\frac{1}{5}$ **19.** $19\frac{1}{4}$

21. $\frac{1}{7}$ **22.** $\frac{1}{5}$ **23.** $\frac{1}{10}$ **25.** $\frac{36}{385}$ **26.** $39\frac{3}{8}$

27. 11   **29.** 242   **30.** $15\frac{99}{160}$ **31.** $11\frac{3}{8}$ **33.** 17

34. $1\frac{3}{10}$ **35.** $2\frac{1}{10}$ **37.** $1827\frac{1}{10}$ **38.** $1291\frac{5}{42}$

39. $3016\frac{26}{75}$ **41.** 40   **42.** 20   **43.** 20

45. 75   **46.** 100   **47.** $\frac{5}{16}$ **49.** $\frac{9}{14}$

50. $\frac{7}{10}$ **51.** 10 ft, 22 ft   **53.** 177 mi

54. $1012\frac{7}{8}$ ft   **55.** 90 pages; 18 hr

EXERCISES 3.8   PAGE 101

1. $\frac{8}{9}$ **2.** $\frac{4}{15}$ **3.** $\frac{5}{7}$ **5.** $\frac{7}{5} = 1\frac{2}{5}$

6. $\frac{11}{3} = 3\frac{2}{3}$ **7.** $\frac{1}{3}$ **9.** $\frac{3}{4}$ **10.** $\frac{7}{16}$

11. $\frac{4}{9}$ **13.** $\frac{5}{8}$ **14.** $\frac{24}{35}$ **15.** $\frac{39}{32} = 1\frac{7}{32}$

17. $\frac{9}{10}$ **18.** $\frac{10}{9} = 1\frac{1}{9}$ **19.** 1   **21.** $\frac{32}{21} = 1\frac{11}{21}$

22. $\frac{8}{3} = 2\frac{2}{3}$ **23.** $\frac{10}{3} = 3\frac{1}{3}$ **25.** $\frac{16}{21}$ **26.** $\frac{10}{39}$

27. $\frac{7}{60}$ **29.** $\frac{63}{50} = 1\frac{13}{50}$ **30.** $\frac{28}{17} = 1\frac{11}{17}$ **31.** $\frac{62}{39} = 1\frac{23}{39}$

33. $\frac{41}{12} = 3\frac{5}{12}$ **34.** $\frac{7}{5} = 1\frac{2}{5}$ **35.** $\frac{41}{3} = 13\frac{2}{3}$ **37.** $\frac{37}{2} = 18\frac{1}{2}$

38. $\frac{9}{32}$ **39.** $\frac{20}{11} = 1\frac{9}{11}$ **41.** $\frac{2}{3}$ **42.** $\frac{1}{5}$

43. 0   **45.** $\frac{43}{450}$ **46.** $\frac{172}{9} = 19\frac{1}{9}$ **47.** $\frac{111}{20} = 5\frac{11}{20}$

49. $\dfrac{3}{2} = 1\dfrac{1}{2}$ **50.** $\dfrac{3}{5}$ **51.** $\dfrac{353}{630}$ **53.** $\dfrac{50}{27} = 1\dfrac{23}{27}$

54. $\dfrac{62}{43} = 1\dfrac{19}{43}$ **55.** $\dfrac{27}{46}$ **57.** $48   **58.** $135,000

EXERCISES 3.9   PAGE 104

1. $\dfrac{15}{7}$ **2.** $\dfrac{44}{25}$ **3.** $\dfrac{11}{70}$ **5.** $\dfrac{5}{7}$

6. $\dfrac{12}{7}$ **7.** $\dfrac{1}{5}$ **9.** $\dfrac{111}{31}$ **10.** $\dfrac{5}{2}$

11. $\dfrac{25}{21}$ **13.** $\dfrac{33}{4}$ **14.** $\dfrac{328}{133}$ **15.** $\dfrac{76}{135}$

17. $\dfrac{429}{185}$ **18.** $\dfrac{27}{40}$ **19.** $\dfrac{63}{19}$ **21.** 140

22. $\dfrac{69}{35}$ **23.** $\dfrac{60}{167}$ **25.** $\dfrac{377}{373}$ **26.** 1

27. 0   **29.** 1   **30.** $\dfrac{211}{1211}$

REVIEW QUESTIONS: CHAPTER 3   PAGE 108

1. 0   **2.** undefined   **3.** $\dfrac{3}{2}, \dfrac{2}{3}$ **4.** associative

5.   **6.** $\dfrac{1}{30}$ **7.** $\dfrac{3}{49}$

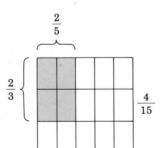

8. $\dfrac{1}{12}$ **9.** 2   **10.** 60   **11.** 65

12. $\dfrac{1}{2}$ **13.** $\dfrac{9}{8}$ **14.** 0   **15.** $\dfrac{5}{4}$

16. $\dfrac{5}{7}$ **17.** $\dfrac{2}{3}$ **18.** $\dfrac{1}{4}$ **19.** $\dfrac{49}{72}$

20. $\dfrac{7}{22}$ **21.** $\dfrac{25}{54}$ **22.** $7\dfrac{5}{6}$ **23.** $3\dfrac{21}{50}$

24. $\dfrac{51}{10}$ **25.** $\dfrac{41}{3}$ **26.** $30\dfrac{3}{4}$ **27.** $6\dfrac{7}{12}$

28. $14\dfrac{1}{10}$ **29.** $6\dfrac{47}{56}$ **30.** $\dfrac{1}{5}$ **31.** 104

32. $\dfrac{2}{9}$ **33.** $\dfrac{3}{4}$ **34.** (a) 45; (b) 64   **35.** $\dfrac{91}{120}$

36. $\dfrac{4}{5}$ by $\dfrac{2}{15}$ **37.** $\dfrac{25}{112}$ **38.** $\dfrac{33}{4}$ **39.** $31\dfrac{7}{8}$ ft

40. $122\dfrac{5}{12}$ ft   **41.** $\dfrac{39}{124}$ **42.** $1000

CHAPTER TEST: CHAPTER 3   PAGE 111

1. 0   **2.** reciprocals

3.   **4.** commutative

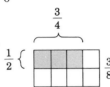

5. $\dfrac{27}{100}$; 75   **6.** $\dfrac{2}{105}$ **7.** $\dfrac{1}{2}$ **8.** $\dfrac{4}{5}$

9. $\dfrac{3}{4}$ **10.** $\dfrac{9}{11}$ **11.** $\dfrac{1}{10}$ **12.** $\dfrac{8}{9}$

13. $\dfrac{29}{66}$ **14.** $\dfrac{1}{8}$ **15.** $\dfrac{16}{169}$ **16.** $\dfrac{39}{1} = 39$

17. $\dfrac{7}{12}$ **18.** $\dfrac{1}{15}$ **19.** $\dfrac{19}{80}$ **20.** $\dfrac{131}{80}$

21. $\dfrac{74}{15}$ or $4\dfrac{14}{15}$ **22.** $\dfrac{10}{3}$ or $3\dfrac{1}{3}$ **23.** $42   **24.** $36\dfrac{1}{3}$ ft

CHAPTER 4

EXERCISES 4.1   PAGE 116

1. 37.498   **2.** 18.76   **3.** 4.11   **5.** 87.003

6. 95.2   **7.** 62.7   **9.** 100.38   **10.** 250.623

11. $82\dfrac{56}{100}$ **13.** $10\dfrac{576}{1000}$ **14.** $100\dfrac{6}{10}$ **15.** $65\dfrac{3}{1000}$

17. 0.014   **18.** 0.17   **19.** 6.28   **21.** 72.392

22. 850.0036   **23.** 700.77   **25.** 600,500.402

26. five tenths   **27.** ninety-three hundredths

29. thirty-two and fifty-eight hundredths

30. seventy-one and six hundredths

31. thirty-five and seventy-eight thousandths

33. eighteen and one hundred two thousandths

34. fifty and eight thousandths

35. six hundred seven and six hundred seven thousandths
37. five hundred ninety-three and eighty-six hundredths
38. four thousand, seven hundred and six hundred seventeen thousandths
39. five thousand and five thousandths
41. nine hundred and four thousand, six hundred thirty-eight ten-thousandths

EXERCISES 4.2   PAGE 119

1.	4.8	**2.**	5.0	**3.**	76.3	**5.**	89.0
6.	7.6	**7.**	18.0	**9.**	14.3	**10.**	0.0
11.	0.39	**13.**	5.72	**14.**	8.99	**15.**	7.00
17.	0.08	**18.**	6.00	**19.**	5.71	**21.**	0.067
22.	0.056	**23.**	0.634	**25.**	32.479	**26.**	9.430
27.	17.364	**29.**	0.002	**30.**	20.770	**31.**	479.
33.	18.	**34.**	20.	**35.**	382.	**37.**	440.
38.	701.	**39.**	6333.	**41.**	5160.	**42.**	6480.
43.	500.	**45.**	1000.	**46.**	380.	**47.**	5480.
49.	92,540.	**50.**	7010.	**51.**	7000.	**53.**	48,000.
54.	103,000.	**55.**	217,000.	**57.**	380,000.	**58.**	4,501,000.
59.	7,305,000.	**61.**	0.00058	**62.**	0.54	**63.**	470.
65.	500.	**66.**	6000.	**67.**	3.230	**69.**	80,000.
70.	78,420.						

EXERCISES 4.3   PAGE 122

1.	2.3	**2.**	11.5	**3.**	7.55	**5.**	72.31
6.	4.6926	**7.**	276.096	**9.**	44.6516	**10.**	481.25
11.	118.333	**13.**	7.148	**14.**	7.4914	**15.**	93.877
17.	103.429	**18.**	46.943	**19.**	137.150	**21.**	1.44
22.	8.93	**23.**	15.89	**25.**	64.947	**26.**	4.895
27.	4.7974	**29.**	2.9434	**30.**	34.186	**31.**	$95.50; $4.50
33.	$94.85; $5.15		**34.**	12.28 in.		**35.**	3.0284; 95.08; 98.1084

EXERCISES 4.4   PAGE 125

1.	0.42	**2.**	7.5	**3.**	0.28	**5.**	18.6
6.	0.004	**7.**	0.025	**9.**	0.246	**10.**	0.07
11.	0.0006	**13.**	0.094	**14.**	0.0663	**15.**	10.79
17.	3.	**18.**	14.	**19.**	1.	**21.**	346.
22.	2057.	**23.**	782.	**25.**	4.35	**26.**	4178.2
27.	0.38	**29.**	71,200	**30.**	251,480.	**31.**	0.000045
33.	0.1484	**34.**	0.00092481	**35.**	7.905642	**37.**	$240.90
38.	$4209.50	**39.**	$3057.60	**41.**	382.4 mi	**42.**	7.8; 8.0; no

EXERCISES 4.5   PAGE 129

1.	2.34	**2.**	0.57	**3.**	9.9
5.	0.08	**6.**	0.9	**7.**	2056.
9.	20.	**10.**	5.	**11.**	0.002
13.	$0.9428 \approx 0.943$	**14.**	0.13	**15.**	$0.8444 \approx 0.844$
17.	$0.1666 \approx 0.167$	**18.**	1.3	**19.**	3.03
21.	$0.0706 \approx 0.071$	**22.**	$12.6567 \approx 12.657$	**23.**	310.6

25. $20.9876 \approx 20.988$ **26.** $511.1111 \approx 511.111$

27.	$7.3239 \approx 7.324$	**29.**	28.4	**30.**	$4.2463 \approx 4.246$
31.	0.784	**33.**	0.0005036	**34.**	0.45621
35.	7.682	**37.**	0.000001826	**38.**	91.112
39.	0.0006122	**41.**	70.4	**42.**	442.8 mi
43.	226.8 mi	**45.**	19.0 mph	**46.**	$1.25
47.	495.9 mi	**49.**	$23.85	**50.**	$295.
51.	$693.25				

EXERCISES 4.6   PAGE 133

1. $\dfrac{2 \text{ nickels}}{3 \text{ nickels}}$ **2.** $\dfrac{6 \text{ chairs}}{5 \text{ students}}$ **3.** $\dfrac{6 \text{ feet}}{5 \text{ feet}}$

5. $\dfrac{1 \text{ bookshelf}}{6 \text{ feet}}$ **6.** $\dfrac{10 \text{ cm}}{10 \text{ cm}} = 1$ **7.** $\dfrac{100 \text{ cm}}{100 \text{ cm}} = 1$

9. $\dfrac{21 \text{ dollars}}{4 \text{ stocks}}$ **10.** $\dfrac{19 \text{ miles}}{1 \text{ gallon}}$

11.	true	**13.**	true	**14.**	true	**15.**	false	**17.**	true
18.	true	**19.**	true	**21.**	true	**22.**	true	**23.**	true
25.	true	**26.**	false	**27.**	false	**29.**	false	**30.**	false
31.	true	**33.**	false	**34.**	true	**35.**	true	**37.**	true
38.	false	**39.**	true						

EXERCISES 4.7   PAGE 136

1.	$x = 12$	**2.**	$y = 2$	**3.**	$z = 20$	**5.**	$B = 40$
6.	$B = 21$	**7.**	$x = 50$	**9.**	$A = \dfrac{21}{2}$	**10.**	$x = 5$
11.	$D = 100$	**13.**	$x = 1$	**14.**	$y = 28\dfrac{2}{9}$	**15.**	$x = \dfrac{3}{5}$
17.	$w = 24$	**18.**	$w = 150$	**19.**	$y = 6$	**21.**	$x = \dfrac{3}{2}$
22.	$R = 40$	**23.**	$R = 60$	**25.**	$A = 2$	**26.**	$B = 80$
27.	$B = 120$	**29.**	$R = \dfrac{100}{3}$	**30.**	$R = \dfrac{200}{3}$	**31.**	$x = 22$
33.	$x = 1$	**34.**	$x = 60$	**35.**	$x = 15$	**37.**	$B = 7.8$
38.	$B = 8.2$	**39.**	$x = 1.56$					

EXERCISES 4.8   PAGE 139

1. 45 mi	**2.** $13.50	**3.** 11 gal
5. $180	**6.** $3.27	**7.** 20 yd
9. 437.5 g	**10.** $120	**11.** $12
13. $1275	**14.** $160,000	**15.** $34
17. 42.86 ft or $42\frac{6}{7}$ ft	**18.** $7\frac{1}{2}$	**19.** 7.5 mph; 52.5 mph
21. 3360 mi	**22.** $11\frac{1}{4}$ hr	**23.** 9 hr
25. 22.3 gal	**26.** 259,200 revolutions	**27.** $648
29. $1\frac{7}{8}$ in.	**30.** 30.48 cm	**31.** e.g., $\frac{10}{4}$
33. $1\frac{1}{2}$ cups	**34.** 18 biscuits	**35.** 4700 g
37. 2.7 kg	**38.** 0.004 mg	**39.** 500 mg
41. 4 drams	**42.** 3 drams	**43.** 1920 minims
45. $8\frac{1}{3}$ grains	**46.** 60 g	**47.** 75 minims
49. 0.5 ounce	**50.** 750 mL	

REVIEW QUESTIONS: CHAPTER 4   PAGE 143

1. 10   **2.** four tenths   **3.** seven and eight hundredths

4. ninety-two and one hundred thirty-seven thousandths

5. eighteen and five thousand, five hundred twenty-six ten-thousandths

6. $81\frac{47}{100}$	**7.** $100\frac{3}{100}$	**8.** $9\frac{592}{1000}$	**9.** $200\frac{5}{10}$
10. 2.17	**11.** 84.075	**12.** 3003.003	**13.** 5900
14. 7.6	**15.** 0.039	**16.** 2.06988	**17.** 26.82
18. 64.151	**19.** 93.418	**20.** 17.79	**21.** 9.02
22. 3.9623	**23.** 104.272	**24.** 22.708	**25.** 0.72
26. 0.02	**27.** 0.0064	**28.** 235.	**29.** 1.7632
30. 5964.1	**31.** 0.12	**32.** 0.1728	**33.** 171.55
34. 0.71	**35.** 880.53	**36.** 200.	**37.** 2.961
38. 0.00567	**39.** 1.9435	**40.** 23.5	

41. extremes—2 and y; means—3 and x

42. true	**43.** true	**44.** false
45. $x = 5$	**46.** $y = 300$	**47.** $w = 5\frac{1}{16}$
48. 149.8 mi	**49.** 4.6855	**50.** 166.85 mi
51. $700	**52.** 40,000 hairpins	**53.** 45 mph
54. $26\frac{2}{3}$ ft	**55.** $266\frac{2}{3}$ mi	

CHAPTER TEST: CHAPTER 4 PAGE 146

1. five thousand, six hundred ninety-two and four tenths; $5692\dfrac{4}{10}$

2. eight and three hundred fifty-seven thousandths; $8\dfrac{357}{1000}$

3. 75.00003	4. 9400	5. 71.36	6. 99.385
7. 41.77	8. 15.92	9. 13.934	10. 13,850
11. 0.14	12. 5.57708	13. 0.325	14. 7000.
15. 0.0839	16. 84.24	17. true	18. false

19. $x = 2\dfrac{4}{7}$ 20. $z = 75$ 21. $x = 1.122$ 22. $1\dfrac{1}{5}$ mi

23. 24.236 mi 24. $108 25. 1440 revolutions

CHAPTER 5

EXERCISES 5.1 PAGE 150

1. 30%	2. 20%	3. 40%	5. 7%
6. 8%	7. 9%	9. 25%	10. 35%
11. 45%	13. 75%	14. 42%	15. 53%
17. 77%	18. 48%	19. 125%	21. 150%
22. 175%	23. 200%	25. 236%	26. 120%
27. 16.3%	29. 13.4%	30. 38.6%	31. 20.25%
33. 0.5%	34. 1.5%	35. 0.25%	37. $10\dfrac{1}{4}$%

38. $1\dfrac{1}{4}$% 39. $24\dfrac{1}{2}$%

EXERCISES 5.2 PAGE 152

1. 2%	2. 9%	3. 10%	5. 70%
6. 90%	7. 36%	9. 83%	10. 75%
11. 25%	13. 40%	14. 65%	15. 2.5%
17. 4.6%	18. 5.5%	19. 0.3%	21. 110%
22. 130%	23. 125%	25. 200%	26. 108%
27. 105%	29. 230%	30. 215%	31. 0.02
33. 0.1	34. 0.18	35. 0.15	37. 0.25
38. 0.3	39. 0.35	41. 0.101	42. 0.115
43. 0.132	45. 0.0525	46. 0.065	47. 0.1375
49. 0.2025	50. 0.185	51. 0.0025	53. 0.0017
54. 0.005	55. 1.25	57. 1.3	58. 1.2
59. 2.22			

EXERCISES 5.3 PAGE 155

1. 3%	2. 16%	3. 7%	5. 50%
6. 75%	7. 25%	9. 55%	10. 70%
11. 30%	13. 20%	14. 40%	15. 80%

17. 26%   **18.** 4%   **19.** 48%   **21.** 12.5%

22. 62.5%   **23.** 87.5%   **25.** $55\frac{5}{9}\%$ **26.** $28\frac{5}{7}\%$

27. $42\frac{6}{7}\%$ **29.** $63\frac{7}{11}\%$ **30.** $45\frac{5}{11}\%$ **31.** $107\frac{2}{14}\%$

33. 105%   **34.** 125%   **35.** 175%   **37.** 137.5%

38. 250%   **39.** 210%   **41.** $\frac{1}{10}$ **42.** $\frac{1}{20}$

43. $\frac{3}{20}$ **45.** $\frac{1}{4}$ **46.** $\frac{3}{10}$ **47.** $\frac{1}{2}$

49. $\frac{3}{8}$ **50.** $\frac{1}{6}$ **51.** $\frac{1}{3}$ **53.** $\frac{33}{100}$

54. $\frac{1}{200}$ **55.** $\frac{1}{400}$ **57.** 1   **58.** $1\frac{1}{4}$

59. $1\frac{1}{5}$ **61.** $\frac{3}{1000}$ **62.** $\frac{1}{40}$ **63.** $\frac{5}{8}$

65. $\frac{3}{400}$

EXERCISES 5.4   PAGE 159

1. 7   **2.** 3.1   **3.** 9   **5.** 9   **6.** 18

7. 36   **9.** 150   **10.** 85   **11.** 700   **13.** 75

14. 84   **15.** 42   **17.** 150%   **18.** 40%   **19.** 20%

21. 50%   **22.** 20%   **23.** $33\frac{1}{3}\%$ **25.** 12.5   **26.** 23.56

27. 110   **29.** 130%   **30.** 150%   **31.** 72   **33.** 520

34. 38   **35.** 200%   **37.** 16.32   **38.** 26.88   **39.** 58.5

EXERCISES 5.5   PAGE 164

1. $5700   **2.** $195; $6305   **3.** $710   **5.** $12,000

6. 0.03   **7.** $4.30

9. 1st salesman—$405; 2nd salesman—$525

10. 20 problems   **11.** 153 free throws   **13.** $750; $600; $636

14. $1.82; $32.02

15. $400; $80; $33\frac{1}{3}\%$ on cost; 25% on selling price

17. rent—32%; food—35%; taxes—12%

18. 88%; 250 pages; 30 pages   **19.** $10.53; 60%

21. $1875; $2025

22. 10%; $11\frac{1}{9}\%$; 20 lb are 10% of 200 lb but $11\frac{1}{9}\%$ of 180 lb

23. $656.40   **25.** $22,018   **26.** the $26 book; $1.26

REVIEW QUESTIONS: CHAPTER 5   PAGE 167

1. hundredths
2. 85%
3. 18%
4. 37%
5. $16\frac{1}{2}\%$
6. 15.2%
7. 115%
8. 6%
9. 30%
10. 67%
11. 2.7%
12. 300%
13. 120%
14. 0.35
15. 0.04
16. 0.0025
17. 0.0025
18. 0.071
19. 1.32
20. 60%
21. 15%
22. 16%
23. 37.5%
24. $41\frac{2}{3}\%$
25. $126\frac{2}{3}\%$
26. $\frac{7}{50}$
27. $\frac{2}{5}$
28. $\frac{33}{50}$
29. $\frac{1}{8}$
30. 4
31. $\frac{67}{200}$
32. 15.6
33. 2.55
34. $233\frac{1}{3}$
35. $42\frac{6}{7}$
36. 20%
37. $33\frac{1}{3}\%$
38. 25%
39. 1.095
40. 50
41. $254\frac{6}{11}$
42. 0.975
43. 200%
44. $11.93
45. 50% on cost; $33\frac{1}{3}\%$ on selling price
46. 8 problems
47. $1950
48. 0.012
49. $5\frac{5}{9}\%$—movie; $38\frac{8}{9}\%$—clothes; $22\frac{2}{9}\%$—anniv.
50. $10,000; $9150

CHAPTER TEST: CHAPTER 5   PAGE 170

1. hundredths
2. 63%
3. 4.5%
4. 90%
5. 24%
6. 137.5%
7. 3.6%
8. 54%
9. 70%
10. 260%
11. 12.5%
12. 0.16
13. 0.081
14. 0.952
15. 1.83
16. 0.1125
17. $\frac{9}{100}$
18. $\frac{11}{20}$
19. $\frac{3}{8}$
20. 3.5
21. $\frac{31}{500}$
22. 13.2
23. 540
24. 5%
25. 35%
26. 72
27. 97
28. $755.25
29. 50% on cost; $33\frac{1}{3}\%$ on selling price
30. $3570

CHAPTER 6

EXERCISES 6.1   PAGE 175

1. $48
2. $60
3. $31.50
5. $11.25

6. $10   **7.** $48   **9.** $730   **10.** $1030

11. $463.50; $36.50   **13.** $37,500   **14.** 72 days   **15.** 10%

17. (a) interest—$7.50   **18.** $1000; 270 days or $\frac{3}{4}$ year
    (b) time—60 days
    (c) rate—18%
    (d) principal—$100

19. 19.2%; $562.50

EXERCISES 6.2   PAGE 179

1. $13,791.70; $14,631.61   **2.** $9459.52

3. $306.83; $5306.83; $6.83 more

5. $370.80 yearly; $376.54 semiannually; $379.52 quarterly

6. $19,201.32   **7.** $800; $1441.96

9. $19,965; no ($17,395.43 @ 5% semiannually); no ($2569.57 more for
    10% annually); no $20,173.40 @ 5% semiannually for 6 years is $208.40
    more than 10% annually for 3 years)

10. $74.18 more compounded monthly   **11.** $1251.39; $900 interest

13. Double in 12 years;

YEAR	PRINCIPAL	INTEREST	TOTAL
1	$10,000.00	$ 600.00	$10,600.00
2	10,600.00	636.00	11,236.00
3	11,236.00	674.16	11,910.16
4	11,910.16	714.61	12,624.77
5	12,624.77	754.49	13,382.26
6	13,382.26	802.93	14,185.19
7	14,185.19	851.11	15,036.30
8	15,036.30	902.20	15,938.50
9	15,938.50	956.29	16,894.79
10	16,894.79	1,003.68	17,908.47
11	17,908.47	1,074.51	18,982.98
12	18,982.98	1,138.99	20,121.96

14. Almost double in 6 years;

YEAR	PRINCIPAL	INTEREST	TOTAL
1	$10,000.00	$1200.00	$11,200.00
2	11,200.00	1344.00	12,544.00
3	12,544.00	1505.28	14,049.28
4	14,049.28	1685.92	15,735.20
5	15,735.20	1888.23	17,623.42
6	17,623.42	2114.81	19,738.23

EXERCISES 6.3   PAGE 183

1. paid $1358.67; 11 monthly payments; $86.67 interest

2. paid $966.40; 11 monthly payments

3. $257 monthly payment; 144%

5. $310. monthly payment; Plan A is more expensive by $526.37
6. A = $2057.22.; B = $1977.54; B had lower plan
7. A = $260.59 monthly payment; B = $333.18 monthly payment; Plan A is cheaper by $38.44; A = 216% last payment; B = 96% last payment
9. 11 payments; $1671.59 total price; $81.59 interest
10. 21 payments; $1852.65 interest; $15,072.65 total cost

EXERCISES 6.4   PAGE 189

1. seller receives $27,050.; buyer owes $19,010.
2. seller receives $13,305.; buyer owes $20,677.50
3. seller receives $34,280.; buyer owes $13,000.
5. seller receives $47,244.; buyer owes $28,535.
6. seller receives $72,230.; buyer owes $46,105.
7. seller receives $16,385.; buyer owes $5,929.

EXERCISES 6.5   PAGE 193

1. Soc. Sci.; Chem. & Phys. and Humanities; 3300, 21.2%
2. fac. sal.—$5,625,000, admin. sal.—$2,500,000, nonteach. sal.—$1,625,000, mainten.—$1,250,000, stud. prog.—$625,000, sav.—$500,000, supplies—$375,000; $3,125,000; 22%; $2,750,000
3. 1977; 18 inches 1979 and 1980; 16 in. average
5. news—300 min, commerc.—180 min, soaps—180 min, sitcoms—156 min, drama—144 min, movies—120 min, children's prog.—120 min
6. wheat—$1500 loss; steel—$1000 gain; steel—$2000 gain; wheat—$1500 loss
7. City B: 15 runners; 30%; City C; 12 runners; 70.59%
9. highest—July; lowest—March; average—16.58
10. The scales represent two different types of quantities; Sue; Bob and Sue; 54.5% studying; 75% studying; yes, longer work hours seem to produce lower GPAs and vice versa.
11. highest—August; $3.5 million; $0.3 million or $300,000; $5.3 million; Dec. 1980—29.41%; Dec. 1981—31.94%

REVIEW QUESTIONS: CHAPTER 6   PAGE 199

1. $97.50   2. $807   3. $7.50   4. $2000
5. Principal   Rate   Time   Interest   6. annually—$803.40
                              $12   semiann.—$815.82
                       18 mo.   quartly.—$822.25

$8\frac{1}{2}\%$

$2000

7.

YEAR	PRINCIPAL	INTEREST	TOTAL
1	$10,000.00	$ 800.00	$10,800.00
2	10,800.00	864.00	11,664.00
3	11,664.00	933.12	12,597.12
4	12,597.12	1,007.77	13,604.89
5	13,604.89	1,088.39	14,693.28
6	14,693.28	1,175.46	15,868.74
7	15,868.74	1,269.50	17,138.24
8	17,138.24	1,371.06	18,509.30
9	18,509.30	1,480.74	19,990.04

8. 16 payments; $10,419.78 total cost

9. seller receives $40,480.; buyer owes $16,610.

10. housing—$10,500; food—$7000; taxes—$3500; transp.—$3500; clothing—$3500; entertainment—$2800; education—$2450; savings—$1750.

CHAPTER TEST: CHAPTER 6 PAGE 201

1. $2825

2. $42.65; 65¢ more monthly

3. $9387.99 total cost

4. seller receives $58,820.; buyer owes $43,790.

5. Sept.—$1.5 million; $1 million difference; $16\frac{2}{3}$%; 100% growth; 50% drop

CHAPTER 7

EXERCISES 7.1 PAGE 209

1. milli, centi, deci, deka, hecto, kilo

2. 100 cm   **3.** 500 cm   **5.** 600 cm   **6.** 2000 mm

7. 300 mm   **9.** 1400 mm   **10.** 16 mm   **11.** 18 mm

13. 350 mm   **14.** 40 dm   **15.** 160 dm   **17.** 210 cm

18. 3000 m   **19.** 5000 m   **21.** 6400 m   **22.** 1.1 cm

23. 2.6 cm   **25.** 4.8 cm   **26.** 0.06 dm   **27.** 0.12 dm

29. 0.03 m   **30.** 0.145 m   **31.** 0.256 m   **33.** 0.32 m

34. 1.5 m   **35.** 1.7 m   **37.** 2.4 km   **38.** 0.5 km

39. 0.4 km   **41.** 462 cm   **42.** 0.063 m   **43.** 0.052 m

45. 6410 mm   **46.** 0.3 cm   **47.** 0.5 cm   **49.** 0.057 m

50. 20 km   **51.** 35 km   **53.** 2300 m   **54.** 0.0005 km

55. 0.0003 km   **57.** 104 mm   **58.** 53.2 m   **59.** 31.4 cm

61. 19.468 cm   **62.** 110 m   **63.** 13.5 cm   **65.** 16 000 m

EXERCISES 7.2 PAGE 213

1. 7000 mg   **2.** 2000 g   **3.** 0.0345 g

5.	4 t	**6.**	5.6 kg	**7.**	73 000 000 mg
9.	540 mg	**10.**	700 mg	**11.**	5000 kg
13.	2000 kg	**14.**	0.896 g	**15.**	896 000 mg
17.	75 kg	**18.**	3 g	**19.**	7 000 000 g
21.	0.00034 kg	**22.**	780 mg	**23.**	0.016 g
25.	3940 mg	**26.**	0.0923 kg	**27.**	5600 kg
29.	3.547 t	**30.**	2.963 t		

EXERCISES 7.3 PAGE 219

1. 300 mm^2 **2.** 560 mm^2 **3.** 870 mm^2 **5.** 6 cm^2
6. 0.28 cm^2 **7.** 14 cm^2 **9.** $400 \text{ cm}^2 = 40\,000 \text{ mm}^2$
10. $730 \text{ cm}^2 = 73\,000 \text{ mm}^2$ **11.** $5700 \text{ cm}^2 = 570\,000 \text{ mm}^2$
13. $1700 \text{ dm}^2 = 170\,000 \text{ cm}^2 = 17\,000\,000 \text{ mm}^2$
14. $290 \text{ dm}^2 = 29\,000 \text{ cm}^2 = 2\,900\,000 \text{ mm}^2$
15. $3 \text{ dm}^2 = 300 \text{ cm}^2 = 30\,000 \text{ mm}^2$ **17.** 1.42 cm^2
18. 58 cm^2 **19.** 2 m^2 **21.** 780 m^2 **22.** $30\,000 \text{ m}^2$
23. 4 m^2 **25.** $869 \text{ a} = 86\,900 \text{ m}^2$ **26.** $781 \text{ a} = 78\,100 \text{ m}^2$
27. $16 \text{ a} = 1600 \text{ m}^2$ **29.** 0.01 ha **30.** 0.15 ha
31. $50\,000 \text{ a} = 500 \text{ ha}$ **33.** $3000 \text{ a} = 30 \text{ ha}$
34. $5320 \text{ a} = 53.2 \text{ ha}$ **35.** 875 cm^2 **37.** 20 mm^2
38. 78.5 m^2 **39.** 7.065 cm^2 **41.** 38.5 mm^2
42. $5 \text{ cm}^2 \text{ or } 500 \text{ mm}^2$ **43.** 6 cm^2 **45.** 57.12 mm^2
46. 21.195 m^2 **47.** 32.28 dm^2 **49.** 7536 m^2 **50.** 15.8 km^2
51. 3.14 m^2 **53.** $1.2 \text{ m}^2 = 12\,000 \text{ cm}^2$ **54.** $2500 \text{ cm}^2 = 0.25 \text{ m}^2$
55. $1750 \text{ cm}^2 = 0.175 \text{ m}^2$

EXERCISES 7.4 PAGE 225

1.	1000 mm^3	**2.**	10 cm	**3.**	10 dm
	1000 cm^3		100 mm		100 cm
	1000 dm^3		100 cm^2		100 dm^2
	$1\,000\,000\,000 \text{ m}^3$		$10\,000 \text{ mm}^2$		$10\,000 \text{ cm}^2$
			1000 cm^3		1000 dm^3
			$1\,000\,000 \text{ mm}^3$		$1\,000\,000 \text{ cm}^3$
5.	73 000 L	**6.**	900 L	**7.**	0.4 L
9.	63 000 mL	**10.**	8700 mL	**11.**	0.5 kL
13.	19 cm^3	**14.**	5000 mm^3	**15.**	2000 cm^3
17.	5300 mL	**18.**	30 mL	**19.**	0.03 L
21.	48 000 L	**22.**	72 kL	**23.**	0.32 hL
25.	0.29 kL	**26.**	0.569 L	**27.**	7 200 000 mL
29.	9500 L	**30.**	0.72 hL	**31.**	70 dm^3
33.	381.51 cm^3	**34.**	904.32 dm^3	**35.**	12.56 dm^3
37.	224 cm^3	**38.**	9106 dm^3	**39.**	282.6 cm^3

EXERCISES 7.5 PAGE 231

1. $77°\text{F}$ **2.** $176°\text{F}$ **3.** $50°\text{F}$ **5.** $10°\text{C}$

6. $37\frac{7}{9}\,^{\circ}\mathrm{C}$ 7. $0\,^{\circ}\mathrm{C}$ 9. 157.48 cm   10. 190.5 cm

11. 2.745 m	13. 96.6 km	14. 161 km	15. 644 cm
17. 124 mi	18. 40.3 mi	19. 21.7 mi	21. 197 in.
22. 39.4 in.	23. 19.35 cm^2	25. 55.8 m^2	26. 27.9 m^2
27. 83.6 m^2	29. 405 ha	30. 101.25 ha	31. 741 acres
33. 53.82 ft^2	34. 11.96 yd^2	35. 4.65 in.^{2}	37. 9.46 L
38. 18.92 L	39. 94.625 L	41. 10.6 qt	42. 26.5 qt
43. 11.088 gal	45. 3277.4 cm^3	46. 353.15 ft^3	47. 4.54 kg
49. 1102.5 lb	50. 3.5 oz		

REVIEW QUESTIONS: CHAPTER 7   PAGE 233

1. 1500 cm   2. 0.35 dm   3. 3700 mm^2
4. 0.17 cm^2 5. 300 a   6. 30 000 m^2
7. 5000 cm^3 8. 36 000 mL   9. 13 000 cm^3
10. 68 000 mm^3 11. 5000 g   12. 3400 mg
13. 6710 kg   14. 0.019 g   15. 8000 g
16. 4.29 kg   17. 12 cm; 3 cm^2 18. 212 mm; 2280 mm^2
19. 25.42 m; 44.13 m^2 20. 33 mm; 63 mm^2 21. 294 L
22. 0.02931 L (rounded off)   23. (a) 10°C; (b) 50°C
24. (a) 7.62 cm; (b) 3.22 km; (c) 1.215 ha; (d) 3.784 L; (e) 2.27 kg

CHAPTER TEST: CHAPTER 7   PAGE 235

1. 20 cm by 180 cm   2. 10 kg by 9990 g
3. volume is the same   4. 0.37 m   5. 2300 cm
6. 2000 cm^3 7. 1.2 kg   8. 5600 kg
9. 7500 m^2 10. 11 m   11. 4000 mm^3
12. 9.6 cm^2 13. 0.0835 g
14. 251.2 cm—perimeter; 5024 cm^2—area
15. 32 m—perimeter; 40 m^2—area
16. 0.33912 L   17. 176°F   18. (a) 2.00152 in.; (b) 2.2472 qt

CHAPTER 8

EXERCISES 8.1   PAGE 239

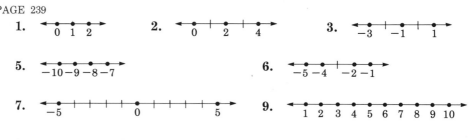

11. (number line) 1 3 5 7 ...

13. (number line) ... −5 −4 −3

14. (number line) ... −10 −9 −8

15. (number line) −3 0 3 6 9

17. (number line) −7 −2 0 2 7

18. (number line) −3 0 1 3

19. (number line) −5 −3 0 3 5

21. 10	**22.** 9	**23.** −14	**25.** 6	**26.** 3
27. −30	**29.** 0	**30.** 7	**31.** 6	**33.** 24
34. 16	**35.** 20	**37.** 0	**38.** 27	**39.** 13
41. −13	**42.** −18	**43.** 30	**45.** −9	**46.** −6
47. 9	**49.** −11	**50.** −12		

EXERCISES 8.2   PAGE 243

1. 2	**2.** 1	**3.** 10	**5.** 19	**6.** −10
7. −9	**9.** 0	**10.** 25	**11.** −4	**13.** 2
14. −3	**15.** −16	**17.** 17	**18.** −4	**19.** 0
21. 4	**22.** −6	**23.** 14	**25.** −1	**26.** 4
27. −15	**29.** 41	**30.** 6	**31.** 13	**33.** 10
34. 8	**35.** −12	**37.** −10	**38.** 0	**39.** 0
41. −235	**42.** −165	**43.** 120	**45.** −121	

EXERCISES 8.3   PAGE 245

1. 3	**2.** 13	**3.** 11	**5.** −7	**6.** −13
7. −9	**9.** 4	**10.** 10	**11.** −10	**13.** 1
14. 3	**15.** 18	**17.** 17	**18.** 23	**19.** −31
21. −5	**22.** −4	**23.** 0	**25.** −5	**26.** 30
27. 8	**29.** 5	**30.** −9	**31.** 80	**33.** −12
34. −29	**35.** 43			

EXERCISES 8.4   PAGE 247

1. 8	**2.** 12	**3.** 6	**5.** 10	**6.** 17
7. −4	**9.** 6	**10.** 6	**11.** −9	**13.** −1
14. −15	**15.** −8	**17.** −18	**18.** −10	**19.** 3
21. 1	**22.** 2	**23.** 0	**25.** 15	**26.** 4
27. 4	**29.** −3	**30.** 22	**31.** −27	**33.** 24
34. 0	**35.** 0	**37.** 5	**38.** 0	**39.** −5

EXERCISES 8.5   PAGE 250

1. −15	**2.** −24	**3.** 24	**5.** −20	**6.** −24

7. 28	**9.** −50	**10.** −33	**11.** −21	**13.** −48
14. −36	**15.** 63	**17.** 0	**18.** 0	**19.** 90
21. 24	**22.** 30	**23.** 60	**25.** −42	**26.** −64
27. −45	**29.** −60	**30.** −70	**31.** 0	**33.** −1
34. −27	**35.** 16	**37.** −84	**38.** −50	**39.** 300
41. −108	**42.** 320	**43.** 280	**45.** 108	**46.** −125
47. 1152	**49.** 2916	**50.** 1400		

EXERCISES 8.6   PAGE 253

1. −3	**2.** −9	**3.** −2	**5.** 4	**6.** 10	**7.** 5
9. −5	**10.** −5	**11.** −3	**13.** 2	**14.** 3	**15.** 4
17. −3	**18.** −3	**19.** −8	**21.** 2	**22.** 4	**23.** 1
25. undefined	**26.** undefined	**27.** 0		**29.** undefined
30. 0	**31.** positive	**33.** positive		**34.** positive
35. negative	**37.** 0	**38.** 0		**39.** undefined

EXERCISES 8.7   PAGE 255

1. 1.7   **2.** 54.9   **3.** 4.0901

5. −1.845   **6.** 1.898   **7.** −5.5

9. 900   **10.** −0.35   **11.** 1.4

13. $-\dfrac{29}{24}$ **14.** $-\dfrac{5}{24}$ **15.** $-\dfrac{41}{24}$

17. $-\dfrac{161}{3}$ or $-53\dfrac{2}{3}$ **18.** $-\dfrac{1271}{12}$ or $-105\dfrac{11}{12}$ **19.** $-\dfrac{11}{30}$

21. −57   **22.** $-\dfrac{21}{2}$ **23.** 0

25. −23   **26.** 33   **27.** −37

29. −8   **30.** −48   **31.** −43

33. −12   **34.** 11   **35.** −100

37. 108   **38.** 1   **39.** 23

REVIEW QUESTIONS: CHAPTER 8   PAGE 258

1. (number line with points at −5, −4, and 0)

2. (number line with points at −1, 1, 3, 4)

3. 7	**4.** 19	**5.** −56	**6.** 6	**7.** 9
8. 19	**9.** −1	**10.** −3	**11.** 0	**12.** 0
13. −82	**14.** 11	**15.** 11	**16.** 37	**17.** −23
18. 0	**19.** 6	**20.** 10	**21.** 6	**22.** 5
23. 2	**24.** 2	**25.** −135	**26.** −78	**27.** −63
28. 40	**29.** −3	**30.** −60	**31.** 16	**32.** −125
33. 120	**34.** 27	**35.** −3	**36.** 2	**37.** −6
38. −5	**39.** 3.65	**40.** $-\dfrac{31}{72}$	**41.** 0	**42.** −9

43. true	**44.** true	**45.** false	**46.** false	**47.** false
48. false	**49.** false	**50.** true	**51.** true	**52.** false

CHAPTER TEST: CHAPTER 8   PAGE 260

1.

$-3\ -2\quad 0\ \ 1$

2. 6	**3.** 25	**4.** -29	**5.** 8
6. 10	**7.** 21	**8.** -1	**9.** 4
10. 0	**11.** -20	**12.** -1	**13.** 93
14. 3	**15.** -46	**16.** -220	**17.** -80
18. 65	**19.** 5	**20.** -252	**21.** -14

22. -17 **23.** 2 **24.** 5.85 **25.** $-\dfrac{421}{72}$ or $-5\dfrac{61}{72}$

26. -0.15 **27.** 21 **28.** 10

29. True; the sum of two opposites is always 0.

30. False; the sum of two positive numbers is a positive number and 0 is not positive.

CHAPTER 9

EXERCISES 9.1   PAGE 264

1. $8x$	**2.** x	**3.** $6x$	**5.** $-7a$
6. $-7y$	**7.** $-12y$	**9.** $-9x$	**10.** $-3x$
11. $-8x$	**13.** 0	**14.** $-p$	**15.** $-c$
17. $-9a$	**18.** $-2c$	**19.** 0	**21.** $5x - 7$
22. $-x + 2$	**23.** $-x + 5$	**25.** $-11a - 2$	**26.** $-3x - 2$
27. $3x + 1$	**29.** $4x + 7$	**30.** $4y - 1$	**31.** -5
33. 0	**34.** -3	**35.** 9	**37.** 22
38. 6	**39.** -11	**41.** -3	**42.** -10
43. 12	**45.** -8		

EXERCISES 9.2   PAGE 266

1. $x + 5$	**2.** $x + 6$	**3.** $x + 10$	**5.** $x - 8$
6. $x - 4$	**7.** $x - 14$	**9.** $x + 11$	**10.** $6 - x$
11. $2x$	**13.** $\dfrac{x}{-7}$	**14.** $\dfrac{-18}{x}$	**15.** $4x - 3$
17. $4(x - 3)$	**18.** $5(x - 7)$	**19.** $-2(x + 5)$	**21.** $13 + 2x$
22. $x + 3x$	**23.** $6(x + 1)$	**25.** $-3 + 3x$	**26.** $8x + 5$
27. $9(x - 3)$	**29.** $7x - 3$	**30.** $16 + 2x$	

31. a number increased by 6 **33.** the product of a number and 5

34. a number divided by 2 **35.** 5 divided by a number

37. 20 minus a number **38.** twice a number plus 5

39. 3 less than twice a number

41. 5 minus 3 times a number

42. 10 times the quantity $x + 3$

43. -4 times the quantity $x - 7$

45. 3 times the difference between a number and 11

46. $17 + d$

47. $y + 4$

49. $\dfrac{T}{3}$

50. $60h$

51. $w + 7$

53. $0.75p$

54. $\dfrac{d}{3.2}$

55. $2.5r$

EXERCISES 9.3   PAGE 270

1. $x = 6$ **2.** $x = 7$ **3.** $y = 22$ **5.** $y = -5$

6. $x = -3$ **7.** $x = -2$ **9.** $y = 2$ **10.** $x = 7$

11. $x = 6$ **13.** $y = -4$ **14.** $x = -8$ **15.** $x = -6$

17. $y = 5$ **18.** $y = 3$ **19.** $x = 13$ **21.** $x = 23$

22. $x = 19$ **23.** $y = -22$ **25.** $y = 7$ **26.** $y = 7$

27. $x = -5$ **29.** $y = 5$ **30.** $y = 10$

EXERCISES 9.4   PAGE 272

1. $x = 1$ **2.** $x = 4$ **3.** $y = 2$ **5.** $x = -3$

6. $y = 1$ **7.** $y = 2$ **9.** $x = -6$ **10.** $x = -1$

11. $y = -3$ **13.** $x = 9$ **14.** $x = 7$ **15.** $y = -3$

17. $x = -1$ **18.** $x = 3$ **19.** $x = -3$ **21.** $x = -5$

22. $y = 4$ **23.** $y = 5$ **25.** $x = 0$ **26.** $x = 0$

27. $x = 3$ **29.** $x = 3$ **30.** $x = 8$ **31.** $x = -7$

33. $x = 5$ **34.** $y = 3$ **35.** $y = 4$ **37.** $x = 5$

38. $x = 4$ **39.** $x = -7$ **41.** $x = -12$ **42.** $y = 75$

43. $x = -\dfrac{5}{2}$ **45.** $x = \dfrac{220}{3}$ or $73\dfrac{1}{3}$

EXERCISES 9.5   PAGE 276

1. 19   **2.** 12   **3.** 54

5. 15   **6.** -1 **7.** -4

9. 4   **10.** -24 **11.** 15

13. 5   **14.** 18, 19   **15.** $-13, -14, -15$

17. 25, 27, 29   **18.** 8, 10, 12   **19.** 14, 16, 18, 20

21. 2   **22.** 5   **23.** -3

25. $-3, -2$ **26.** 8, 10   **27.** 1

29. -28 **30.** 3

REVIEW QUESTIONS: CHAPTER 9   PAGE 279

1. $15x$ **2.** $12y$ **3.** $3x$ **4.** $-6x$

5. $14w$ **6.** 0   **7.** $-2y$ **8.** $-34p$

9. $-8a + 6$ **10.** $-2x + 14$ **11.** $14y - 7$ **12.** $-10x - 10$

13. 29   **14.** 16   **15.** 41   **16.** $x + 8$

17. $5x - 3$ **18.** $\dfrac{x}{9}$ **19.** $-3(x + 2)$ **20.** $10 + 4x$

21. $18 - 2x$ **22.** -4 times a number
23. 5 plus seven times a number
24. -2 times the sum of a number and 6

25. $\dfrac{m}{60}$ hours **26.** $l - 3$ meters **27.** rt miles **28.** $x = 8$

29. $x = -3$ **30.** $y = 10$ **31.** $y = 27$ **32.** $x = 7$
33. $y = 14$ **34.** $x = 3$ **35.** $x = -2$ **36.** $y = 5$
37. $x = 1$ **38.** $x = 14$ **39.** $x = -8$ **40.** $x = -5$
41. $x = -12$ **42.** $x = -1$ **43.** -2 **44.** -1

45. 25, 26, 27 **46.** 21 **47.** 5 **48.** $-\dfrac{2}{3}$

CHAPTER TEST: CHAPTER 9 PAGE 281

1. $-x$ **2.** $16y$ **3.** $-15x$
4. $14p - 3$ **5.** -4 **6.** $-7 - 3x$
7. -11 **8.** -1 **9.** $\dfrac{3}{4}x - 5$
10. $20x + 6$ **11.** $-3(x - 7)$ **12.** $1 + 2(x - 9)$
13. 6 minus four times a number **14.** $4w - 3$
15. $d + .15d$ or $1.15d$ **16.** $x = 40$
17. $x = 6$ **18.** $x = -8$ **19.** $y = -12$ **20.** $x = -3$
21. $x = 5$ **22.** $x = 20$ **23.** $y = -15$ **24.** -16
25. 82, 84, 86 **26.** 36

CHAPTER 10

EXERCISES 10.1 PAGE 285

1. perfect square **2.** perfect square
3. perfect square **5.** perfect square
6. perfect square **7.** not a perfect square
9. not a perfect square **10.** not a perfect square
11. $1.732 \cdot 1.732 = 2.999824$ and $1.733 \cdot 1.733 = 3.003289$

13. -10 **14.** $\sqrt{3}$ **15.** $-10, \dfrac{3}{4}, 7.2, \sqrt{3}$
17. $2\sqrt{7}$ **18.** $2\sqrt{6}$ **19.** $4\sqrt{2}$
21. $12\sqrt{2}$ **22.** $11\sqrt{3}$ **23.** $11\sqrt{2}$
25. $10\sqrt{3}$ **26.** $8\sqrt{2}$ **27.** $5\sqrt{5}$
29. $7\sqrt{2}$ **30.** $11\sqrt{5}$ **31.** $5\sqrt{6}$
33. 14 **34.** $20\sqrt{2}$ **35.** $4\sqrt{5}$
37. $2\sqrt{10}$ **38.** 16 **39.** 19

EXERCISES 10.2 PAGE 290

1. true **2.** true **3.** false **5.** false
6. true **7.** false **8.** false **9.** false

10. true

11. $-1 < x < 2$, open interval

13. $x < 0$, open interval

14. $0 \le x \le 3$, closed interval

15. $-1 \le x$, half-open interval

17. open interval

18. closed interval

19. closed interval

21. half-open interval

22. closed interval

23. open interval

25. half-open interval

26. open interval

27. half-open interval

29. $x > 4$

30. $y \ge 2$

31. $y \le -3$

33. $y > -\dfrac{6}{5}$

34. $y > \dfrac{2}{3}$

35. $x < \dfrac{11}{7}$

37. $-5 > x$

38. $x \ge 7$

39. $x \le 5$

41. $x > -6$

42. $5 < x$

43. $x \le -1$

45. $x > -1$

46. $0 \ge x$

47. $x < -5$

49. $1 \leq x \leq 4$ 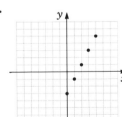 **50.** $-1 \leq y \leq -\dfrac{1}{2}$

51. $\dfrac{1}{3} \leq x \leq 2$ **53.** $-3 \leq y \leq 1$

54. $-5 \geq x > -6$ **55.** $3 < x \leq 5$

EXERCISES 10.3   PAGE 295

1. $\{(-5, 1), (-3, 2), (-1, 1), (1, 2), (2, 1)\}$
2. $\{(-3, 1), (-1, 4), (0, 5), (1, 4), (3, -3)\}$
3. $\{(-3, 2), (-2, -1), (-2, 1), (0, 0), (2, 1), (3, 0)\}$
5. $\{(-3, -4), (-3, 3), (-1, -1), (-1, 1), (1, 0)\}$
6. $\{(-1, -4), (-1, -3), (-1, 0), (-1, 2), (-1, 5)\}$
7. $\{(-1, -5), (0, -4), (1, -3), (2, -2), (3, -1), (4, 0)\}$
9. $\{(-1, 4), (0, 3), (1, 2), (2, 1), (3, 0), (4, -1), (5, -2)\}$
10. $\{(-3, 0), (-2, -1), (-1, -2), (0, -3), (0, 0), (1, -2), (2, -1), (3, 0)\}$

11. 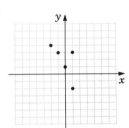   **13.** 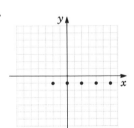   **14.**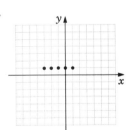

15. 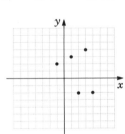   **17.** 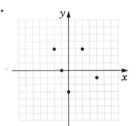   **18.**

19. 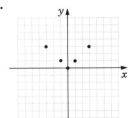   **21.** 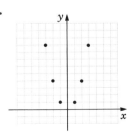   **22.**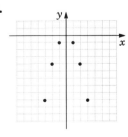

23.

25.

26.

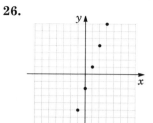

27.

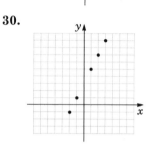

29.

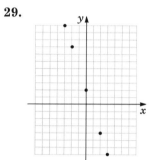

30.

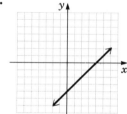

EXERCISES 10.4   PAGE 300

1.

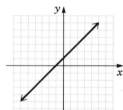

2.

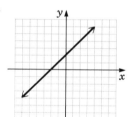

3.

5.

6.

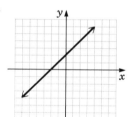

7.

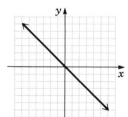

9.

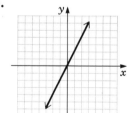

10.

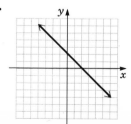

11.

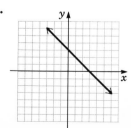

13.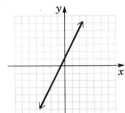

14.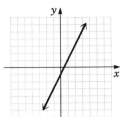

15.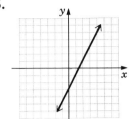

17.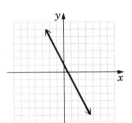

18.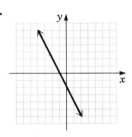

19.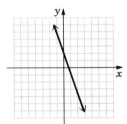

21.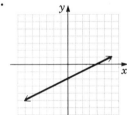

22.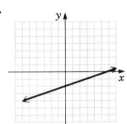

23.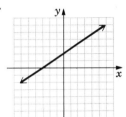

25.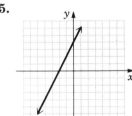

26.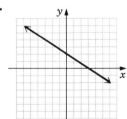

27.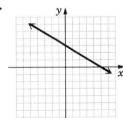

29.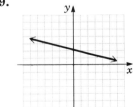

30.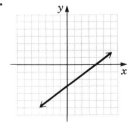

31.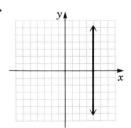

33. 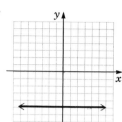 **34.** 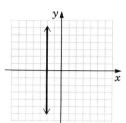 **35.**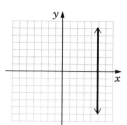

37. 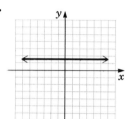 **38.** 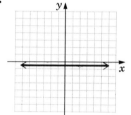 **39.**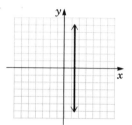

EXERCISES 10.5 PAGE 305

1.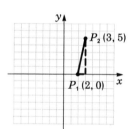

$$d = \sqrt{(3-2)^2 + (5-0)^2}$$
$$= \sqrt{26}$$

2.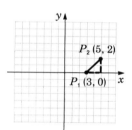

$$d = \sqrt{(5-3)^2 + (2-0)^2}$$
$$= \sqrt{8} = 2\sqrt{2}$$

3.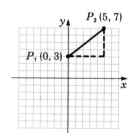

$$d = \sqrt{(5-0)^2 + (7-3)^2}$$
$$= \sqrt{41}$$

5.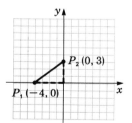

$$d = \sqrt{(0-(-4))^2 + (3-0)^4}$$
$$= 5$$

6.

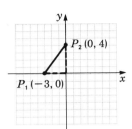

$$d = \sqrt{(0 - (-3))^2 + (4 - 0)^2}$$
$$= 5$$

7.

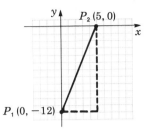

$$d = \sqrt{(5 - 0)^2 + (0 - (-12))^2}$$
$$= 13$$

9.

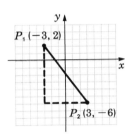

$$d = \sqrt{(3 - (-3))^2 + (-6 - 2)^2}$$
$$= 10$$

10.

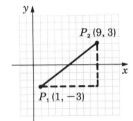

$$d = \sqrt{(9 - 1)^2 + (3 - (-3))^2}$$
$$= 10$$

11.

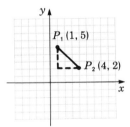

$$d = \sqrt{(4 - 1)^2 + (2 - 5)^2}$$
$$= \sqrt{18} = 3\sqrt{2}$$

13.

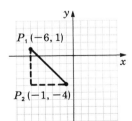

$$d = \sqrt{(-1 - (-6))^2 + (-4 - 1)^2}$$
$$= \sqrt{50} = 5\sqrt{2}$$

14.

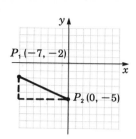

$$d = \sqrt{(0 - (-7))^2 + (-5 - (-2))^2}$$
$$= \sqrt{58}$$

15.

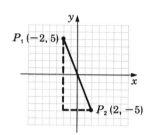

$$d = \sqrt{(2 - (-2))^2 + (-5 - 5)^2}$$
$$= \sqrt{116} = 2\sqrt{29}$$

17.

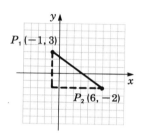

$$d = \sqrt{(6 - (-1))^2 + (-2 - 3)^2}$$
$$= \sqrt{74}$$

18.

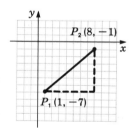

$$d = \sqrt{(8 - 1)^2 + (-1 - (-7))^2}$$
$$= \sqrt{85}$$

19.

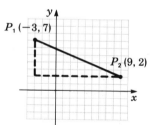

$$d = \sqrt{(9 - (-3))^2 + (2 - 7)^2}$$
$$= 13$$

21.

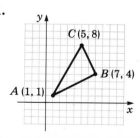

$\overline{AB} = \sqrt{45}, \ \overline{AC} = \sqrt{65}, \ \overline{BC} = \sqrt{20}$

The triangle ABC is a right triangle since $(\overline{AB})^2 + (\overline{BC})^2 = (\overline{AC})^2$ or $45 + 20 = 65$.

22.

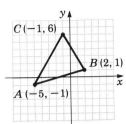

$\overline{AB} = \sqrt{53}, \overline{AC} = \sqrt{65}, \overline{BC} = \sqrt{34}$
The triangle ABC is not a right triangle since $(\overline{AB})^2 + (\overline{BC})^2 \neq (\overline{AC})^2$ or $53 + 34 \neq 65$.

REVIEW QUESTIONS: CHAPTER 10   PAGE 308

1. $(2.645)^2 = 6.996025$
$(2.646)^2 = 7.001316$

2. 0 is an integer.

3. $-\dfrac{1}{2}$, 0, 6.13 are rational numbers.

4. $-\sqrt{5}$, $-\dfrac{1}{2}$, 0, 6.13 are real numbers.

5. 13

6. 15

7. $8\sqrt{2}$

8. $3\sqrt{7}$

9. $3\sqrt{6}$

10. $11\sqrt{2}$

11. $10\sqrt{3}$

12. $15\sqrt{2}$

13. open interval
$-\dfrac{1}{2}$ $\dfrac{3}{4}$

14. half-open interval
0   5

15. half-open interval
$\sqrt{2}$

16. closed interval
-3 3.1

17. open interval
$\dfrac{1}{3}$

18. half-open interval
14   15

19. $-2 < x \leq 3$, half-open interval

20. $x < \dfrac{1}{2}$, open interval

21. $x < -4$
-4

22. $y \leq 7$
7

23. $y \geq \dfrac{8}{3}$
$\dfrac{8}{3}$

24. $-3 > x$
-3

25. $-2 < x$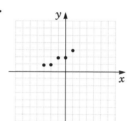

26. $2 \le x \le 3$

27. $\{(-4, 1)\ (-3, 0),\ (-2, -1),\ (0, 0),\ (1, 1),\ (2, 0)\}$

28. $\{(-1, 1),\ (0, 1),\ (1, 1),\ (2, 1)\}$

29. $\{(-1, 0),\ (0, 2),\ (1, 4),\ (2, 2),\ (3, 0)\}$

30.
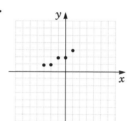

31.
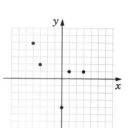

32.
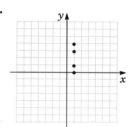

33.
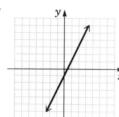

34.
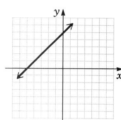

35.
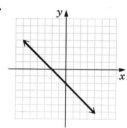

36.
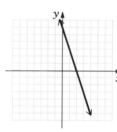

37.
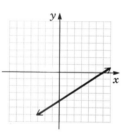

38.
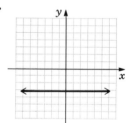

39. $d = \sqrt{(4 - 0)^2 + (6 - 3)^2} = 5$

40. $d = \sqrt{(5 - (-2)^2 + (-1 - 1)^2} = \sqrt{53}$

41.
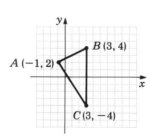

$\overline{AB} = \sqrt{20},\ \overline{AC} = \sqrt{52},\ \overline{BC} = 8$

$(\overline{AB})^2 + (\overline{AC})^2 \ne (\overline{BC})^2$ or $20 + 52 \ne 64$.

So, the triangle is not a right triangle.

CHAPTER TEST: CHAPTER 10   PAGE 310

1. $(2.828)^2 = 7.997584$
$(2.829)^2 = 8.003241$

2. -7 is an integer.

3. $-7, -3.14, \dfrac{1}{4}, 2\dfrac{1}{2}$ are rational numbers.

4. $\sqrt{6}$ is an irrational number.

5. $-7, -3.14, \dfrac{1}{4}, \sqrt{6}, 2\dfrac{1}{2}$ are all real numbers.

6. 16   **7.** $7\sqrt{2}$ **8.** $10\sqrt{5}$ **9.** $4\sqrt{7}$

10. 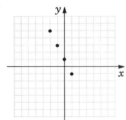 closed interval

11. open interval

12. open interval

13. $1 < x < 3.6$, open interval

14. $-\dfrac{2}{3} \le x \le 1\dfrac{2}{5}$, closed interval

15. $x \ge 4$ **16.** $y > 1$

17. $-1 \le x$ **18.** $-1 < x < 1$

19. $\{(-3, -4), (-2, -3), (-1, -2), (0, -1), (1, 0), (2, 1)\}$

20. 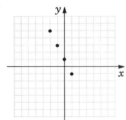   **21.**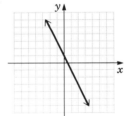

22. 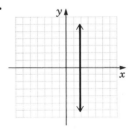   **23.**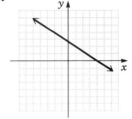

24. $d = \sqrt{(1 - (-3))^2 + (8 - 4)^2} = \sqrt{32} = 4\sqrt{2}$

25. $\overline{AB} = 5, \quad \overline{AC} = \sqrt{61}, \quad \overline{BC} = 6$ and $(\overline{AB})^2 + (\overline{BC})^2 = (\overline{AC})^2$ or $25 + 36 = 61$.

So, the triangle is a right triangle.

APPENDIX I

EXERCISES I.1   PAGE 315

1. 254	**2.** 163,041	**3.** 21,327	**5.** 140	**6.** 256
7. 53	**9.** 2362	**10.** 97	**11.** 744	**13.** 978
14. (64)		**15.** (532)	**17.** (846)	

14. (64)

(a) ∩∩∩ / ∩∩∩ IIII

(b) ∴∴

(c) ΓΔΔIIII

(d) LXIV

15. (532)

(a) ??? / ?? ∩∩∩ II

(b) ⁑

(c) ΓᴴΔΔΔII

(d) DXXXII

17. (846)

(a) ???? / ???? ∩∩∩ III

(b) ∸

(c) ΓᴴHHHΔΔΔΔΓI

(d) DCCCXLVI

EXERCISES I.2   PAGE 318

1. 4983	**2.** 32	**3.** 1802	**5.** 47,900
6. 222	**7.** 46	**9.** 999	

10. (a) VVVV VVV   <<<< << VV

(b) $\overline{\text{ϒOB}}$

(c) 罒
  万
  七
  十
  二

11. (a) VVVVV VVVV   <<< << < VVV VVV

(b) $\overline{\text{φCF}}$

(c) 五
  万
  九
  十
  六

13. (a) V V VVV VV

(b) $\overline{/ΓXΞE}$

(c) 三
  千
  六
  万
  六
  十
  五

14. (a) VV V   <<< VVV

(b) $\overline{/ZΣCΓ}$

(c) 七
  千
  二
  万
  九
  十
  三

15. (a) VVV $\underset{\sim}{\leq}\underset{\sim}{\leq}{}^{<}$VV

       A

    (b) /MΩNB

    (c) +

       Ŧ

       八

       ᛒ

       ⴱ

       +

       =

APPENDIX II

EXERCISES II.1   PAGE 322

1. $35 = 3(10^1) + 5(10^0)$ 2. $761 = 7(10^2) + 6(10^1) + 1(10^0)$

3. $8469 = 8(10^3) + 4(10^2) + 6(10^1) + 9(10^0)$

5. $62{,}322 = 6(10^4) + 2(10^3) + 3(10^2) + 2(10^1) + 2(10^0)$

6. $11_{(2)} = 1(2^1) + 1(2^0) = 1(2) + 1(1) = 2 + 1 = 3$

7. $101_{(2)} = 1(2^2) + 0(2^1) + 1(2^0) = 1(4) + 0(2) + 1(1) = 4 + 0 + 1 = 5$

9. $1011_{(2)} = 1(2^3) + 0(2^2) + 1(2^1) + 1(2^0) = 1(8) + 0(4) + 1(2) + 1(1)$
$$= 8 + 0 + 2 + 1 = 11$$

10. $1101_{(2)} = 1(2^3) + 1(2^2) + 0(2^1) + 1(2^0) = 1(8) + 1(4) + 0(2) + 1(1)$
$$= 8 + 4 + 0 + 1 = 13$$

11. $110{,}111_{(2)} = 1(2^5) + 1(2^4) + 0(2^3) + 1(2^2) + 1(2^1) + 1(2^0)$
$$= 1(32) + 1(16)\ 0(8) + 1(4) + 1(2) + 1(1)$$
$$= 32 + 16 + 0 + 4 + 2 + 1 = 55$$

13. $101{,}011_{(2)} = 1(2^5) + 0(2^4) + 1(2^3) + 0(2^2) + 1(2^1) + 1(2^0)$
$$= 1(32) + 0(16) + 1(8) + 0(4) + 1(2) + 1(1)$$
$$= 32 + 0 + 8 + 0 + 2 + 1 = 43$$

14. $11{,}010_{(2)} = 1(2^4) + 1(2^3) + 0(2^2) + 1(2^1) + 0(2^0)$
$$= 1(16) + 1(8) + 0(4) + 1(2) + 0(1) = 16 + 8 + 0 + 2 + 0$$
$$= 26$$

15. $1000_{(2)} = 1(2^3) + 0(2^2) + 0(2^1) + 0(2^0) = 1(8) + 0(4) + 0(2) + 0(1)$
$$= 8 + 0 + 0 + 0 = 8$$

17. $11{,}101_{(2)} = 1(2^4) + 1(2^3) + 1(2^2) + 0(2^1) + 1(2^0)$
$$= 1(16) + 1(8) + 1(4) + 0(2) + 1(1) = 16 + 8 + 4 + 0 + 1$$
$$= 29$$

18. $10{,}110_{(2)} = 1(2^4) + 0(2^3) + 1(2^2) + 1(2^1) + 0(2^0)$
$$= 1(16) + 0(8) + 1(4) + 1(2) + 0(1) = 16 + 0 + 4 + 2 + 0$$
$$= 22$$

19. $111{,}111_{(2)} = 1(2^5) + 1(2^4) + 1(2^3) + 1(2^2) + 1(2^1) + 1(2^0)$
$$= 1(32) + 1(16) + 1(8) + 1(4) + 1(2) + 1(1)$$
$$= 32 + 16 + 8 + 4 + 2 + 1 = 63$$

21.   111

EXERCISES II.2   PAGE 324

1. $24_{(5)} = 2(5^1) + 4(5^0) = 2(5) + 4(1) = 10 + 4 = 14$
2. $13_{(5)} = 1(5^1) + 3(5^0) = 1(5) + 3(1) = 5 + 3 = 8$
3. $10_{(5)} = 1(5^1) + 0(5^0) = 1(5) + 0(1) = 5 + 0 = 5$
5. $104_{(5)} = 1(5^2) + 0(5^1) + 4(5^0) = 1(25) + 0(5) + 4(1) = 25 + 0 + 4 = 29$
6. $312_{(5)} = 3(5^2) + 1(5^1) + 2(5^0) = 3(25) + 1(5) + 2(1) = 75 + 5 + 2$
$= 82$
7. $32_{(5)} = 3(5^1) + 2(5^0) = 3(5) + 2(1) = 15 + 2 = 17$
9. $423_{(5)} = 4(5^2) + 2(5^1) + 3(5^0) = 4(25) + 2(5) + 3(1) = 100 + 10 + 3$
$= 113$
10. $444_{(5)} = 4(5^2) + 4(5^1) + 4(5^0) = 4(25) + 4(5) + 4(1) = 100 + 20 + 4$
$= 124$
11. $1034_{(5)} = 1(5^3) + 0(5^2) + 3(5^1) + 4(5^0) = 1(125) + 0(25) + 3(5) + 4(1)$
$= 125 + 0 + 15 + 4 = 144$
13. $244_{(5)} = 2(5^2) + 4(5^1) + 4(5^0) = 2(25) + 4(5) + 4(1) = 50 + 20 + 4$
$= 74$
14. $3204_{(5)} = 3(5^3) + 2(5^2) + 0(5^1) + 4(5^0) = 3(125) + 2(25) + 0(5) + 4(1)$
$= 375 + 50 + 0 + 4 = 429$
15. $13,042_{(5)} = 1(5^4) + 3(5^3) + 0(5^2) + 4(5^1) + 2(5^0)$
$= 1(625) + 3(125) + 0(25) + 4(5) + 2(1)$
$= 625 + 375 + 0 + 20 + 2 = 1022$
17.   yes; 13   **18.**   yes; 28   **19.** $\{0, 1, 2, 3, 4, 5, 6, 7\}$

EXERCISES II.3   PAGE 327

1. $1000_{(2)} = 8$ **2.** $311_{(5)} = 81$ **3.** $11,000_{(2)} = 24$
5. $432_{(5)} = 117$ **6.** $1000_{(2)} = 8$ **7.** $1011_{(2)} = 11$
9. $433_{(5)} = 118$ **10.** $1302_{(5)} = 202$ **11.** $10,100_{(2)} = 20$
13. $10,100_{(2)} = 20$ **14.** $1222_{(5)} = 187$ **15.** $2241_{(5)} = 321$
17. $110,111_{(2)} = 55$ **18.** $23,240_{(5)} = 1695$ **19.** $3002_{(5)} = 377$
21. $30,141_{(5)} = 1921$ **22.** $1,101,001_{(2)} = 105$ **23.** $110,001_{(2)} = 49$
25. $10,001,010_{(2)} = 138$

APPENDIX III

EXERCISES   PAGE 331

1. 4	**2.** 4	**3.** 17	**5.** 10	**6.** 6
7. 3	**9.** 6	**10.** 11	**11.** 2	**13.** 8
14. 11	**15.** 4	**17.** 12	**18.** 11	**19.** 25
21. 1	**22.** 20			
23. 1	**26.** relatively prime			

27. relatively prime
29. relatively prime
30. not rel. prime; GCD is 7
31. relatively prime
33. relatively prime
34. not rel. prime; GCD is 11
35. relatively prime

APPENDIX IV

EXERCISES   PAGE 333

1. 22.36	2. 19.24	3. 1.41	5. 170.00
6. 150.00	7. 15.81	9. 9.75	10. 8.37
11. 9.27	13. 25.10	14. 4.24	15. 50

INDEX

1 2 3 4 5 6 7 8 9 0

POWERS, ROOTS, AND PRIME FACTORIZATIONS

NO.	SQUARE	SQUARE ROOT	CUBE	CUBE ROOT	PRIME FACTORIZATION
1	1	1.0000	1	1.0000	—
2	4	1.4142	8	1.2599	prime
3	9	1.7321	27	1.4423	prime
4	16	2.0000	64	1.5874	2 · 2
5	25	2.2361	125	1.7100	prime
6	36	2.4495	216	1.8171	2 · 3
7	49	2.6458	343	1.9129	prime
8	64	2.8284	512	2.0000	2 · 2 · 2
9	81	3.0000	729	2.0801	3 · 3
10	100	3.1623	1000	2.1544	2 · 5
11	121	3.3166	1331	2.2240	prime
12	144	3.4641	1728	2.2894	2 · 2 · 3
13	169	3.6056	2197	2.3513	prime
14	196	3.7417	2744	2.4101	2 · 7
15	225	3.8730	3375	2.4662	3 · 5
16	256	4.0000	4096	2.5198	2 · 2 · 2 · 2
17	289	4.1231	4913	2.5713	prime
18	324	4.2426	5832	2.6207	2 · 3 · 3
19	361	4.3589	6859	2.6684	prime
20	400	4.4721	8000	2.7144	2 · 2 · 5
21	441	4.5826	9261	2.7589	3 · 7
22	484	4.6904	10,648	2.8020	2 · 11
23	529	4.7958	12,167	2.8439	prime
24	576	4.8990	13,824	2.8845	2 · 2 · 2 · 3
25	625	5.0000	15,625	2.9240	5 · 5
26	676	5.0990	17,576	2.9625	2 · 13
27	729	5.1962	19,683	3.0000	3 · 3 · 3
28	784	5.2915	21,952	3.0366	2 · 2 · 7
29	841	5.3852	24,389	3.0723	prime
30	900	5.4772	27,000	3.1072	2 · 3 · 5
31	961	5.5678	29,791	3.1414	prime
32	1024	5.6569	32,768	3.1748	2 · 2 · 2 · 2 · 2
33	1089	5.7446	35,937	3.2075	3 · 11
34	1156	5.8310	39,304	3.2396	2 · 17
35	1225	5.9161	42,875	3.2711	5 · 7
36	1296	6.0000	46,656	3.3019	2 · 2 · 3 · 3
37	1369	6.0828	50,653	3.3322	prime
38	1444	6.1644	54,872	3.3620	2 · 19
39	1521	6.2450	59,319	3.3912	3 · 13
40	1600	6.3246	64,000	3.4200	2 · 2 · 2 · 5
41	1681	6.4031	68,921	3.4482	prime
42	1764	6.4807	74,088	3.4760	2 · 3 · 7
43	1849	6.5574	79,507	3.5034	prime
44	1936	6.6333	85,184	3.5303	2 · 2 · 11
45	2025	6.7082	91,125	3.5569	3 · 3 · 5
46	2116	6.7823	97,336	3.5830	2 · 23
47	2209	6.8557	103,823	3.6088	prime
48	2304	6.9282	110,592	3.6342	2 · 2 · 2 · 2 · 3
49	2401	7.0000	117,649	3.6593	7 · 7
50	2500	7.0711	125,000	3.6840	2 · 5 · 5